MODIFIED GRAVITY

Progresses and Outlook of
Theories, Numerical Techniques
and Observational Tests

MODIFIED GRAVITY

Progresses and Outlook of Theories, Numerical Techniques and Observational Tests

Editors

Baojiu Li
Durham University, UK

Kazuya Koyama
University of Portsmouth, UK

World Scientific

NEW JERSEY · LONDON · SINGAPORE · BEIJING · SHANGHAI · HONG KONG · TAIPEI · CHENNAI · TOKYO

Published by

World Scientific Publishing Co. Pte. Ltd.

5 Toh Tuck Link, Singapore 596224

USA office: 27 Warren Street, Suite 401-402, Hackensack, NJ 07601

UK office: 57 Shelton Street, Covent Garden, London WC2H 9HE

Library of Congress Cataloging-in-Publication Data

Names: Li, Baojiu, editor. | Koyama, Kazuya, editor.

Title: Modified gravity : progresses and outlook of theories, numerical techniques and
     observational tests / edited by Baojiu Li (Durham University, UK),
     Kazuya Koyama (University of Portsmouth, UK).

Description: [Hackensack] : World Scientific, [2019]. | Includes bibliographical references.

Identifiers: LCCN 2019023491 | ISBN 9789813273993 (hardcover)

Subjects: LCSH: Gravitation. | Cosmology.

Classification: LCC QC178 .M633 2019 | DDC 521/.1--dc23

LC record available at https://lccn.loc.gov/2019023491

British Library Cataloguing-in-Publication Data

A catalogue record for this book is available from the British Library.

The articles in this book were previously published in *International Journal of Modern Physics D*, Vol. 27, No. 15.

Cover image credit: UCL Mathematical and Physical Sciences

For photocopying of material in this volume, please pay a copying fee through the Copyright Clearance Center, Inc., 222 Rosewood Drive, Danvers, MA 01923, USA. In this case permission to photocopy is not required from the publisher.

For any available supplementary material, please visit
https://www.worldscientific.com/worldscibooks/10.1142/11090#t=suppl

Desk Editor: Ng Kah Fee

Contents

Chapter 1

Gravity Beyond General Relativity

Kazuya Koyama

*Institute of Cosmology & Gravitation, University of Portsmouth,
Dennis Sciama Building, Portsmouth, PO1 3FX, UK
Kazuya.Koyama@port.ac.uk*

We introduce the standard model of cosmology based on General Relativity (GR) and discuss its successes and problems. We then discuss motivations to consider gravitational theories beyond GR and summarise observational and theoretical constraints that these theories need to satisfy. A special focus is laid on screening mechanisms, which hide deviations from GR in the Solar System and enable large modifications to GR on astrophysical and cosmological scales. Finally, several modified gravity models are introduced, which satisfy the Solar System constrains as well as the constraint on the speed of gravitational waves obtained from almost simultaneous detections of gravitational waves and gamma ray bursts from a neutron star merger (GW170817/GRB 170817A).

Keywords: Cosmology; general relativity; modified gravity.

1. Introduction

Cosmology has entered the era of precision science. We now have the standard model of cosmology, Lambda Cold Dark Matter (ΛCDM) model, which explains various observations from the Cosmic Microwave Background to large-scale structure (LSS) with only 6 cosmological parameters.[1] At the same time, in the ΛCDM model, only 5% of the energy density of the Universe is made of known matter. 25% of the energy density is made of dark matter while 70% is made of the cosmological constant. The cosmological constant is required to explain the accelerated expansion of the Universe but its observed value is many orders of magnitude smaller than what we expect from the standard model of particle physics. This cosmological constant problem motivates us to consider alternatives to the ΛCDM model. In this chapter, we focus on modifications to General Relativity (GR) and discuss motivations to consider theories beyond GR and explain various constraints that these theories need to satisfy. We pay particular attention to screening mechanisms to evade stringent Solar System constraints while having large modifications on astrophysical and cosmological scales. We then discuss several examples of models that satisfy the Solar System constrains as well as the constraint on the speed of gravitational waves obtained by almost simultaneous detections of gravitational waves and gamma ray burst from a neutron star merger (GW170817/GRB 170817A).[2]

Throughout this chapter, we use the metric signature $(-, +, +, +)$.

2. Lambda Cold Dark Matter (ΛCDM) Model

This section introduces the standard ΛCDM model, emphasising the basic assumptions of the model, and summarises observational constraints on the model mainly from the Planck 2018 results.[3]

2.1. *ΛCDM model*

The standard model of cosmology is based on three assumptions:

(1) Our Universe is homogeneous and isotropic on average on large scales.
(2) Gravity is described by GR on all scales.
(3) The matter content of our Universe is given by Cold Dark Matter (CDM), baryons and radiation.

The first assumption implies that the metric describing the background homogeneous and isotropic Universe is given by

$$ds^2 = -dt^2 + a(t)^2\left[d\chi^2 + f_K(\chi)(d\theta^2 + \sin^2\theta d\phi^2)\right], \quad f_K = \frac{1}{\sqrt{-K}}\sinh(\sqrt{-K}\chi), \quad (1)$$

where $a(t)$ is the scale factor and $K = 0, 1$ and -1 correspond to flat, closed and open geometry of the three-dimensional constant time hypersurface, respectively. The comoving distance is defined as

$$\chi = -\int_{t_0}^{t}\frac{1}{a(t')}dt' = \frac{1}{H_0}\int_0^z\frac{dz'}{E(z')}, \quad E(z) = \frac{H(z)}{H_0}, \quad (2)$$

where quantities with the subscript 0 indicate that they are evaluated today at $t = t_0$, $z = 1/a - 1$ is the redshift and $H = \dot{a}/a$ is the Hubble expansion rate, with an overdot denoting the derivative with respect to the physical time. The luminosity and angular diameter distances are defined as

$$d_L = f_K(\chi)(1+z), \quad d_A = \frac{d_L}{(1+z)^2}. \quad (3)$$

The Hubble parameter today H_0 plays a key role in dark energy and modified gravity models. In the natural unit, $c = \hbar = k_B = 1$, it is given as $H_0 = 2.13 \times 10^{-42}h$ GeV where h is the dimensionless Hubble parameter, $h \equiv H_0/(100\text{kms}^{-1}\text{Mpc}^{-1})$, which needs to be determined by observations.

The second assumption implies that gravity is described by the Einstein-Hilbert action

$$S = \frac{1}{16\pi G}\int d^4x\sqrt{-g}(R + 2\Lambda) + \int d^4x \mathcal{L}_m[g_{\mu\nu}], \quad (4)$$

where G is the Newton constant, R is the Ricci curvature scalar and Λ is the cosmological constant. The Einstein equations are given by

$$G_{\mu\nu} + \Lambda g_{\mu\nu} = 8\pi G\, T_{\mu\nu}, \tag{5}$$

where $g_{\mu\nu}$ is the metric tensor, the energy-momentum tensor is defined as $T^{\mu\nu} = (2/\sqrt{-g})\delta S_m/\delta g_{\mu\nu}$. The Planck energy is defined as $M_{\rm pl} \equiv \sqrt{1/8\pi G} = 2.43 \times 10^{18}$ GeV. As we will discuss later, we do not expect that GR is valid at energy scales higher than $M_{\rm pl}$ but this is not relevant as long as we consider the late time evolution of the Universe. Applying to the metric given by (1), we obtain the Friedmann equations

$$H^2 = \left(\frac{\dot{a}}{a}\right)^2 = \frac{8\pi G}{3}\rho - \frac{K}{a^2} + \frac{\Lambda}{3}, \tag{6}$$

$$\frac{\ddot{a}}{a} = -\frac{4\pi G}{3}(\rho + 3P) + \frac{\Lambda}{3}, \tag{7}$$

where $T^\mu_\nu = \mathrm{diag}(-\rho, P, P, P)$, ρ is the energy density and P is the pressure of matter. The Ricci curvature scalar is given by $R = 6(a\ddot{a} + \dot{a}^2 + K)/a^2$.

Now we use the final assumption to specify the matter content of the Universe. The Friedmann equation gives

$$E(z) = \Omega_m(1 + z)^3 + \Omega_r(1 + z)^4 + \Omega_K(1 + z)^2 + \Omega_\Lambda, \tag{8}$$

where $\Omega_m = \Omega_b + \Omega_c, \Omega_r, \Omega_K$ and Ω_Λ are the contributions from cold dark matter and baryons, radiation, curvature and the cosmological constant, respectively. The contribution from radiation, Ω_r, is precisely known from Cosmic Microwave Background (CMB) as $\Omega_r h^2 = 2.47 \times 10^{-5}$. The minimum model that fits the CMB data is the flat ΛCDM model, which has 6 parameters, $\Omega_b h^2$, $\Omega_c h^2$, θ_*, τ, A_s and n_s, where θ_* is the angular scale of the sound horizon at the last scattering, τ is the optical depth, A_s is the amplitude of the primordial fluctuations and n_s is the spectrum index of the power spectrum of the primordial curvature perturbation. The Planck 2018 result[3] gave the constraint $\Omega_m = 0.315 \pm 0.007, \Omega_\Lambda = 0.685 \pm 0.007$ and $H_0 = 67.4 \pm 0.5$ km/s/Mpc, assuming $\Omega_K = 0$. Note that the constraint on H_0 is an indirect one derived from the 6 parameters by marginalising over other parameters. Therefore it highly depends on the assumption of the cosmological model, i.e. the flat ΛCDM model.[4] The curvature is constrained as $\Omega_K = 0.001 \pm 0.002$ by the joint constraint with baryon acoustic oscillations (BAO) measurements from galaxy surveys.

The most notable feature of the ΛCDM model is the large contribution from the cosmological constant. This is required because the expansion of the Universe is accelerating rather than decelerating, $\ddot{a} > 0$. This was first found by the observations of Type Ia Supernovae (SNe Ia),[5,6] which measure the luminosity distance d_L to SNe. On the other hand, the acoustic peaks of CMB anisotropies are determined by the sound horizon at radiation drag and the measurements of these acoustic peaks give the information on the angular diameter distance to the last scattering surface.

The same acoustic oscillation features are encoded in the distribution of galaxies and provide the measurement of angular diameter distances at low redshifts from galaxy surveys.[7,8] These measurements are complementary in the constraints of Ω_m and Ω_Λ. The ΛCDM model is currently consistent with all these measurements.

Finally, we mention some tension in the current data with ΛCDM. The late-time observations of galaxy clusters such as Sunyaev-Zel'dovich effects and weak lensing prefer lower amplitudes of the matter fluctuations (σ_8, the rms density fluctuation at a smoothing scale of $8h^{-1}$Mpc) compared with the Planck measurements of the initial amplitude at last scattering in ΛCDM.[1,9–11] The most notable tension is in the measurements of the Hubble constant H_0. The local measurements using the distance ladder prefer higher H_0 than the Planck measurements. For example, the Hubble Space Telescope (HST) measurements give $H_0 = 73.45 \pm 1.66$ km/s/Mpc,[12] which is inconsistent at the 3.5 sigma level with Planck ΛCDM. Note that the recent gravitational wave detection with almost simultaneous detections of short gamma ray bursts from a neutron star merger gave an independent measurement of H_0, $H_0 = 70^{+12.0}_{-8.0}$ km/s/Mpc.[13]

It still needs to be seen whether these discrepancies are due to unknown systematics or not. However, if they persist in the future measurements, this may be an indication of new physics beyond ΛCDM.

2.2. *Cosmic acceleration*

Although the current observational data is consistently explained by the ΛCDM model, it requires the existence of the cosmological constant. The energy scale of the cosmological constant is given by

$$\rho_\Lambda \equiv \frac{\Lambda}{8\pi G} \sim (10^{-3}\text{eV})^4. \tag{9}$$

Quantum field theory predicts that all particles give rise to vacuum energy $T_{\mu\nu} = -V_{\text{vac}}g_{\mu\nu}$, which contributes to the cosmological constant. The Standard Model of particle physics includes particles up to the TeV scale and we would estimate the vacuum energy to be $V_{\text{vac}} > (\text{TeV})^4$. This is 60 orders of magnitude larger than the cosmological constant that we need to explain the current observations. This is known as the (old) cosmological constant problem, i.e. why the vacuum energy does not gravitate as expected according to Einstein's theory of gravity.[14] This problem existed even before the discovery of the accelerated expansion of the Universe. The latter has created a new cosmological constant problem, i.e. the coincidence problem - why does the expansion of the Universe start to accelerate just now? This requires that the cosmological constant is fine-tuned to be given by (9).

There have been many attempts to explain the observed accelerated expansion of the Universe. Note that in most of these approaches, the old cosmological constant problem is assumed to be solved by some mechanism. There are three main approaches depending on which assumption of the ΛCDM model to abandon.

The first possibility is to consider an inhomogeneous universe. Indeed, our Universe is inhomogeneous, otherwise we do not exist. If the formation of inhomogeneous structures causes a back reaction to the expansion of the Universe and it causes the expansion of the Universe to accelerate, we can solve the coincidence problem as the onset of the acceleration coincides with the formation of first nonlinear structures. The back reaction is a notoriously difficult problem in GR due to the non-linear nature of Einstein's equations. There are still on-going debates on the magnitude of back-reaction[15, 16] but it is a general consensus that it is likely not large enough to explain the current acceleration of the Universe. Another related possibility is to abandon the Copernican principle and assume that we are living in a special place in the Universe such as a centre of a large void. Although it is relatively easy to explain supernovae observations in this model, it is not easy to reconcile it with all other observations such as kinetic Sunyaev-Zel'dovich effects.[17] Note that even if inhomogeneity cannot explain the accelerated expansion, it can still play an important role in precise cosmological measurements.

The second possibility is that general relativity is modified on large scales. See reviews.[18–20] As we will see below, GR has been tested in the Solar System to high accuracy. It successfully predicts the orbital decay of the Hulse-Taylor binary pulsar.[21] Recently the prediction of gravitational waves has been confirmed by direct detections of gravitational waves from binary black holes[22] as well as a binary neutron star.[23] However, all these constraints on deviations from GR apply to scales that are vastly different from scales accessible by cosmological and astrophysical measurements. Thus it is still possible that large distance modifications of gravity can account for the accelerated expansion of the Universe.

Finally, the acceleration can be caused by unknown dark energy. The distinction between dark energy and modified gravity is generally ambiguous. Indeed, in the background, any modification of gravity or dark energy can be described as the existence of an effective energy density ρ_{dark} and pressure P_{dark}. The only parameters required to describe the background expansion are the equation of state $w_{\text{dark}}(z) = P_{\text{dark}}/\rho_{\text{dark}}$ and the present-day density parameter Ω_{dark}. The background expansion can be parametrised as

$$E(z) = \Omega_m(1+z)^3 + \Omega_r(1+z)^4 + \Omega_K(1+z)^2 + \Omega_{\text{dark}} \exp\left(3\int_0^z dz' \frac{1+w_{\text{dark}}(z')}{1+z'}\right).$$

(10)

The Planck 2018 gave a constraint on a constant equation of state as $w_{\text{dark}} = -1.028 \pm 0.032$ by combining CMB, BAO and SNe measurements. A time dependence of $w_{\text{dark}}(z)$ would be a clear evidence that the acceleration is not caused by the cosmological constant. A recent attempt to combine various data sets indicates that the evolving w_{dark} model is preferred at a 3.5σ significance level based on the improvement in the fit, although the Bayesian evidence for the dynamical w_{dark} model is insufficient to favour it over ΛCDM.[24]

If we consider the formation of structure, there is a difference between simple dark energy models and modified gravity models. Here simple dark energy models mean that dark energy does not cluster and it has no effects on structure formation other than through a modified expansion history. This is the case in quintessence models where dark energy is described by a scalar field with the standard kinetic term.[25, 26] In these models, the inhomogeneity of the scalar field can be ignored below horizon scales. In these models, the only parameters of the model are the equation of state w_{dark} and the density parameter Ω_{dark}. Thus structure formation is completely determined by the background evolution of the Universe, while in modified gravity models, this connection is generally lost as the force law is modified. Thus the combination of measurements of the background expansion and structure formation offers a way to distinguish between these two scenarios. See Ref. 27 for discussions on the distinction between dark energy and modified gravity models.

3. Modified Gravity Models

In this chapter, we mainly focus on modified gravity models. We will first discuss motivations to consider modifications to general relativity. Then we discuss various observational constraints as well as theoretical constraints that any modified gravity models need to satisfy.

3.1. *Motivations*

The discovery of cosmic acceleration has renewed interests to study theories of modified gravity. There are mainly four motivations to consider such theories.

(1) Quantum gravity

GR is not a renormalisable theory as is clear from the fact that the coupling constant G has a mass dimension of -2. Thus it is thought as an effective theory that is valid up to the Planck scale M_{pl} where quantum gravity effects become important. There are many attempts to construct quantum theory of gravity such as string theory but the complete quantum theory of gravity is still out of our reach. An important lesson that we have learnt is that the Planck scale M_{pl} is not necessary the scale at which GR is modified and the quantum gravity scale can be much lower. For example, there has been an attempt to construct a model where the fundamental scale of gravity is TeV to solve the hierarchy between the electroweak scale and the Planck scale.[28]

(2) The cosmological constant problem

One of the reasons to modify gravity is to explain why we do not experience the effect of large vacuum energy expected from the Standard Model of particle physics. One possible solution is that the vacuum energy does not gravitate. This is known as degravitation or self-tuning. There is a no-go theorem by Weinberg that forbids self-tuning solutions under several assumptions[14] and any attempts need to evade this no-go theorem. A simple example is a braneworld model in six dimensions (see Ref. 29 for a review). In this model, the standard model particles are confined to a four dimensional brane in six dimensions with two extra spatial dimensions. The cosmological constant in the four dimensional brane does not change the geometry of the four dimensional spacetime and it only curves the extra two dimensional space. The challenge is that it is difficult to modify gravity only for the cosmological constant while reproducing GR for normal matter. See Ref. 30 for a review and references therein.

(3) Cosmic acceleration

Modified gravity models provide a possibility to realise the accelerated expansion of the Universe without the cosmological constant. A simply example is provided by a five dimensional braneworld model proposed by Dvali, Gabadadze and Porrati (DGP).[31] In this model the Friedman equation on a four-dimensional brane is given by $H^2 = H/r_c + 8\pi G\rho/3$ where r_c is defined as the ratio between the four-dimensional and five-dimensional Newton constants. Even if the energy density ρ does not contain the cosmological constant, the expansion of the Universe accelerates as the Hubble parameter approaches constant, $H \to 1/r_c$, at late times. This is called self-acceleration. Unfortunately, this particular solution suffers from instabilities[32, 33] but the idea of self-acceleration has been studied intensively in many other modified gravity models.

(4) Tests of general relativity

General relativity has been tested exquisitely in the Solar System and by binary pulsars.[34] However, this does not imply that it is valid on all scales in any environments. It is worth pursuing the tests of GR on different scales and in different environments to understand its regime of validity. For example, the recent detections of gravitational waves have opened a new window to test gravity in strong gravity regime as well as to test the propagation of gravitational waves. In the next decade, a number of cosmological surveys aiming to reveal the nature of cosmic acceleration will produce measurements that can be used to test GR on cosmological scales. To satisfy stringent Solar System constraints, modified gravity theories often incorporate screening mechanisms to hide modifications of gravity in dense environments. These screening mechanisms provide novel ways to test deviations from GR. For example, screening mechanisms can introduce environment-dependent modifications of gravity, which can be tested using astrophysical measurements. Individual atoms inside a large enough high-vacuum chamber can be unscreened, giving a possibility to detect modifications of gravity in laboratory tests. These novel tests of gravity are one of the main focuses of this book.

3.2. *Observational constraints*

Modified gravity models need to satisfy various observational constraints. Here we list several important constraints that will be used later. See Ref. 34 for a comprehensive review.

(1) Solar system tests

The first conditions come from the Solar System tests. The Parametrised Post Newtonian (PPN) metric is given by

$$ds^2 = -\left(1 - 2U + 2\beta U^2\right) dt^2 + (1 - 2\gamma U)\, \delta_{ij} dx^i dx^j, \tag{11}$$

where $U = \int d^3x' \rho(x')/|\mathbf{x} - \mathbf{x}'|$. The time dilation due to the effect of the Sun's gravitational field was measured very accurately using the signal from Cassini satellite. This gives the constraint on γ as

$$\gamma - 1 < (2.1 \pm 2.3) \times 10^{-5}. \tag{12}$$

The constraint on β comes from the perihelion shift of Mercury. Assuming the Cassini bound, the constraint on β is given by $\beta - 1 = (-4.1 \pm 7.8) \times 10^{-5}$.

(2) Lunar Laser Ranging

The Lunar Laser Ranging (LLR) experiment gives various constraints.[35] It gives tight constraints on the deviations of gravitational potential from the GR prediction. The anomalous perihelion angular advance of the Moon is constrained as

$$|\delta\theta| = \left| \pi r \frac{d}{dr} \left[r^2 \frac{d}{dr} \left(\frac{\varepsilon}{r} \right) \right] \right| < 2.4 \times 10^{-11}, \quad \varepsilon = \frac{\delta \Psi}{\Psi}, \tag{13}$$

where ε is the radial dependent deviation of the gravitation potential $\Psi = -GM/r$. The LLR experiments also constrain the time variation of the Newton constant

$$\frac{\dot{G}}{G} = (2 \pm 7) \times 10^{-13} \text{ per year.} \tag{14}$$

Finally, the LLR provides precision tests of the weak equivalence principle. The difference between the accelerations of the Earth and the Moon is constrained as

$$\eta \equiv \frac{2|a_{\text{earth}} - a_{\text{moon}}|}{a_{\text{earth}} + a_{\text{moon}}} < 10^{-13}. \tag{15}$$

(3) Hulse-Taylor binary pulsar

The orbital decay of the Hulse-Taylor binary pulsar due to gravitational wave emissions is consistent with the GR prediction.[36] After correcting the effect of a relative acceleration between the binary pulsar system and the Solar System caused by the differential rotation of the galaxy, the observed rate of change of orbital period compared with the GR prediction is given by

$$\frac{\dot{P}}{\dot{P}_{\text{GR}}} = 0.997 \pm 0.002. \tag{16}$$

(4) Gravitational wave propagation

On 17 August 2017, gravitational waves from a neutron star merger were detected by LIGO.[23] Almost simultaneously, short gamma ray bursts were detected. The observed time delay was $(+1.74 \pm 0.05)$ s. This put stringent constraints on the difference between the speed of gravitational waves c_{GW} and the speed of light c as[2]

$$-3 \times 10^{-15} < \frac{c_{\text{GW}}}{c} - 1 < 7 \times 10^{-16}. \tag{17}$$

The lower bound was obtained assuming that the short gamma ray bursts signal was emitted 10 s after the GW signal. This constraint applies to the local Universe. The distance to the neutron star merger is estimated as $43.8^{+2.9}_{-6.9}$Mpc. In terms of redshift, this corresponds to $z \sim 0.01$. Note that there is also a lower bound $-2 \times 10^{-15} < c_{\text{GW}}/c - 1$ from the gravitational Cherenkov radiation emitted by high energy cosmic rays if the speed of gravitational waves is smaller than the speed of light.[37] This constraint has a significant implication for many modified gravity models[38–44] as we will see later.

3.3. *Theoretical conditions*

Another condition that modified gravity models need to satisfy is the theoretical consistency of the model. Modified gravity models often introduce additional degrees of freedom as we discuss below and these additional modes can lead to instabilities. To illustrate several types of instabilities, let us consider a simple scalar field described by the action

$$S = \int dt d^3 x (K_t \dot{\phi}^2 - K_x (\partial_i \phi)(\partial^i \phi) - m^2 \phi^2). \tag{18}$$

The tachyonic instability arises when the scalar field has a negative mass squared $m^2 < 0$. Whether this instability is catastrophic or not depends on the instability time scale determined by the mass given by $|m^2|^{-1/2}$. Another instability arises when the gradient term has a negative coefficient $K_x/K_t < 0$. In this case, the instability time scale is determined by the wave number of the mode, so the time scale becomes shorter on smaller scales. Finally, the ghost instability arises when the time kinetic term of the scalar field has a wrong sign $K_t < 0$. At the quantum level, the vacuum is unstable as negative energy particles can be created from vacuum and it decays instantaneously. To avoid the instability, it is required to introduce a non-Lorentz-invariant cut-off in the theory (see[45] for a review).

Another problem is known as the strong coupling problem, which arises from non-linear interaction terms. An example of the non-linear term that appears later is[32]

$$S_{\text{non-linear}} = -\int d^4 x \frac{1}{\Lambda_3^3} \Box \phi (\partial \phi)^2. \tag{19}$$

This non-linear interaction is suppressed by Λ_3. Quantum corrections generate other terms suppressed by Λ_3 such as $S \propto -(1/\Lambda_3^6) \int d^4 x \phi \Box^4 \phi$, and we lose control of the theory beyond Λ_3. This strong coupling scale is often associated with the energy scale related to the accelerated expansion of the Universe, H_0, which is extremely small compared with the scale of gravity M_{Pl}. Thus we will often find that the strong coupling scale is rather low in modified gravity models. This means that we need to treat these theories as an effective theory, which is valid only at energy scales lower than Λ_3. It is an interesting question whether there exists a standard (i.e. local, unitary, analytic and Lorentz-invariant) Ultra-Violet (UV) completion of these theories.[46] The condition to have a standard UV completion restricts the parameter space of effective theory (see for example, Ref. 47).

3.4. *Classifications*

There are numerous modified gravity models proposed in the literature. To classify these models in a general framework, Lovelock's theorem plays an important role. Lovelock's theorem proves that Einstein's equations are the only *second-order local* equations of motion for a *single metric* derivable from the *covariant action* in *four-dimensional* spacetime. This indicates that if we modify GR we need to violate one or more of the assumptions in Lovelock theorem.

(1) Higher derivatives

We normally require that the equations of motion contain up to second time derivatives. The Ostrogradsky theorem states that higher order time derivatives introduce additional degrees of freedom, which makes the Hamiltonian unbounded from below. This can be shown by introducing a new variable for a higher derivative term. This theorem applies if all the conjugate momenta including those associated with higher derivatives can be expressed in terms of velocities and variables. This is called the non-degeneracy condition. If this non-degeneracy condition is not satisfied, it is possible to have higher derivative theories without introducing instabilities.[48] A typical example is $f(R)$ gravity models, which contain fourth order time derivatives of metric in the equations of motion but do not propagate the Ostrogradsky ghost. Another example is degenerate higher order scalar tensor theories.[49] We will review these theories below.

(2) Non-locality

An example of non-local gravity is to modify Einstein-Hilbert action to $Rf(\Box^{-1}R)$ where $\Box^{-1}$ is the inverse Laplacian operator.[50, 51] The main advantage of this theory is that $\Box^{-1}R$ has no dimension so in principle it is possible to avoid the introduction of the fine-tuned mass dimension. Another example is $m^2R\Box^{-2}R$.[52] The meaning of the non-local operator $\Box^{-1}$ needs to be defined carefully. In the equations of motion, these operators are understood as the retarded Green function in order to preserve causality. This cannot be obtained from a classical action, thus these actions should be considered as a quantum effective action. These theories can be localised by introducing scalar fields. In fact, the $m^2R\Box^{-2}R$ theory can be written as a bi-scalar tensor theory. One of the scalar field has a ghost-like kinetic term but it is argued that there is no quanta associated with this ghost field. See Ref. 53 for details.

(3) Higher dimensional spacetime

Another radical idea is to consider higher dimensional spacetime. There are stringent constraints on the existence of extra-dimensions for standard model particle interactions. These constraints are normally avoided by compactifying the extra-dimensions. In this case, modifications of gravity only appears on length scales shorter than the size of extra-dimensions. Another possibility is to consider a brane on which standard model particles are confined. We will review a braneworld model that admits long distance modifications of gravity.

(4) Lorentz violation

A typical example of Lorentz non-invariant theory is Einstein-Aether theory. In this theory, a vacuum expectation value of a vector field breaks Lorentz invariance. Lorentz non-invariant theory plays a role in the construction of renormalisable theory of gravity. By admitting higher spatial derivatives, it is possible to improve the UV behaviour of gravity as in Horava gravity. These theories are tightly constrained by various tests of gravity, most notably the constraints on the deviation of the gravitational wave speed from the speed of light. See Refs. 54 and 55 for details.

(5) Extra degrees of freedom

In additional to metric, additional degrees of freedom such as scalar fields, vector fields or tensor fields can be introduced. Scalar tensor theories are one of the most well studied modified gravity models where a scalar field is non-minimally coupled to gravity. Other well studied theories include massive gravity and bi-gravity theories. In massive gravity, a fiducial metric is introduced while in bi-gravity models, an additional dynamical metric is introduced. See Ref. 56 for a review.

4. Screening Mechanisms

In the previous section, we classify various modified gravity models. Many of these models introduce an additional scalar degree of freedom. For example, five dimensional gravitons and massive gravitons in four-dimensions both have five degrees of freedom. One of them is a helicity-0 mode, which behaves as a scalar field on small scales.

If this scalar degree of freedom couples to matter, it is subject to stringent constraints from the Solar System tests. Screening mechanisms provide a way to evade this condition and they are included in many well-studied modified gravity models. See Refs. 19 and 20 for reviews and references therein.

4.1. *Motivations*

In order to understand the need for screening mechanisms, let us consider Brans-Dicke gravity described by the action

$$S = \frac{1}{16\pi G} \int d^4x \sqrt{-g} \left(\psi R - \frac{\omega_{\rm BD}}{\psi} (\nabla \psi)^2 \right) + S_m[g_{\mu\nu}]. \tag{20}$$

We are interested in forces generated by a non-relativistic source $T_0^0 = -\rho$. Using the quasi-static approximations to neglect time derivatives, the perturbations of the

Minkowski metric

$$ds^2 = -(1 + 2\Psi)dt^2 + (1 - 2\Phi)\delta_{ij}dx^i dx^j, \tag{21}$$

and the scalar field perturbation $\psi = 1 + \varphi$ obey the following equations

$$\nabla^2 \Psi = 4\pi G\rho - \frac{1}{2}\nabla^2 \varphi, \tag{22}$$

$$(3 + 2\omega_{\mathrm{BD}})\nabla^2 \varphi = -8\pi G\rho, \tag{23}$$

$$\Phi - \Psi = \varphi. \tag{24}$$

From these equations, we observe that the scalar field perturbation φ gives an additional contribution to the Poisson equation. This introduces an additional force, often called the fifth force. The scalar field perturbation also changes the relation between the two metric perturbations Φ and Ψ. The solutions for the metric perturbations are given by

$$\nabla^2 \Psi = 4\pi G\mu\rho, \quad \Psi = \gamma^{-1}\Phi, \tag{25}$$

where

$$\mu = \frac{4 + 2\omega_{\mathrm{BD}}}{3 + 2\omega_{\mathrm{BD}}}, \quad \gamma = \frac{1 + \omega_{\mathrm{BD}}}{2 + \omega_{\mathrm{BD}}}. \tag{26}$$

We recover GR in the large ω_{BD} limit. Indeed, imposing the Solar System constraints (12), we obtain $|\gamma - 1| = (2.1 \pm 2.3) \times 10^{-5}$. Then the constraint on ω_{BD} is given by $\omega_{\mathrm{BD}} > 40,000$.[34] Once we impose this constraint on the model parameter ω_{BD}, the deviations from GR is suppressed, i.e. $|\mu - 1| < 10^{-5}$ and $|\gamma - 1| < 10^{-5}$, on all scales.

4.2. *Classifications*

In order to avoid this blanket suppression of deviations from GR on all scales, the Brans-Dicke theory needs to be extended to include non-linear functions of the scale field. Schematically, the generalised action is given by

$$S = \frac{1}{16\pi G}\int d^4x \sqrt{-g}\left(\psi R - \frac{\omega(\psi)}{\psi}(\nabla\phi)^2 + \mathcal{K}\left[(\nabla\psi)^2, (\nabla^2\psi)\right] - 2U(\psi)\right) + S_m[g_{\mu\nu}], \tag{27}$$

where $\mathcal{K}$ is a function of first and second derivatives of the scalar field. This function needs to be chosen carefully not to introduce the Ostrogradsky ghost. We will give examples of these functions later.

The screening mechanisms can be classified formally depending on which non-linear function in Eq. (27) is used to suppress the fifth force.

(1) Chameleon mechanism[57,58]

In this mechanism, the potential $U(\psi)$ is introduced. The scalar field takes a background value ψ_{BG} determined by the background[a] density. The potential introduces a mass term for the scalar field perturbation around this background

$$(3 + 2\omega_{\mathrm{BD}})\nabla^2\varphi + m(\psi_{\mathrm{BG}})^2\varphi = -8\pi G\rho, \qquad (28)$$

where the mass depends on the background field. The scalar field does not propagate beyond the Compton wavelength m^{-1} as the solution decays as $\exp(-mr)$ where r is the distance from the source. Since the mass depends on the background density, it is possible to realise a situation where the mass is large in dense environments such as the Solar System while it is small on a cosmological background, modifying gravity non-trivially.

(2) Dilaton[59] and Symmetron mechanism[60]

The coupling of the scalar field to matter is determined by the function $\omega(\psi)$, which is determined by the background scalar field ψ_{BG} thus depends on a background density. It is possible to consider a model where $\omega(\psi_{\mathrm{BG}})$ is large in dense environments so that the scalar field decouples from matter while ω is $\mathcal{O}(1)$ on cosmological background.

(3) K-mouflage[61] and Vainshtein mechanism[62]

In this case, the non-linear kinetic term $\mathcal{K}\left[(\nabla\psi)^2, (\nabla^2\psi)\right]$ plays a role to effectively suppress the coupling between the scalar field and matter. K-mouflage mechanism uses the non-linearity of the first derivative of the scalar field while the Vainshtein mechanism relies on the non-linearity of the second derivative of the scalar field in the action.

These screening mechanisms can be classified in another way depending on how it operates to suppress the fifth force for a spherically symmetric object with a mass M and size R.[19]

(1) Thin-shell screening

In this category of models, screening is determined by the gravitational potential of the object $|\Psi| = GM/R$. The condition for screening is given by the thin-shell condition,[57,58] which requires that the gravitational potential is larger than a critical value set by the background scalar field ψ_{BG}, $|\Psi| > \chi_{\mathrm{BG}}(\psi_{\mathrm{BG}})$i, where χ_{BG} is defined later. Chameleon, Dilaton and Symmetron screening belong to this type of screening mechanism.[63]

(2) K-mouflage mechanism

In this case, screening is determined by the first derivative of the gravitational potential $|\partial\Psi| \sim GM/R^2$. The screening happens if this exceeds a critical mass scale Λ_c set by a model parameter, $|\partial\Psi| > \Lambda_c$.

(3) Vainshtein mechanism

Screening is determined by the second derivative of the gravitational potential $|\partial^2\Psi| \sim GM/R^3$. This is the three-dimensional curvature generated by the

---

[a]Here 'background' means some environment where the system under consideration resides, and not necessarily the cosmological background.

object. Screening operates if the curvature of the object exceeds a critical mass scale squared set by a model parameter, $|\partial^2 \Psi| > \Lambda_c^2$.

4.3. *Frame transformation*

The screening mechanisms that utilise non-linear functions of the scalar field, $\omega(\psi)$ and $U(\psi)$, are described by generalised Brans-Dicke theory described by the action

$$S = \frac{1}{16\pi G} \int d^4 x \sqrt{-g} \left(\psi R - \frac{\omega(\psi)}{\psi}(\partial\psi)^2 - 2U(\psi) \right) + S_m[g_{\mu\nu}]. \tag{29}$$

Chameleon, dilaton and symmetron mechanisms can be implemented in this theory. This action is formulated in the Jordan frame where matter is coupled to metric $g_{\mu\nu}$ minimally. In order to understand the dynamics of the scalar field, it is useful to define the Einstein frame in which the scalar field is minimally coupled to gravity instead. This can be achieved by a conformal transformation

$$g_{\mu\nu}^E = \psi g_{\mu\nu}. \tag{30}$$

By this transformation, the action becomes

$$S^E = \int d^4 x \sqrt{-g^E} \left[\frac{1}{16\pi G} R^E - \frac{1}{2}(\nabla\phi)^2 - V(\phi) \right] + S_m[A^2(\phi)g_{\mu\nu}^E], \tag{31}$$

where

$$\left(\frac{d\phi}{d\psi} \right)^2 = \frac{1}{16\pi G} \frac{3+2\omega}{\psi^2}, \quad A(\phi) = \psi^{-1/2}, \quad V(\phi) = \frac{U(\psi)}{8\pi G\psi^2}. \tag{32}$$

In this frame, matter is coupled to the scalar field through the effective metric $A^2(\phi)g_{\mu\nu}^E$. This changes the geodesic equation.

Let us consider the motion of non-relativistic particles. These particles follow the geodesic of the effective metric $g_{\mu\nu} = A^2(\phi)g_{\mu\nu}^E$. We consider the weak field limit and expand the scalar filed as $\phi = \bar{\phi} + \delta\phi$ where $A(\bar{\phi}) = 1$. By defining the metric perturbations in the Einstein frame

$$ds_E^2 = -(1 + 2\Psi^E)dt^2 + (1 - 2\Phi^E)\delta_{ij}dx^i dx^j, \tag{33}$$

the geodesic equation is given by

$$\ddot{\vec{x}} = -\vec{\nabla}\Psi^E - \frac{\beta}{M_{\rm pl}}\vec{\nabla}\delta\phi, \quad \beta \equiv M_{\rm pl}\frac{d\ln A}{d\phi} = \sqrt{\frac{1}{2(3+2\omega)}}, \tag{34}$$

where Ψ^E satisfies the standard Poisson equation

$$\nabla^2\Psi^E = 4\pi G\rho^E, \tag{35}$$

where ρ^E is the energy density in the Einstein frame. In the weak field limit, this coincides with the energy density in the Jordan frame since $A(\bar{\phi}) = 1$. See Ref. 64 for discussions on various definitions of the density. In the following we replace ρ^E

by ρ. On the other hand, in the Jordan frame, the geodesic equation is unmodified but the Poisson equation is modified (see Eq. (24))

$$\ddot{\vec{x}} = -\vec{\nabla}\Psi, \quad \nabla^2\Psi = 4\pi G\rho - \frac{1}{2}\nabla^2\varphi. \tag{36}$$

By using the relation between the gravitational potential in the Jordan frame Ψ and that in the Einstein frame Ψ^E, $\vec{\nabla}\Psi = \vec{\nabla}\Psi^E + (\beta/M_{\rm pl})\vec{\nabla}\phi$, and the relation between φ and ϕ, $\vec{\nabla}\phi = -(M_{\rm pl}/2\beta)\vec{\nabla}\varphi$, we can easily check that the trajectories of non-relativistic particles in the weak field limit are the same in the Jordan and Einstein frame. In general, it is possible to show that observational predictions are exactly the same in both frames (see Ref. 65 and references therein).

The dynamics of the scalar field is easier to analyse in the Einstein frame. In the following, we will omit the superscript E unless there is an ambiguity. The equation of motion for the scalar field is given by

$$\Box\phi = \frac{dV}{d\phi} + \frac{\beta(\phi)\rho}{M_{\rm pl}} \equiv \frac{dV_{\rm eff}}{d\phi}, \tag{37}$$

where we consider non-relativistic matter and the effective potential is defined as

$$V_{\rm eff}(\phi) = V(\phi) + \rho\ln A(\phi). \tag{38}$$

We see that the function β determines the coupling between the scalar field and matter.

So far we have assumed the weak equivalence principle, i.e. the scalar field is coupled universally to all matter species. It is possible to break this principle and couple the scalar field differently to different species:

$$S_m = S_{\rm cdm}[A^2_{\rm cdm}g^E_{\mu\nu}] + S_{\rm baryon}[A^2_{\rm baryon}g^E_{\mu\nu}] + \dots \tag{39}$$

The stringent constraints coming from the Solar System tests can be avoided by assuming $\beta_{\rm baryon} = 0$. These models are often called interacting dark energy models[66] if we identify the scalar field as dark energy. In this case, dark energy is interacting with dark matter but not baryons, and there is no need to introduce screening mechanisms.

4.4. *Thin shell screening*

In this section, we review screening mechanisms in which screening is determined by the gravitational potential $|\Psi|$. These include chameleon, symmetron and dilaton mechanisms. Chameleons screen the fifth force by having a large mass of the scalar field in dense environments while symmetrons and dilatons screen the fifth force by having a small coupling to matter in dense environments. Despite this difference, the condition under which screening operates is given by the same thin shell condition as we will see below. A comprehensive review on chameleon mechanisms can be found in Ref. 64.

4.4.1. Chameleon mechanism

A typical choice of $V(\phi)$ and $A(\phi)$ to realise the chameleon mechanism is

$$V(\phi) = V_0 + \frac{\Lambda_{\text{ch}}^{\ell+4}}{\phi^\ell}, \quad A(\phi) = \exp\left(\frac{\beta\phi}{M_{\text{pl}}}\right), \tag{40}$$

where V_0 is a bare cosmological constant, Λ_{ch} is a parameter of the potential with mass dimension one and ℓ is a dimensionless parameter. The minimum of the effective potential depends on the density as

$$\phi_{\text{BG}} = \left(\frac{\ell M_{\text{pl}} \Lambda_{\text{ch}}^{\ell+4}}{\beta\rho}\right)^{\frac{1}{\ell+1}}. \tag{41}$$

Around this minimum, the scalar field acquires a mass

$$m(\phi_{\text{BG}}) = \frac{\mathrm{d}^2 V_{\text{eff}}(\phi_{\text{BG}})}{\mathrm{d}\phi^2} = \ell(\ell+1)\Lambda_{\text{ch}}^{\ell+4}\left(\frac{\beta\rho}{\ell M_{\text{pl}} \Lambda_{\text{ch}}^{\ell+4}}\right)^{\frac{\ell+2}{\ell+1}}. \tag{42}$$

If $\ell > -1$, in high density backgrounds, ϕ_{BG} becomes small and $m(\phi_{\text{BG}})$ becomes large, satisfying the condition to realise the chameleon mechanism. Figure 1 shows the behaviour of the effective potential.

In order to understand how the chameleon mechanism operates, we consider a spherical object with density ρ_s of size R embedded in a constant density $\rho_{\text{BG}} \ll \rho_s$. At infinity, we set the value of the scalar field at the minimum of the potential,

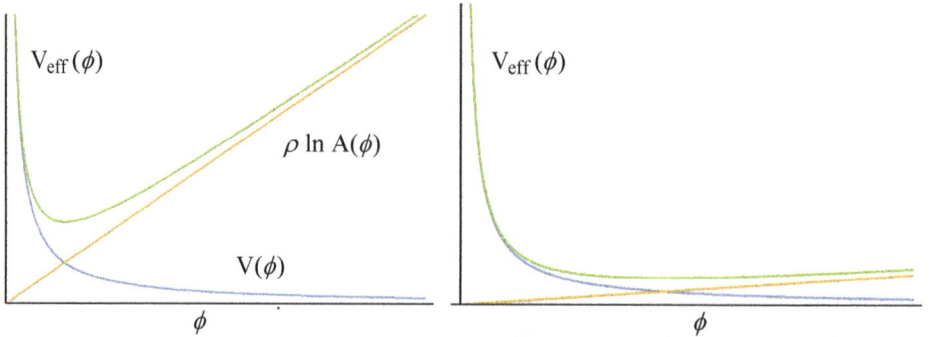

Fig. 1: An example of the potential $V(\phi)$, the contribution from matter $\rho \ln A(\phi)$ and the effective potential $V_{\text{eff}}(\phi)$ in chameleon models. The left panel shows the situation in high density environment while the right panel shows the low density case.

$\phi = \phi_{\text{BG}}$ set by ρ_{BG}. Away from the source, perturbations around the background $\phi = \phi_{\text{BG}} + \varphi$ satisfies

$$\frac{1}{r^2}\frac{d}{dr}\left(r^2\frac{d\varphi}{dr}\right) = m(\phi_{\text{BG}})^2\varphi. \tag{43}$$

The solution for this equation is given by $\varphi(r) = -(C/r)\exp[-m(\phi_{\text{BG}})r]$ where C is an integration constant. On the other hand, inside the source, the scalar field is trapped at the minimum of effective potential $\phi = \phi_s$ determined by ρ_s between $r = 0$ and $r = r_{\text{scr}}$. Between r_{scr} and R, the scalar field rolls from ϕ_s toward ϕ_{BG}. By matching these solutions, we can determine the integration constants C and r_{scr}. The solution outside the source $r > R$ is given by[58]

$$\frac{\phi(r)}{M_{\text{pl}}} = -\left(\frac{3\Delta R}{R}\right)\frac{2GM\beta}{r}e^{-m(\phi_{\text{BG}})r} + \frac{\phi_{\text{BG}}}{M_{\text{pl}}}, \tag{44}$$

where

$$\frac{3\Delta R}{R} \equiv \frac{3(R - r_{\text{scr}})}{R} = \frac{\phi_{\text{BG}} - \phi_s}{2\beta M_{\text{pl}}|\Psi|}, \tag{45}$$

where $|\Psi| = GM/R$. This solution exists only when the thin shell condition $3\Delta R/R \ll 1$ is satisfied. If this condition is not satisfied, the scalar field does not reach ϕ_s inside the source and the solution is given by Eq. (44) with $3\Delta R/R = 1$.

This result can be understood as follows. Due to the large mass of the scalar field, it is trapped at the minimum of the potential between $r = 0$ and $r = r_{\text{scr}}$ and the matter inside this region does not generate a scalar force if the thin shell condition is satisfied. Only the mass in a shell between $r = r_{\text{scr}}$ and $r = R$

$$\frac{M - M(r_{\text{scr}})}{M} \sim 3\frac{\Delta R}{R} \tag{46}$$

generate the scalar force where $M(r_{\text{scr}})$ is an enclosed mass in the region $0 < r < r_{\text{scr}}$ (see Fig. 2). This is the reason why r_{scr} is called the screening radius.

The thin shell condition can be expressed in terms of the screening parameter

$$\chi_{\text{BG}} \equiv \frac{\phi_{\text{BG}} - \phi_s}{2\beta M_{\text{pl}}} \approx \frac{\phi_{\text{BG}}}{2\beta M_{\text{pl}}}, \tag{47}$$

as $\chi_{\text{BG}} < |\Psi|$ given that $\phi_s \ll \phi_{\text{BG}}$. Once this condition is satisfied, the scalar field generated by an object with mass M outside the object is given by

$$\frac{\phi(r)}{M_{\text{pl}}} = -\left(\frac{\chi_{\text{BG}}}{|\Psi|}\right)\frac{2GM\beta}{r}e^{-m(\phi_{\text{BG}})r} + \frac{\phi_{\text{BG}}}{M_{\text{pl}}}. \tag{48}$$

Recalling that the geodesic equation for a non-relativistic test particle is given by

$$\ddot{\vec{x}} = -\vec{\nabla}\Psi - \frac{\beta}{M_{\text{pl}}}\vec{\nabla}\phi, \tag{49}$$

we define the normal Newtonian force and the fifth force as

$$\vec{F}_N = -\vec{\nabla}\Psi, \quad \vec{F}_\phi = -\frac{\beta}{M_{\text{pl}}}\vec{\nabla}\phi. \tag{50}$$

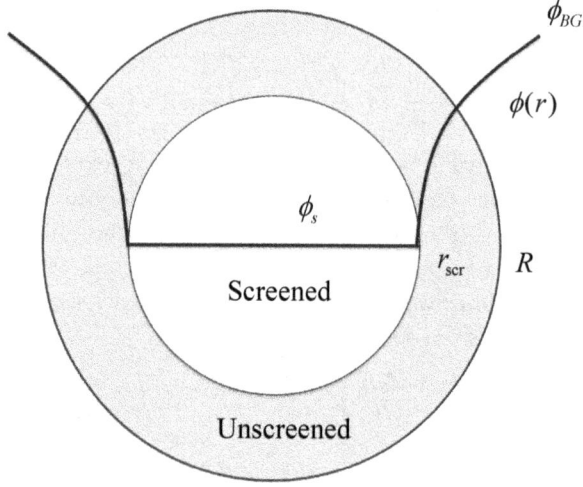

Fig. 2: This figure shows how the thin shell screening works for an object with size R. The scalar field is trapped at the minimum of the effective potential determined by the density of the object, ϕ_s, which is much smaller than the background scalar filed, ϕ_{BG}. The mass inside $r = r_{scr}$ does not produce the scalar force and only the shell between $r = r_{scr}$ and $r = R$ contribute to the fifth force, suppressing the ratio of the scalar to Newtonian force.

Using the solution for the scalar field, the ratio of the scalar to Newtonian force outside a source (but within the Compton wavelength) is given by

$$\frac{F_\phi}{F_N} = 2\beta^2 \min\left\{\left(\frac{\chi_{BG}}{|\Psi|}\right), 1\right\}. \tag{51}$$

4.4.2. *Symmetron/dilaton mechanism*

A typical choice of $A(\phi)$ and $V(\phi)$ for the symmetron mechanism is given by

$$V(\phi) = V_0 - \frac{1}{2}\mu^2\phi^2 + \frac{\lambda}{4}\phi^4, \quad A(\phi) = 1 + \frac{\phi^2}{2M_s^2}, \tag{52}$$

where λ is a dimensionless parameter, V_0 is the bare cosmological constant and μ, M_s are model parameters of mass dimension one, so that the effective potential becomes

$$V_{\text{eff}} = -\frac{1}{2}\mu^2\left(1 - \frac{\rho}{\mu^2 M_s^2}\right)\phi^2 + \frac{\lambda}{4}\phi^4. \tag{53}$$

It is easy to see that there is a critical density $\rho_* = \mu^2 M_s^2$. For high densities, $\rho > \rho_*$, the symmetry is restored and the scalar field stays at the minimum $\phi_{\min} = 0$ (see Fig. 3). At this minimum, $\beta(\phi_{\min} = 0) = 0$ and the scalar field does not couple

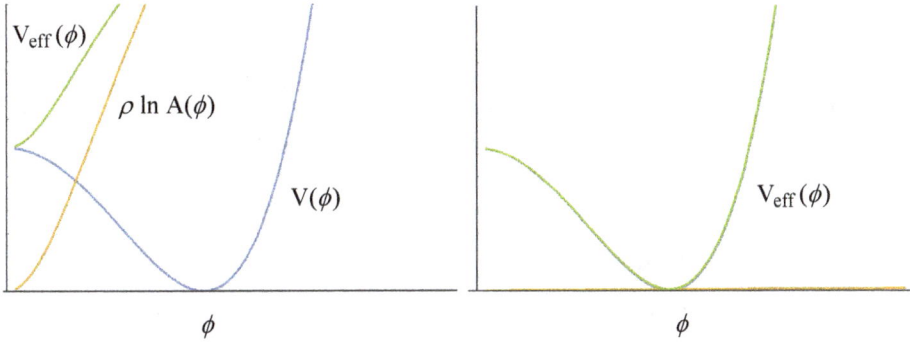

Fig. 3: An example of the potential $V(\phi)$, the contribution from matter $\rho \ln A(\phi)$ and the effective potential $V_{\text{eff}}(\phi)$ in symmetron models. The left panel shows the case with high density, in which the symmetry is restored by the matter contribution and the minimum is given by $\phi_{\min} = 0$. The right panel shows the case with low density.

to matter. On the other hand, for low densities, $\rho < \rho_*$, the scalar field evolves to the minimum

$$\phi_{\min} \sim \pm \frac{\mu}{\sqrt{\lambda}}. \tag{54}$$

Around this minimum, the coupling to matter is given by

$$\beta(\phi_{\min}) = \left| \frac{M_{\text{pl}}\phi_{\min}}{M_s^2} \right| \sim \frac{\mu M_{\text{pl}}}{\sqrt{\lambda}M_s^2}. \tag{55}$$

Another related mechanism is dilaton. A typical choice of $V(\phi)$ and $A(\phi)$ is

$$V(\phi) = V_0 \exp\left(-\frac{\lambda\phi}{M_{\text{pl}}}\right), \quad A(\phi) = 1 + \frac{A_2}{2M_{\text{pl}}^2}\phi^2, \tag{56}$$

where λ and $A_2 \ll 1$ are dimensionless model parameters. The minimum of the effective potential is given by

$$\phi_{\min} = \frac{\lambda V_0 M_{\text{pl}}}{A_2 \rho}, \tag{57}$$

where we assumed $\lambda\phi/M_{\text{pl}} \ll 1$. At the minimum, the coupling constant is given by $\beta = A_2\phi/M_{\text{pl}} = \lambda(V_0/\rho)$. As in the symmetron case, for high densities, the minimum of the potential is driven to $\phi_{\min} = 0$ thus the coupling of the scalar to matter vanishes, $\beta = 0$, while the coupling can be $O(1)$ when the density becomes low compared to λV_0.

Although the mechanism to suppress the fifth force is different from chameleons, it was shown that the screening condition in these mechanisms is similar.[63] If we consider a spherically symmetric object, the ratio of the scalar to Newtonian force is given in terms of the screening parameter χ_{BG} Eq. (47) as in chameleons, where β is given by $\beta_{BG} \equiv \beta(\phi_{BG})$.

In the case of symmetrons, domain walls can be formed after the phase transition. This requires the study of full time dependent dynamics of the scalar field. This was performed with N-body simulations and it was shown that domain walls do form but their effects on the distribution of matter are small.[67]

4.4.3. *Thin shell screening*

In all above screening mechanisms, the thin shell condition can be expressed in terms of the *self-screening* parameter

$$\chi_{BG} \equiv \frac{\phi_{BG}}{2\beta_{BG} M_{pl}}, \tag{58}$$

where $\beta_{BG} = \beta(\phi_{BG})$. The thin shell condition is simply given by $\chi_{BG} < |\Psi|$. This is called self-screening because this condition only takes into account the screening of the scalar field by an isolated object. The ratio of the scalar to Newtonian force outside a source acting on a non-relativistic particle is given by

$$\frac{F_\phi}{F_N} = 2\beta_{BG}^2 \min\left\{\left(\frac{\chi_{BG}}{|\Psi|}\right), 1\right\}. \tag{59}$$

The above formula applies to the force acting on a test particle. The attraction due to the scalar field between two objects separated by distance r is in general given by

$$F_{1,2} = 2Q_1 Q_2 \frac{GM_1 M_2}{r^2}, \tag{60}$$

where

$$Q_i = \beta_{BG} \min\left\{\left(\frac{\chi_{BG}}{|\Psi|_i}\right), 1\right\}, \tag{61}$$

and $|\Psi|_i$ is the gravitational potential of the object i.

Table 1[64] summarises the gravitational potential of several classes of astrophysical objects, which are useful to estimate the self-screening of these objects.

It was found that screening also depends on environments. For example, if an object with a deep potential exists nearby, this can screen an object that does not satisfy the self-screening condition. To identify the environmental screening, the full partial differential equation needs to be solved with a given distribution of matter. This was done in the context of N-body simulations.[68] It was found that environmental screening can be estimated by computing the gravitational potential of nearby objects within the Compton wavelength

$$\Psi_{ext} = \sum_{d_i < m(\phi_{BG})^{-1} + R_i} \frac{GM_i}{d_i}, \tag{62}$$

Table 1: Objects commonly considered as probes of chameleon/symmetron/dilaton screening. The second column shows surface Newtonian potential $|\Psi| \equiv GM/R$. Here v_c is the circular velocity, $v_c^2 = GM/R^2$. From Ref. 64.

Object	Newtonian Potential $|\Psi|$
Earth	10^{-9}
Moon	10^{-11}
Main-sequence star (Sun-like)	10^{-6}
Post-main-sequence star ($\mathcal{M} = 1\text{--}10 M_\odot$, $R = 10\text{--}100 R_\odot$)	$10^{-7}\text{--}10^{-8}$
Spiral galaxy (Milky Way-like, $v_c \sim 200$ km/s)	10^{-6}
Dwarf galaxy ($v_c \sim 50$ km/s)	10^{-8}

where d_i is the distance to the neighbouring object with mass M_i and radius R_i. The environmental screening condition is given by $\chi_{\text{BG}} < \Psi_{\text{ext}}$. This condition was verified by studying screening of dark matter halos in N-body simulations and applied to SDSS galaxies within 200Mpc to create a screening map, which identifies the regions where environmental screening does not apply.[69] A more sophisticated method has been developed to include the effect of missing mass and it was concluded that it is in principle possible to test the thin shell condition to the level of $\chi_{\text{BG}} \sim 10^{-7}$.[70]

4.4.4. *Constraints*

There are various constraints on thin shell screened models.

(1) Cassini experiment

The constraint from the Cassini experiment, Eq. (12), imposes the condition that the modification of the unscreened Cassini satellite in the vicinity of the Sun should be suppressed as

$$\beta_{\text{galaxy}}^2 \left(\frac{\chi_{\text{galaxy}}}{|\Psi_{\text{sun}}|} \right) < 10^{-5}, \tag{63}$$

where the background is assumed to be determined by the Milky Way galaxy.

(2) Lunar Laser Ranging experiments

There is a strong constraint on the difference between the acceleration of the Earth and the Moon due to the Sun, Eq. (15). If the Earth, Moon and Sun are thin-shell screened, the accelerations are given by

$$a_{\text{earth}} = a_N(1 + 2Q_{\text{earth}}Q_{\text{sun}}), \quad a_{\text{moon}} = a_N(1 + 2Q_{\text{moon}}Q_{\text{sun}}). \tag{64}$$

Using the gravitation potential of the Earth, Moon and Sun in Table 1, the constraint can be written as

$$\beta_{\text{galaxy}}^2 \left(\frac{\chi_{\text{galaxy}}}{|\Psi_{\text{earth}}|} \right)^2 < 10^{-14}, \tag{65}$$

where again we assumed that the background is determined by the Milky Way galaxy.

(3) Milky Way galaxy constraint

The above constraints assume that the scalar field is at the minimum of its effective potential in the Milky Way galaxy. To ensure that the Milky Way galaxy is screened we need to impose the condition

$$\beta_{\text{cosmology}}^2 \left(\frac{\chi_{\text{cosmology}}}{|\Psi_{\text{galaxy}}|} \right) < 1, \tag{66}$$

where the background is assumed to be determined by the cosmological background.

In addition, there are constraints from laboratory experiments and astrophysics tests,[64] which will be discussed in details in the last two chapters of this book.

Note that the Solar System constraints and the LLR constraint impose the condition on the scalar field with the galaxy density. To convert these constraints to those on model parameters, we need to specify the potential and coupling function. The galaxy density is given by $\rho_{\text{galaxy}} \sim 10^5 \rho_{\text{cosmology}}$. The background scalar field scales as $\phi_{\text{BG}} \propto \rho^{-1/(\ell+1)}$ (see Eq. (41)). If $1 \ll 1/(\ell+1)$, the condition for the Milky Way galaxy screening (66) typically gives the strongest constraint on $\phi_{\text{cosmology}}$ as ϕ_{BG} is highly suppressed in denser environments.[71] This is the case in $f(R)$ gravity models discussed in Sec. 5.1 where $2 \leq 1/(\ell+1)$.[72]

4.5. K-mouflage mechanism

This mechanism can be described by the action

$$S = \int d^4x \sqrt{-g} \left(\frac{1}{16\pi G} R + \Lambda_2^4 K \left(\frac{X}{\Lambda_2^4} \right) \right) + S_m[A(\phi)^2 g_{\mu\nu}], \quad X = -\frac{1}{2}\partial^\mu \phi \partial_\mu \phi. \tag{67}$$

The equation of motion for the scalar field is given by

$$\frac{1}{\sqrt{-g}}\partial_\mu \left(\sqrt{-g}\partial^\mu \phi K' \right) = \frac{d\ln A}{d\phi}\rho. \tag{68}$$

We consider perturbations around a background $\phi = \phi_{\text{BG}} + \varphi(x)$ and focus on quasi-static perturbations with a non-relativistic source. The equation of motion for φ can be integrated once[73]

$$K'\nabla\varphi = 2\beta M_{\text{pl}}\nabla\Psi, \tag{69}$$

where the gravitational potential satisfies the normal Poisson equation

$$\nabla^2\Psi = 4\pi G\delta\rho, \tag{70}$$

and

$$K' \equiv \frac{dK(\chi)}{d\chi}, \quad \chi = -\frac{1}{2\Lambda_2^4}(\nabla\phi)^2. \tag{71}$$

Note that in principle the energy density fluctuation $\delta\rho$ contains the energy density of the scalar field itself but this is sub-dominant compared with the matter density and we will ignore it. For a spherically symmetric object with mass M, we can define the K-mouflage radius as

$$r_K = \left(\frac{2\beta G M M_{\rm pl}}{\Lambda_2^2}\right)^{1/2}. \tag{72}$$

For $r < r_K$, $|\chi|$ becomes large and if $K'(\chi)$ becomes large, the fifth force is suppressed:

$$\frac{F_\phi}{F_N} = \frac{2\beta^2}{K'}. \tag{73}$$

The condition for the screening, $r < r_K$, can be written as

$$\frac{GM}{r^2} > \Lambda_c \equiv \frac{\Lambda_2^2}{2\beta M_{\rm pl}}. \tag{74}$$

Thus in this screening mechanism, screening is determined by the first derivative of the gravitational potential.

4.5.1. *Constraints*

The ratio of the scalar to Newtonian force is constrained by the Cassini experiment, Eq. (12)

$$\frac{F_\phi}{F_N} = \frac{2\beta^2}{K'} < 10^{-5}, \tag{75}$$

at distances of order one AU from the Sun. Assuming Λ_2 is associated with dark energy energy scale, i.e. $\Lambda^2/M_{\rm pl} \sim H_0$, the K-mouflage radius r_K is estimated as[73]

$$r_K \sim 3500 \text{ AU} \sqrt{\frac{\beta M}{M_{\rm sun}}}. \tag{76}$$

The correction to the gravitational potential is the radial dependent and this introduces an anomalous perihelion for the Moon orbiting around the Earth given by Eq. (13) where the ratio between the scalar and Newtonian potential is given by

$$\varepsilon = \frac{\beta\varphi}{M_{\rm pl}\Psi}. \tag{77}$$

Using the equation of motion, this can be written as[73]

$$\delta\theta = -8\pi \frac{\beta^2}{K'} \frac{\chi K''}{K'} \frac{1}{c_s^2}, \quad c_s^2 = \frac{K' + 2\chi K''}{K'}. \tag{78}$$

The Earth-Moon system gives a constraint $|\delta\theta| < 2.4 \times 10^{-11}$.

4.6. Vainshtein mechanism

The Vainshtein mechanism is the oldest idea of the screening mechanism. This mechanism was identified in the context of massive gravity. Massive graviton has five degrees of freedom. In the linear theory, the helicity-0 mode does not decouple even in the massless limit. This is known as the van Dam-Veltman-Zakharov discontinuity.[74, 75] This problem is cured by non-linear interactions of the helicity-0 mode, which decouple this mode. This is known as the Vainshtein mechanism. The massive gravity models considered by Vainshtein contain a Boulware-Deser ghost[76] but the same mechanism applies to a braneworld model[31] and also a ghost-free massive gravity model.[77]

4.6.1. Vainshtein mechanism

The simplest model to describe the Vainshtein mechanism is given by the action

$$S = \int d^4x \sqrt{-g} \left(\frac{1}{16\pi G} R - \frac{1}{2}(\partial\phi)^2 - \frac{1}{2\Lambda_3^3}\Box\phi(\partial\phi)^2 \right) + S_m[A(\phi)^2 g_{\mu\nu}], \quad (79)$$

where Λ_3 is a model parameter of mass dimension one. We consider static perturbations around the Minkowski background for simplicity. The equation of motion for the scalar field in the quasi-static limit is

$$\nabla^2\phi + \frac{1}{\Lambda_3^3}\left[(\nabla^2\phi)^2 - (\nabla_i\nabla_j\phi)(\nabla^i\nabla^j\phi)\right] = \frac{\beta\rho}{M_{\rm pl}}. \quad (80)$$

Again we consider a spherically symmetric source and compute quasi-static perturbations. Outside the matter source, the solution is obtained as

$$\frac{1}{M_{\rm pl}}\frac{d\phi}{dr} = \frac{\beta GM}{r^2}\left(\frac{r}{r_V}\right)^3 \left(\sqrt{1 + 4\left(\frac{r_V}{r}\right)^3} - 1\right), \quad (81)$$

where the Vainshtein radius is given by

$$r_V = \left(\frac{4\beta GMM_{\rm pl}}{\Lambda_3^3}\right)^{1/3}. \quad (82)$$

At large distance, we recover the linear solution. On the other hand, inside the Vainshtein radius $r \ll r_V$, we obtain

$$\frac{1}{M_{\rm pl}}\frac{d\varphi}{dr} = \frac{2\beta GM}{r^2}\left(\frac{r}{r_V}\right)^{3/2}, \quad (83)$$

and the ratio of the scalar to Newtonian force is given by

$$\frac{F_\phi}{F_N} = 2\beta^2\left(\frac{r}{r_V}\right)^{3/2}. \quad (84)$$

The condition for the screening $r < r_V$ can be written as

$$\frac{GM}{r^3} > \Lambda_c^2 \equiv \frac{\Lambda_3^3}{4\beta M_{\rm pl}}. \quad (85)$$

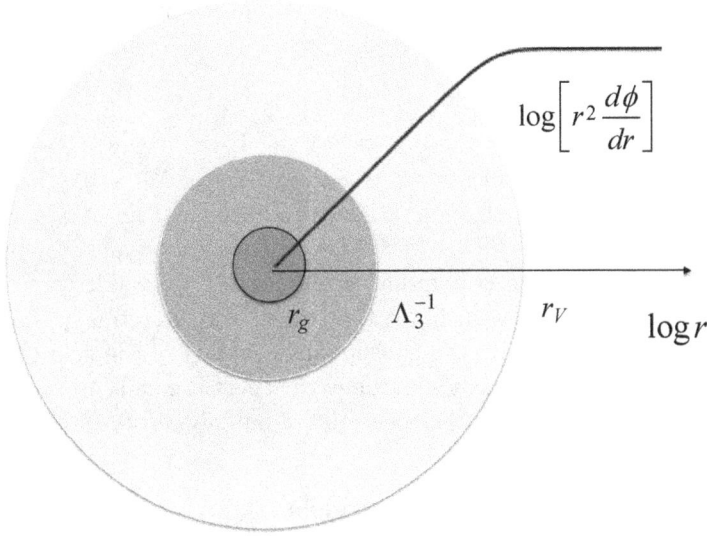

Fig. 4: The scalar field profile for a spherically symmetric solution with the Vainshtein mechanism. $r_g = 2GM$ is the Schwarzchild radius.

Table 2: The radii R (physical size), $r_g = 2GM$ (Schwarzschild radius) and r_V (Vainshtein radius) of some typical objects. Unit is meter. For estimation we have used $\beta = 2\sqrt{2}/3$ and $r_c = H_0^{-1} \sim 4000\text{Mpc} \sim 1.2 \times 10^{26}\text{m}$.

Object	R	r_g	r_V
Milky Way	$\sim 0.9 \times 10^{21}$	$\sim 2 \times 10^{15}$	$\sim 3 \times 10^{22}$
Sun	$\sim 0.7 \times 10^{9}$	$\sim 3 \times 10^{3}$	$\sim 3.5 \times 10^{18}$
Earth	$\sim 6 \times 10^{6}$	$\sim 9 \times 10^{-3}$	$\sim 5 \times 10^{16}$
Atom	$\sim 5 \times 10^{-11}$	$\sim 1.8 \times 10^{-54}$	$\sim 3 \times 10^{-1}$

The length scale associated with Λ_c, $r_c \equiv \Lambda_c^{-1}$, is typically assumed to be $r_c \sim H_0^{-1}$. Table 2 summarises the Vainshtein radius and the Schwarzschild radius $r_g = 2GM$ for several objects with this choice of the parameter.[78]

4.6.2. *Constraints*

The Vainshtein radius r_V is much larger than the Schwarzschild radius for $r_c \sim H_0$. For the Sun, r_V is significantly larger than the size of the Solar System. Thus inside the Solar System, the fifth fore contribution is well approximated by Eq. (83). The ratio between the scalar and gravitation potential is given by

$$\varepsilon = -4\beta^2 \left(\frac{r}{r_V} \right)^{3/2}. \tag{86}$$

Note that the scalar field is normalised so that it vanishes at infinity. The perihelion precession per orbit, Eq. (13), is then calculated as[79]

$$\delta\theta = \frac{3}{4}\pi\varepsilon. \tag{87}$$

Lunar Laser Ranging experiments put constraints $|\delta\theta| < 2.4 \times 10^{-11}$. Using $r_g = 2GM = 0.886$ cm for the Earth and $r = 3.84 \times 10^{10}$ cm for the Earth-Moon distance we obtain $r_c > 350$ Mpc for $\beta \sim O(1)$.[80] These results are obtained by treating the Moon as a test body. This assumption is not necessarily valid. For example for the Earth-Moon system, it was shown that there is a correction to the universal precession rate Eq. (87) due to the nonsuperimposability of the field that depends on the mass ratio of the two bodies although the effect is small (4%) for the Earth-Moon system.[81] See Ref. 82 for the constraints from laboratory tests.

5. Examples of Modified Gravity Models

In this section, we will review several modified gravity models focusing on those that will be used in later chapters of this book.

5.1. $f(R)$ gravity

One of the most popular modified gravity models is $f(R)$ gravity where the Einstein-Hilbert action is generalised to a nonlinear function of the Ricci curvature scalar[83]

$$S = \frac{1}{16\pi G}\int d^4x\sqrt{-g}F(R) + \int d^4x\sqrt{-g}\mathcal{L}_m. \tag{88}$$

See Refs. 84–86 for reviews. The modified Einstein equations are given by

$$F_R R_{\mu\nu} - \frac{1}{2}g_{\mu\nu}F(R) + (g_{\mu\nu}\Box - \nabla_\mu\nabla_\nu)F_R = 8\pi G T_{\mu\nu}. \tag{89}$$

These equations contain up to fourth order derivatives in terms of $g_{\mu\nu}$. However, this theory is an example where the non-degeneracy condition for the Ostrogradsky theorem is not satisfied. Indeed, it is possible to introduce a scalar field and make the equation of motion second order. The action is equivalent to

$$S = \frac{1}{16\pi G}\int d^4x\sqrt{-g}\Big(F(\phi) + (R-\phi)F'(\phi)\Big). \tag{90}$$

By taking a variation with respect to ϕ, we obtain $(R-\phi)F''(\phi) = 0$. As long as $F''(\phi) \neq 0$, $R = \phi$ and we recover the original action. By defining $\psi = F'(\phi)$ and $2U = -F(\phi) + \phi F'(\phi)$, the action can be written as the same form as Brans-Dicke gravity with a potential

$$S = \frac{1}{16\pi G}\int d^4x\sqrt{-g}\Big(\psi R - 2U(\psi)\Big). \tag{91}$$

Comparing this with the action Eq. (20), we notice that the BD parameter is given by $\omega_{BD} = 0$. As such, if we ignore the potential, this model is already excluded by the Solar System constraints. However, by choosing the potential, i.e. the form of the $F(R)$ function appropriately, it is possible to incorporate a chameleon mechanism to evade the Solar System constraint.[72, 87] It is often more convenient to separate the Einstein-Hilbert action in $F(R)$ so we define

$$F(R) = R + f(R). \tag{92}$$

The freedom to choose the function $f(R)$ leads to many models but the successful models for the late time cosmology share the same features.[72, 88, 89] In the high curvature limit, the function looks like

$$f(R) = -2\Lambda + |f_{R0}| \frac{\bar{R}_0^{n+1}}{R^n}, \tag{93}$$

where $\bar{R}_0$ is the background curvature today. This model requires an effective cosmological constant to explain the observed accelerated expansion of the Universe.

The correction to ΛCDM disappears in the high curvature limit $R \gg \bar{R}$. This suggests the existence of a screening mechanism in this model. In the Einstein frame, the potential and the coupling function take the following form

$$V(\phi) = V_0 - \Lambda^{\frac{3n+4}{1+n}} \phi^{\frac{n}{1+n}}, \quad A(\phi)^2 = e^{\sqrt{2/3}\phi/M_{Pl}}, \tag{94}$$

and the scalar field is related to $f_R \equiv df/dR$ as

$$\phi = -\sqrt{\frac{3}{2}} M_{Pl} \log(1 + f_R). \tag{95}$$

This is a particular case of the chameleon theories where the coupling constant is given by $\beta = d\ln A/d\phi = \sqrt{1/6}$ and the potential given by Eq. (40) with $\ell = -n/(1+n)$.

Under quasi-static approximations, the modified Poisson equation and the perturbations of f_R is given by[72]

$$\nabla^2 \delta f_R = \frac{a^2}{3} [\delta R(f_R) - 8\pi G \delta \rho], \tag{96}$$

$$\nabla^2 \Psi = \frac{16\pi G}{3} a^2 \delta \rho - \frac{a^2}{6} \delta R(f_R), \tag{97}$$

where $\delta f_R = f_R(R) - f_R(\bar{R}), \delta R = R - \bar{R}$ and $\delta \rho = \rho - \bar{\rho}$. If we linearise the equation, $\delta R(f_R)$ is given by

$$\delta R(f_R) = 3m^2 \delta f_R, \quad m^2 = \frac{1}{3f_{RR}}. \tag{98}$$

Note that in order to avoid tachyonic instabilities, we need to impose $f_{RR} > 0$. The solution for the gravitational potential is given by[90]

$$k^2 \Psi = -4\pi G \left(\frac{4 + 3a^2 m^2/k^2}{3 + 3a^2 m^2/k^2} \right) a^2 \delta \rho, \tag{99}$$

where we performed the Fourie transformation and k is the comoving wavenumber. On scales larger than the Compton wavelength $k/a < m$, the scalar field does not propagate and we recover the standard solution. On the other hand, on scales smaller than the Compton wavelength $k/a > m$, gravity is enhanced by the fifth force and the effective Newton constant becomes $4G/3$ as expected from the fact that $f(R)$ gravity has $\omega_{BD} = 0$ (see (26)). For a model with $n = 1$, the scalar field mass is given by the background Ricci curvature as

$$m^2 = \frac{1}{6|f_{R0}|} \frac{\bar{R}^3}{\bar{R}_0^2}.$$
(100)

At early times, the background Ricci curvature is large and the mass is large. Thus deviations from ΛCDM are suppressed. At late times, the Compton wavelength becomes large, modifying gravity on cosmological and astrophysical scales. The compton wavelength today is given by

$$m_0^{-1} = 32\sqrt{\frac{|f_{R0}|}{10^{-4}}} \text{ Mpc.}$$
(101)

For non-linear structures, it is not possible to linearise the equation and we need to solve fully non-linear equations. This requires significant efforts in particular in the context of N-body simulations. This will be discussed in Chapter 3. Note that the self-screening condition is given in this model by

$$\chi_{BG} = \frac{3}{2}|f_R(\bar{R})| < |\Psi|.$$
(102)

The constraint that the Milky Way galaxy is screened, Eq. (66), imposes the condition $|f_{R0}| < 10^{-6}$. This constraint implies that $m_0^{-1} < 3$Mpc. Thus the scalar field dynamics is not important cosmologically and the background expansion history is observationally indistinguishable from that of ΛCDM.[87] This is a general conclusion in models with the chameleon mechanism.[91]

For a general choice of $f(R)$, it is possible to mimic the ΛCDM expansion history or any expansion history by tuning the function $f(R)$ provided that R does not have a minimum in the past (see Ref. 92 for discussions). This is called designer models.[93] In the designer models, linear cosmological perturbations are characterised by the Compton wavelength parameter

$$B(a) = \frac{f_{RR}(\bar{R})}{1 + f_R(\bar{R})} \bar{R}' \frac{H}{H'},$$
(103)

where $\bar{R}$ is the background Ricci curvature, H is the Hubble parameter and the prime denotes the derivative with respect to the scale factor.

Finally, the speed of gravitational waves is the same as that of light in $f(R)$ gravity models, thus the constraint Eq. (17) does not impose any condition on the model.

5.2. *Braneworld gravity*

The idea of braneworld was inspired by the discovery of *D-branes* in string theory and subsequent developments of large extra-dimensions (see Ref. 28 for a review). In this model, we are living on a four-dimensional membrane (brane) in a higher dimensional spacetime (bulk). The standard model particles are confined to the brane but gravity can propagate in the bulk. In these models, it is possible to have a situation where higher dimensional gravity becomes important on large scales. The simplest model is the Dvali-Gabadadze-Porrati (DGP) model described by the action[31]

$$S = \frac{1}{16\pi G_{(5)}} \int d^5x \sqrt{-^{(5)}g}\,^{(5)}R + \int d^4x \sqrt{-g}\left(\frac{1}{16\pi G}R + \mathcal{L}_m\right), \tag{104}$$

where $^{(5)}R$ and $G_{(5)}$ are five-dimensional Ricci curvature and Newton constant, respectively. The ratio between the five-dimensional and four-dimensional Newton constant $r_c = G_{(5)}/2G$ is a parameter of the model and called the cross-over radius. The Friedman equation on the brane is given by[94]

$$H^2 = \pm\frac{H}{r_c} + \frac{8\pi G}{3}\rho. \tag{105}$$

At early times $Hr_c \gg 1$, we recover the usual 4D Friedmann equation. On the other hand, at late times, the Hubble parameter approaches a constant $H \to 1/r_c$ in the upper branch of the solution. This is known as the self-accelerating branch. On the other hand, the lower branch solution, the normal branch solution, requires the cosmological constant or dark energy to realise the accelerated expansion of the Universe. In order to recover the standard cosmology at early times, the cross-over radius needs to be tuned as $r_c \sim H_0^{-1}$.

The perturbations around this background under the quasi-static approximations are described by the Brans-Dicke theory with an additional non-linear interaction term[79, 95, 96]

$$\nabla^2\Psi = 4\pi Ga^2\delta\rho - \frac{1}{2}\nabla^2\varphi, \tag{106}$$

$$(3 + 2\omega_{\mathrm{BD}}(a))\nabla^2\varphi + \frac{r_c^2}{a^2}\left[(\nabla^2\varphi)^2 - (\nabla_i\nabla_j\varphi)^2\right] = -8\pi Ga^2\delta\rho, \tag{107}$$

$$\Phi - \Psi = \varphi, \tag{108}$$

where the BD parameter $\omega_{\mathrm{BD}}(a)$ is given by

$$\omega_{\mathrm{BD}} = \frac{3}{2}(\beta(a) - 1), \quad \beta(a) = 1 \mp 2r_cH\left(1 + \frac{\dot{H}}{3H^2}\right), \tag{109}$$

where "dot" indicates a derivative with respect to time.

Note that the Brans-Dicke parameter is always negative in the self-accelerated branch and the scalar field mediates a repulsive force. This is the manifestation of the problem that the self-accelerating solutions suffer from a ghost instability.[32, 33, 97–99] In addition, at late times $\omega_{\mathrm{BD}} \sim O(1)$ thus by linearising the equation, we again find that this theory would be excluded by the Solar System constraints. However, the coefficient of the non-linear interaction term is very large, $r_c^2 \sim H_0^{-2}$. This is responsible for the Vainshtein mechanism.[100] Indeed, the equation (107) has the same form as Eq. (80) so the constraint from the LLR imples $r_c > $ a few hundreds Mpc.

In the normal branch solution, it is possible to tune the equation of state of dark energy to realise the ΛCDM background expansion history.[101] This is not particularly well-motivated theoretically. However, this model makes it easier to study the effect of the Vainshtein mechanism on structure formation compared with ΛCDM and thus it is frequently used in N-body simulations and phenomenological studies.

On large scales $r > r_c$, gravitational waves propagate in the bulk while light is confined to the brane. Thus there can be a time delay between the arrival of gravitational waves and light as gravitational waves can make a short-cut in the bulk.[102] However, the event GW170817 happens at the distance $\sim$ 40Mpc from us and the frequencies of the gravitational waves are $10 - 100$Hz. Due to the short-wavelength of the gravitational waves ($10^{-16} - 10^{-15}$ Mpc), the time delay will be well suppressed. For example, for a single massive graviton, the constraint Eq. (17) gives only a weak constraint on the graviton mass $m < 10^{-22}$ eV ($m^{-1} > 10^{-7}$Mpc). Thus once we impose the LLR constraint, the constraint Eq. (17) is well satisfied in this model.

5.3. *Horndeski theory and beyond*

Scalar-tensor theories of gravity represent the simplest modification in terms of additional degrees of freedom compared to general relativity, i.e. a single scalar field. In four-dimensional space-time, the most general scalar-tensor theory with second order equations of motion was derived by Horndeski in 1974,[103] and later rediscovered in the context of the so-called (covariant) galileon theories.[104–109] See Ref. 110 for a review. The Horndeski action is given by

$$\mathcal{L}_2 = K(\phi, X), \tag{110}$$

$$\mathcal{L}_3 = G_3(\phi, X)\Box\phi, \tag{111}$$

$$\mathcal{L}_4 = G_4(\phi, X)R - 2G_{4X}(\phi, X)\left[(\Box\phi)^2 - (\nabla_\mu\nabla_\nu\phi)^2\right], \tag{112}$$

$$\mathcal{L}_5 = G_5(\phi, X)G^{\mu\nu}\nabla_\mu\nabla_\nu\phi$$
$$+ \frac{1}{3}G_{5X}(\phi, X)\left[(\Box\phi)^3 - 3\Box\phi(\nabla_\mu\nabla_\nu\phi)^2 + 2(\nabla_\mu\nabla_\nu\phi)^3\right], \tag{113}$$

where X in the above equations is defined as $X \equiv \partial^\mu\phi\partial_\mu\phi$ and K, G_3, G_4 and G_5 are free function of the scalar field and X.

This theory encompasses many models including $f(R)$ gravity and screened models discussed above. These theories have $G_4 = G_4(\phi)$ and $G_5 = 0$. The extension of these theories to non-trivial $G_4(\phi, X)$ and $G_5(\phi, X)$ has a significant implication for the propagation of gravitational waves on the cosmological background. The propagation of gravitational waves

$$ds^2 = -dt^2 + a^2(\delta_{ij} + h_{ij})dx^i dx^j, \quad h_i^i = \partial_i h_j^i = 0, \tag{114}$$

is described by the following equation[111]

$$\ddot{h}_{ij} + (3 + \alpha_M)H\dot{h}_{ij} + c_{GW}^2 \frac{\nabla^2}{a^2} h_{ij} = 0, \tag{115}$$

where the tensor sound speed is given by

$$c_{GW}^2 = 1 + \alpha_T \tag{116}$$

and

$$M_*^2 = 2\left[G_4 - 2XG_{4X} - \frac{1}{2}XG_{5\phi} - \dot\phi HXG_{5X}\right], \tag{117}$$

$$HM_*^2\alpha_M = \frac{dM_*^2}{dt}, \tag{118}$$

$$M_*^2\alpha_T = 2X[2G_{4X} + G_{5\phi} - (\ddot\phi - H\dot\phi)G_{5X}]. \tag{119}$$

The constraint on the speed of gravitational waves (17) gives a strong constraint $|\alpha_T| < 10^{-15}$. If we do not invoke a fine tuning and set $c_{GW} = c$, this forces us to choose

$$G_4 = G_4(\phi), \quad G_5 = 0. \tag{120}$$

The screening mechanisms introduced earlier, i.e. chameleon, symmetron and dilaton models, satisfy these conditions so they are not affected by this constraint.

A theory with non-zero $K(\phi, X)$ and $G_3(\phi, X)$ are called Kinetic Gravity Braiding (KGB) model.[112] One interesting feature of this model is that it is possible to explain the accelerated expansion of the Universe without introducing a potential. A simple example is given by[113]

$$K(X) = \frac{X}{2}, \quad G_3(X) = -M_{Pl}\left(-\frac{r_c^2}{2M_{Pl}^2}X\right)^n. \tag{121}$$

Note that the sign for the kinetic term is opposite to the normal scalar field. However, this does not imply that the perturbations are ghostly when expanding around the non-trivial cosmological background. The strong observational constraints in this model come from the Integrated Sachs-Wolfe (ISW) effects. The cross-correlation between ISW and galaxy clustering has a different sign compared with the ΛCDM model for samll n, which is disfavoured. This gives a condition on n to be larger than a few hundreds.[114, 115]

The requirement of second order equations of motion is sufficient to avoid the Ostrogradsky ghost, but this condition is not necessary.[116, 116–119] In recent years, there have been several attempts to construct healthy theories that relax this condition, exploiting transformations of the metric.[120, 121] This brought initially to the class of beyond Horndeski theories.[121, 122] A breakthrough in the subject came with the works of Refs. 49, 123, which developed a general method to identify the degeneracy conditions that remove the Ostrogradsky ghost despite the appearance of higher derivatives in the equation of motion. A larger class of new degenerate higher order scalar-tensor theories propagating up to 3 degrees of freedom was identified and classified up to cubic order in the second order derivative of the scalar field.[124–126]

Again the constraint on c_{GW} has significant implications to this extended class of scalar tensor theories. If we impose $c_{GW} = c$, we are only left with the following possibility in addition to $K(\phi, X)$ and $G_3(\phi, X)$:[127]

$$\mathcal{L}_4 = \sum_{i=1}^{5} \mathcal{L}_i + \mathcal{L}_R, \tag{122}$$

where

$$\mathcal{L}_1[A_1] = A_1(\phi, X)\phi_{\mu\nu}\phi^{\mu\nu}, \qquad \mathcal{L}_2[A_2] = A_2(\phi, X)(\Box\phi)^2, \tag{123}$$

$$\mathcal{L}_3[A_3] = A_3(\phi, X)(\Box\phi)\phi^\mu\phi_{\mu\nu}\phi^\nu, \qquad \mathcal{L}_4[A_4] = A_4(\phi, X)\phi^\mu\phi_{\mu\rho}\phi^{\rho\nu}\phi_\nu, \tag{124}$$

$$\mathcal{L}_5[A_5] = A_5(\phi, X)(\phi^\mu\phi_{\mu\nu}\phi^\nu)^2, \qquad \mathcal{L}_R[G_4] = G_4(\phi, X)R, \tag{125}$$

and the functions A_I satisfy

$$A_1 = 0 \tag{126}$$

$$A_2 = 0, \qquad A_5 = \frac{A_3}{2G_4}(4G_{4X} + A_3 X), \tag{127}$$

$$A_4 = -\frac{1}{8G}\left[8A_3 G_4 - 48G_{4X}^2 - 8A_3 G_{4X} X + A_3^2 X^2\right],$$

This theory has an interesting implication for the Vainshtein mechanism. We introduce a mass dimension Λ_3 and assume the following scaling for the functions G_4, A_3, A_4 and A_5

$$G_4 \sim M_{pl}^2, \quad X A_3 \sim X A_4 \sim X^2 A_5 \sim M_{pl}\Lambda_3^{-3}, \tag{128}$$

where M_{pl} is the Planck mass and $X \sim M_{pl}\Lambda_3^3$. There appears the Vainshtein radius $r_V = (M/M_{pl}\Lambda_3^3)^{1/3}$ below which the scalar field perturbation becomes non-linear for a spherically symmetric object with mass M. The spherically symmetric solution for the metric perturbations below the Vainshtein radius is given by[128–130]

$$\Psi' = \frac{G_N M}{r^2} + \frac{\Upsilon_1 G_N}{4}M'',$$

$$\Phi' = \frac{G_N M}{r^2} - \frac{5\Upsilon_2 G_N}{4r}M' + \Upsilon_3 G_N M'', \tag{129}$$

where

$$\Upsilon_1 = -\frac{(4G_{4X} - X A_3)^2}{4A_3 G_4},$$

$$\Upsilon_2 = \frac{8G_{4X} X}{5G_4}, \tag{130}$$

$$\Upsilon_3 = -\frac{-16G_{4X}^2 + A_3^2 X^2}{16A_3 G_4},$$

$$G_N = \left[8\pi \left(2G_4 - 2XG_{4X} - 3A_3X^2/2\right)\right]^{-1}. \tag{131}$$

Outside the matter source, M is constant and we recover GR solutions inside the Vainshtein radius. On the other hand, inside a source, gravity is modified so the Vainshtein mechanism is broken. This gives an interesting possibility to test these theories using astrophysical objects.[131] Note that there is a strong constraint from the Hulse-Taylor binary pulsars. The gravitational constant for tensor perturbations G_{GW} is given by G_4, which is different from G_N. This changes the prediction of the orbital decay.[132] From the constraint (16), we obtain the constraint[133]

$$-7.5 \times 10^{-3} < \frac{G_{GW}}{G_N} - 1 < 2.5 \times 10^{-3}, \tag{132}$$

where

$$\frac{G_{GW}}{G_N} - 1 = \frac{2XG_{4X}}{G_4} + \frac{3A_3X^2}{2G_4}. \tag{133}$$

The modification to the gravitational potential Ψ is constrained as

$$-2/3 < \Upsilon_1 < 1.6 \tag{134}$$

from the stellar structure.[134–136]

In addition, these theories admit the self-accelerating solutions where the accelerated expansion can be realised without introducing the cosmological constant or the potential of the scalar field.[137–139]

As discussed in Sec. 3.3, there is a strong coupling problem in these theories due to the presence of the energy scale Λ_3, which is typically given by $\Lambda_3 = (H_0^2 M_{pl})^{1/3}$ to explain the late time acceleration of the Universe. This energy scale is given by

$$\Lambda_3 \sim 10^{-22} \text{ GeV} \sim (10^3 \text{ km})^{-1} \sim 250\text{Hz}. \tag{135}$$

Incidentally, this scale is close to the frequencies of gravitational waves detected by LIGO.[140] This implies that quantum effects cannot be ignored in these theories when discussing gravitational waves detected by LIGO. It was pointed out that gravitational waves decay into the scalar field if $A_3 \neq 0$ at the time scale of Λ_3^{-1}, which effectively imposes the condition $A_3 = 0$ as the gravitational waves would not be detected by LIGO.[141,b] On the other hand, new operators naturally appear near the strong coupling scales and the UV completion of the theory may alleviate some of the constraints obtained from gravitational waves detected by LIGO.[140]

6. Conclusion

This chapter gave an introduction to theories of modified gravity with a particular focus on screening mechanisms. Several examples of modified gravity models were discussed that accommodate the screening mechanisms. These models can evade the

---

[b]This conclusion holds only when the sound speed of the scalar field perturbation is less than one.

stringent constraints from the Solar System tests thanks to the screening mechanism. In addition, they can avoid the stringent constraint on the gravitational wave speed obtained from the almost simultaneous detections of gravitational waves and short gamma ray bursts made on 17 August 2017. Thus these models serve as a test bed to constrain deviations from general relativity using laboratory experiments, astrophysical objects and cosmological observations. The other chapters in this book will discuss methodologies to perform these tests, show the latest results from these novel tests of GR and discuss future prospects.

Acknowledgments

KK is supported by the STFC grant ST/N000668/1. The work of KK has received funding from the European Research Council (ERC) under the European Union's Horizon 2020 research and innovation programme (grant agreement 646702 "CosTesGrav").

Chapter 2

Parametrizations for Tests of Gravity

Lucas Lombriser

Département de Physique Théorique, Université de Genève, 24 quai Ansermet,
CH-1211 Genève 4, Switzerland

Institute for Astronomy, University of Edinburgh, Royal Observatory,
Blackford Hill, Edinburgh, EH9 3HJ, UK
lucas.lombriser@unige.ch

With the increasing wealth of high-quality astronomical and cosmological data
and the manifold departures from General Relativity in principle conceivable,
the development of generalized parameterization frameworks that unify gravitational models and cover a wide range of length scales and a variety of observational probes to enable systematic high-precision tests of gravity has been a
stimulus for intensive research. A review is presented here for some of the formalisms devised for this purpose, covering the cosmological large- and small-
scale structure, the astronomical static weak-field regime as well as emission and
propagation effects for gravitational waves. This includes linear and nonlinear
parametrized post-Friedmannian frameworks, effective field theory approaches,
the parametrized post-Newtonian expansion, the parametrized post-Einsteinian
formalism as well as an inspiral-merger-ringdown waveform model among others.
Connections between the different formalisms are highlighted where they have
been established and a brief outlook is provided for general steps towards a unified global framework for tests of gravity and dark sector models.

1. Introduction

With the great volume of modified gravity, dark energy, and dark sector interaction
models that have been proposed as extensions to General Relativity (GR) and the
Standard Model,[18–20, 26, 27, 142, 143] the development of a more systematic approach
to explore their astronomical and cosmological implications has been and contin-
ues to be of prominent interest. The parametrized post-Newtonian (PPN)[144, 145]
expansion fulfills this purpose for astrophysical phenomena amenable to a low-
energy static approximation, and stringent model-independent constraints have
been inferred using this formalism, particularly from employing observations in
the Solar System.[34] The PPN expansion, however, is not straightforwardly applicable to gravitational modifications endowed with screening mechanisms[19, 27] and
corresponding generalizations have only been developed more recently.[146, 147]

More crucially, the formalism is not suitable for evolving backgrounds such as encountered in cosmology. Inspired by the successes of PPN, the development of an equivalent counterpart formalism on cosmological scales has therefore been the objective of intensive research. The frameworks devised for this purpose are generally referred to as parametrized post-Friedmannian (PPF) formalisms,[148–152] a term that has been coined by Ref. 153, and particularly adopted for the formalisms of Refs. 152 and 154. They aim at providing a consistent generalized description for the gravitational effects in the cosmological background and formation of structure. At the level of linear cosmological perturbations, this comprises both purely phenomenological parametrizations[148–152, 155, 156] as well as theoretically more constrained effective field theory (EFT)[111, 157–164] approaches.

Eventually linear theory fails in describing the cosmological structure below a few tens of Mpc. Together with the screening effects, this severely complicates performing generalized and consistent parametrized tests of gravity. However, a number of approaches have been developed to access the nonlinear scales, including generalized cosmological perturbation theory,[63, 165–172] interpolation functions bridging the modified and screened regimes,[152, 173] parametrizations of screening mechanisms combined with nonlinear structure-formation models,[174, 175] or a unified parametrization of the Lagrangian for a class of modified gravity models.[63]

Further approaches to extending the PPN formalism to both linear and nonlinear cosmological scales have been pursued with patching together post-Newtonian expansions in small regions of spacetime to a parametrized post-Newtonian cosmology (PPNC)[176] or by performing a post-Friedmannian expansion of the cosmological metric in powers of the inverse light speed in vacuum,[177] which however remains to be extended for parametrized gravitational modifications.

With the dawn of gravitational wave astronomy heralded by the LIGO and Virgo detections,[22, 178] powerful new tests of gravity are being facilitated. Effects of modifying gravity can both manifest in the emitted waveforms by modifications at the source[179–184] but also in the observed waveforms entering through effects on the propagation of the wave.[38, 185] Cosmological propagation effects are well described by the PPF or EFT formalisms whereas source modifications can be captured by a parametrized post-Einsteinian (ppE)[179] framework and the generalized inspiral-merger-ringdown waveform model.[180] The gravitational effects in the different formalisms can also be unified,[186] where however an incorporation of screening mechanisms remains to be elaborated.

This chapter reviews the different frameworks that have been developed to parametrize gravitational effects for different astrophysical probes and scales and discusses connections between the formalisms where they have been elaborated, also providing an outlook for how such connections may be established where they are currently missing. Section 2 first reviews the PPF framework. It addresses the evolution of the cosmological background in Sec. 2.1 with parametrizations of dark energy[187, 188] and a classification of genuine self-acceleration from modifying gravity.[38, 39] Section 2.2 discusses parametrization frameworks on linear scales, ranging from the modified large-scale structure with the growth-index parametriza-tion,[155, 156] closure relations for modified linear perturbations,[148–152] and EFT to the modified gravitational wave propagation and the choice of time and scale dependence in the parametrized modifications. Parametrizations of modified gravity effects on the nonlinear cosmological structure are reviewed in Sec. 2.3, separating the weekly and deeply nonlinear regimes. Section 3 is devoted to the PPN formal-ism, briefly reviewing the post-Newtonian expansion in Sec. 3.1, the parametrization of modified gravity effects on this expansion in Sec. 3.2, and the incorporation of screening mechanisms in Sec. 3.3. The parametrization of gravitational waveforms is addressed in Sec. 3.4 and a brief overview of further parametrizations for tests of gravity is given in Sec. 4. Finally, a summary and a brief outlook for the field is presented in Sec. 5.

The following conventions will be adopted: the metric signature is $-+++$, primes denote derivatives with respect to $\ln a$ unless indicated otherwise, scalar perturba-tions of the Friedmann-Lemaître-Robertson-Walker (FLRW) metric are defined in Newtonian gauge as $\Psi \equiv \delta g_{00}/(2g_{00})$ and $\Phi \equiv \delta g_{ii}/(2g_{ii})$, bold mathematical sym-bols denote spatial vectors, and the speed of light in vacuum is set to unity unless otherwise specified.

2. Parametrized Post-Friedmannian Frameworks

In order to enable generalized analyses and observational tests of the cosmological effects of modified gravity and dark energy, extensive efforts have been invested in the development of a cosmological counterpart formalism to the static PPN expan-sion. The different PPF approaches aim at consistently unifying the description of generalized modified gravity and dark energy effects in the cosmological background evolution as well as the linear and nonlinear cosmological structure formation. A further connection to the cosmological effects on the propagation of gravitational waves has also become desirable with the increasing amount of gravitational wave data.

The background evolution in the PPF formalism is addressed first in Sec. 2.1. Section 2.2 then discusses the linear perturbation regime, both for large-scale struc-ture and gravitational wave propagation. Finally, recent progress towards an exten-sion of the linear PPF formalism to the nonlinear regime of cosmological structure formation is reviewed in Sec. 2.3.

2.1. *Background cosmology*

Modifications of gravity or dark energy models that differ substantially from a cos-mological constant can affect the geometry of spacetime in an observable manner. At the background level the two extensions to standard cosmology can however not be distinguished in general. They therefore share a parametrization for their effects on the expansion history, often characterized by the dark energy equation of state $w(t)$. Section 2.1.1 briefly discusses the most common parametrization for $w(t)$.

Much of the motivation for modifications of gravity on cosmological scales has been drawn from offering an alternative explanation to dark energy or the cosmological constant as the driver for the observed late-time accelerated expansion of our Universe. Section 2.1.2 provides a discussion of the state of cosmic self-acceleration due to genuine modifications of gravity from a parametrized perspective.

2.1.1. *Dark energy equation of state*

The evolution of a spatially-flat homogeneous and isotropic cosmological background is to full generality described by only one free function of time, the scale factor $a(t)$ of the FLRW metric, or equivalently by the Hubble parameter $H(t) \equiv a^{-1}da/dt$. The contribution of a general dark energy, modified gravity, or dark sector interaction model on this evolution can be characterized in terms of an effective fluid with energy-momentum tensor $T_{\rm eff}^{\mu\nu} \equiv \kappa^{-2}G^{\mu\nu} - T_{\rm m}^{\mu\nu}$, where $\kappa^2 = 8\pi G$ with bare gravitational constant G. $G^{\mu\nu}$ is the Einstein tensor and $T_{\rm m}^{\mu\nu}$ denotes the matter component including baryonic contributions, cold dark matter, and radiation.

At the background level, the effective fluid is fully specified by its energy density $\bar{\rho}(t) = -T_{\rm eff\ 0}^0$ and pressure $\bar{p}(t) = T_{\rm eff\ i}^i$, which can be parametrized by the equation of state $w(t) = \bar{p}(t)/\bar{\rho}(t)$. For the cosmological constant Λ, the two relate as $\bar{p}_\Lambda = -\bar{\rho}_\Lambda$ with $w = -1$. For generic dark energy models, however, the equation of state departs from that value. For instance, in quintessence models[25, 189] $-1 < w(t) \leq 1$ and for a choice of $w(t)$ remaining in this regime, one can in principle always find a scalar field potential that reproduces the desired equation of state. More exotic dark energy models are required in order to cover the phantom regime $w < -1$. This can for instance be achieved with an additional coupling of the scalar field to the metric or matter components. Generally, one finds

$$\bar{\rho} = \bar{\rho}_0 \exp\left[3 \int_a^1 \frac{1 + w(a')}{a'} da'\right], \tag{1}$$

which follows from energy conservation, where $\bar{\rho}_0$ denotes the energy density today.

The most common approach to parametrizing deviations from ΛCDM in the cosmological background is to adopt the Chevallier-Polarski-Linder[187, 188] (CPL) equation of state

$$w(t) = w_0 + w_a[1 - a(t)]. \tag{2}$$

This relation is tested with geometric probes of the expansion history. It is also directly tested with the growth of structure if restricting to quintessence models. Any observational sign of a departure from $w = -1$ or $w_a = 0$ would provide evidence against ΛCDM. With the parametrization in Eq. (2), Eq. (1) simplifies to

$$\bar{\rho} = \bar{\rho}_0 a^{-3(1+w_0+w_a)} \exp[3w_a(a - 1)]. \tag{3}$$

More generally, the functional dependence of the dark energy equation of state $w(a)$ can be explored within a principal component analysis (PCA).[190] Often, however, in explorations of the effects of parametrized modifications of gravity a cosmological background is assumed that is instead equivalent to concordance cosmology ($w = -1$). This approach is usually observationally motivated and adopted to separate tests of the growth of structure from the background evolution. But it can also be motivated in view of some modified gravity models, for instance, chameleon gravity, where viable models produce $w \approx -1$.[87, 92, 191, 192]

2.1.2. *Genuine cosmic self-acceleration before/after GW170817*

Traditionally, modifications of gravity in the cosmological context have been motivated as alternative to the cosmological constant or dark energy as explanation for the late-time accelerated expansion. Besides introducing a dark energy potential or cosmological constant, the accelerated expansion can instead arise from the kinetic terms of a dark energy field as for instance in k-essence models.[193] Such a kinetic self-acceleration furthermore emerges in Kinetic Gravity Braiding models[112] or the cubic Galileon[104] model, but these can be viewed as imperfect dark energy fluids that, for instance, do not change the propagation of gravitational waves (see Sec. 2.2.4). A modification of this propagation may in contrast be motivated as a requirement for a genuine modification of gravity.[194] We shall therefore briefly explore how a cosmic acceleration can arise in modified gravity that does not rely on kinetic or potential energy contributions of a new field but can genuinely be attributed to an intrinsic property of the modified gravitational interactions.

In Ref. 91, it was argued that for cosmic acceleration to be genuinely attributed to a modified gravity effect there should be no accelerated expansion in the Einstein frame, i.e., where the modified field equations have been transformed into the Einstein field equations (see Chapter 1). Cosmic acceleration should otherwise be ascribed to a dark energy contribution instead. Note that the background evolution in this frame is described by the standard Friedmann equations for some exotic non-minimally coupled matter sector. This classification can be applied to effective field theory (Sec. 2.2.3) with an effective conformal (or pseudo-conformal) transformation $\Omega(t)$ that connects the two frames defined at the cosmological background,[38] absorbing contributions from both conformal and disformal factors at the non-perturbative level.[a] One then finds that genuine self-acceleration should imply

$$\frac{d^2\tilde{a}}{d\tilde{t}^2} = \frac{1}{\sqrt{\Omega}}\left[\left(1 + \frac{1}{2}\frac{d\ln\Omega}{d\ln a}\right)\frac{d^2a}{dt^2} + \frac{aH^2}{2}\frac{d^2\ln\Omega}{d(\ln a)^2}\right] \leq 0, \tag{4}$$

where tildes indicate quantities in the transformed frame. The observed late-time acceleration implies $d^2a/dt^2 > 0$ for $a \gtrsim 0.6$ such that $-d\ln\Omega/d\ln a \sim \mathcal{O}(1)$. It can

---

[a]The transformed frame was dubbed the Einstein-Friedmann frame in Ref. 38, which does not coincide with the Einstein frame unless $c_T = 1$. An equivalent argument can generally be made in the Einstein frame by instead performing the transformations defined in Ref. 195.

furthermore be shown that[38]

$$-\frac{d\ln\Omega}{d\ln a} = \frac{d\ln(G_{\rm eff}/c_{\rm T}^2)}{d\ln a} \sim \mathcal{O}(1), \tag{5}$$

where $G_{\rm eff}$ is an effective gravitational coupling and $c_{\rm T}$ denotes the speed of gravitational waves.[b] These can generally both evolve in time. Hence, genuine cosmic self-acceleration is either due to a change in the strength of gravity or in its speed, assuming that the speed of light in vacuum remains constant.

Since cosmic acceleration is an order unity effect on the cosmological background dynamics, if due to modified gravity, one would naturally expect an effect of similar strength on structure formation. Importantly, however, the modifications in $c_{\rm T}$ and $G_{\rm eff}$ can cancel out in the observed structure. This can for instance occur in Horndeski gravity[103] when[38, 196]

$$1 - c_{\rm T}^2 = \frac{\Delta G_{\rm eff}}{\Delta G_{\rm eff} + \alpha_B G_{\rm eff}} \frac{d\ln G_{\rm eff}}{d\ln a} \tag{6}$$

or when this relation holds approximately within the observational uncertainties. Here, $\Delta G_{\rm eff} \equiv G_{\rm eff} - G$ and α_B describes the braiding effect between the kinetic contributions of the scalar and metric fields. Specifically, the condition (6) follows from imposing a standard Poisson equation and the absence of effective anisotropic stress, which will be discussed in more detail in Sec. 2.2.3 (see Eq. (21) in particular). Importantly, a direct measurement of the gravitational wave speed $c_{\rm T}$ breaks the degeneracy implied by Eq. (6).[38]

The recent gravitational wave observation GW170817[23] with LIGO & Virgo emitted by a neutron star merger in the NGC 4993 galaxy and the wealth of near-simultaneous electromagnetic counterpart measurements constrains the relative deviation between the speeds of gravity and light at $\mathcal{O}(10^{-15})$.[2] With anticipation of this event and bound,[38, 197] implications on cosmic self-acceleration from modified gravity from such a measurement were first analyzed in Refs. 38 and 39 (see Sec. 2.2.4 for further discussions on the speed of gravitational waves). In particular, $c_{\rm T} \simeq 1$ implies that self-acceleration must be due to an evolving $G_{\rm eff}$ from Eq. (5) and that the order unity effect of the gravitational modification can no longer hide in the cosmic structure due to Eq. (6).[38] Ref. 39 then showed that in this case the minimal modification of gravity required for self-acceleration in the Horndeski framework (Sec. 2.2.5) provides a 3σ inferior fit to current cosmological data than the cosmological constant. The tension is mainly attributed to the cross correlations of the integrated Sachs-Wolfe (ISW) effect in the cosmic microwave background (CMB) temperature anisotropies with foreground galaxies and cannot be evaded with screening mechanisms. Moreover, a rescaled constraint is also applicable to theories where a self-acceleration could emerge from a self-interaction in the dark matter, which would not affect the baryonic components and Solar System tests.

---

[b]A transformation into Einstein frame instead yields $d\ln(G_{\rm eff}/c_{\rm T})/d\ln a \sim \mathcal{O}(1)$ in Eq. (5).

It should, however, be noted that dark energy fields embedded in the Horndeski theory are still viable candidates to explain the late-time acceleration. Such fields may also naturally be expected to couple to matter but the corresponding modification of gravity cannot be the main driver for cosmic acceleration. This also removes a preferred scale for the strength of the coupling. However, a modification of gravity with a Hubble scale deviation from GR may also be a remnant of a mechanism that tunes the cosmological constant to its observed value, which would indirectly govern cosmic acceleration. Finally, it is also worth noting that the dark degeneracy reappears in more general theories than Horndeski gravity,[196] which again allows self-acceleration to be hidden in the large-scale structure and evade the tension in galaxy-ISW cross correlations. The minimal gravitational modification that is necessary in general scalar-tensor theories to produce genuine cosmic self-acceleration[39] is, however, expected to leave a 5σ tension in Standard Sirens tests since the gravitational waves are not affected by the degeneracy[38] (Sec. 2.2.4).

2.2. *Linear cosmology*

In addition to the cosmological background evolution, modifications of gravity or exotic dark energy models can manifest in the linear fluctuations around that background. The incurring effects are manifold and extensive efforts have been devoted to establish general parametrization frameworks for the linear cosmological perturbation theory of modified gravity and dark energy models. Different approaches include the growth-index parametrization (Sec. 2.2.1), the PPF formalisms based on defining closure relations for the system of differential equations determining the linear cosmological perturbations (Sec. 2.2.2), the EFT of dark energy and modified gravity (Sec. 2.2.3), and a parametrization for the impact on the cosmological propagation of gravitational waves (Sec. 2.2.4). The parametrizations encountered in these different formalisms are in general free time and scale dependent functions. A few different choices that are frequently adopted shall briefly be discussed in Sec. 2.2.5.

2.2.1. *Growth-index parametrization*

Besides testing for a departure from the cosmological constant in the equation of state $w(a)$ with geometric probes (Sec. 2.1.1), the growth of large-scale structure is a further source of valuable constraints on dark energy and modified gravity. GR predicts that the growth function of matter density fluctuations $D(a)$ should behave as[155]

$$f \equiv \frac{d \ln D(a)}{d \ln a} = \Omega_{\mathrm{m}}(a)^{\tilde{\gamma}}, \tag{7}$$

where $\tilde{\gamma} \approx 6/11$ and $\Omega_{\mathrm{m}}(a) \equiv \kappa^2 \bar{\rho}_{\mathrm{m}}/(3H^2)$. In many modified gravity theories or dark sector interaction models $\tilde{\gamma}$ departs from this value.[156] Measurements of $\tilde{\gamma}$ therefore serve as a *consistency test* of GR, comparing measurements of geometry and growth (also see Refs. 198–200). While observationally useful, it is often not a very natural parametrization to encompass the wealth of possible modified gravity and dark interaction models, and $\tilde{\gamma}$ should generally be a non-trivial time and scale dependent function.

Moreover, the growth index parametrization is incomplete. It does not uniquely specify the linear cosmological fluctuations unless additional assumptions are made for how the perturbations relate. Specifically, while it parametrizes the growth of structure, the weak gravitational lensing caused by that structure can differ between models. Furthermore, the definition made in Eq. (7) does not capture violations of the conservation of comoving curvature (see Eq. (10) in Sec. 2.2.2) encountered at near-horizon scales once departing from ΛCDM. The growth-index parametrization can, however, be completed by connecting it, for instance, to the closure relations discussed in Sec. 2.2.2 with an additional parametrization for the evolution of the comoving curvature ζ.[201] Alternatively, Eq. (7) can be reinterpreted as the velocity-to-density ratio $\mathcal{F} \equiv -k_H V_m / \Delta_m$, which absorbs the evolution of ζ and reduces to f on subhorizon scales $k_H \gg 1$ (see Eq. (10) in Sec. 2.2.2),[202, 203] where $k_H \equiv k/(aH)$.

2.2.2. Closure relations for linear perturbations

Upon revisiting the effective energy-momentum tensor $T_{\text{eff}}^{\mu\nu}$ defined in Sec. 2.1.1, which is associated with the extra terms encountered in a modified Einstein field equation, one can educe that because of the Bianchi identities and the energy-momentum conservation of the matter components, it must also hold that $\nabla_\mu T_{\text{eff}}^{\mu\nu} = 0$. Hence, the usual cosmological perturbation theory can be applied. This implies four fluctuations each in the metric and the energy-momentum tensor. Four of those are fixed by the Einstein and conservation equations and another two are fixed by the adoption of a particular gauge. Hence, one is left with two undetermined relations, which requires the introduction of two closure relations that are specified by the particular modified gravity or dark energy model. This then closes the system of differential equations determining the cosmological perturbations. In practice, the two closure relations are designed such that the effective fluid mimics the relations between the metric and matter fluctuations of the model in question.

This is typically done through the introduction of an effective modification of the Poisson equation and a gravitational slip, or effective anisotropic stress,[148–151]

$$k_H^2 \Psi = -\frac{\kappa^2 \bar{\rho}_m}{2H^2} \mu(a, k) \Delta_m \,, \tag{8}$$

$$\Phi = -\gamma(a, k)\Psi \,, \tag{9}$$

respectively (also see Refs. 204–206), where the comoving gauge is adopted here for the matter perturbations. ΛCDM is recovered for $\mu = \gamma = 1$. Alternatively, a range of combinations of these equations are used to define the two closure relations with a variety of symbols used as notation for the effective modifications (see Tab. 1 in Ref. 205 for relations between some of them). It is also worth noting that in modified gravity theories μ and γ are generally only taking on a simple analytic form in the subhorizon limit. At superhorizon scales ($k_H \ll 1$) the evolution of the perturbations needs to be absorbed into the effective modifications, which may instead more naturally be described as an extra summand in the Poisson equation.[152] In this chapter Eqs. (8) and (9) will be adopted for all scales. The system determining the evolution of the scalar modes is then closed by the energy-momentum conservation equations,

$$\Delta'_m = -k_H V_m - 3\zeta' , \tag{10}$$
$$V'_m = -V_m + k_H \Psi , \tag{11}$$

where $\zeta \equiv \Phi - V_m/(kH)$ denotes the comoving curvature (see Ref. 207 for how a breaking of the conservation equations can be tested independently of μ and γ).

At subhorizon scales ζ' can be neglected in the energy conservation equation and together with momentum conservation and Eq. (8), it follows that

$$\Delta''_m + \left(2 + \frac{H'}{H}\right) \Delta'_m - \frac{3}{2} \Omega_m(a) \mu(a, k) \Delta_m = 0 , \tag{12}$$

which determines the growth of structure. More generally, Eq. (12) can be written as a first-order nonlinear differential equation for $\mathcal{F}$ without adopting a subhorizon approximation,[203] which then reduces to an equation for f when $k_H \gg 1$ (Sec. 2.2.1). The gravitational slip parameter γ quantifies a deviation of the lensing potential

$$\frac{1}{2}(\Phi - \Psi) = -\frac{1}{2}(1 + \gamma)\Psi \tag{13}$$

from the Newtonian potential Ψ governing the gravitational dynamics. This is in analogy to the parametrized post-Newtonian parameter γ in Sec. 3. The gravitational lensing due to a matter distribution is then determined from the Poisson equation of the potential in Eq. (13) with the effective modification replaced by the combination $\Sigma \equiv \frac{1}{2}(1 + \gamma)\mu$. The effective modifications μ and γ are generally time and scale dependent, and some frequently adopted parametrizations shall briefly be inspected in Sec. 2.2.5.

Finally, it is worth noting that at superhorizon scales, the equations of motion also simplify. More specifically, from diffeomorphism invariance, metric gravitational theories with energy-momentum conservation satisfy the adiabatic fluctuations of a flat universe with comoving curvature conservation $\zeta' = 0$ in the limit of $k \to 0$.[208] It then follows from momentum conservation that

$$\zeta'' - \frac{H''}{H'}\zeta' = \Phi'' - \Psi' - \frac{H''}{H'}\Phi' - \left(\frac{H'}{H} - \frac{H''}{H'}\right)\Psi \to 0 \tag{14}$$

such that γ and H determine the evolution of the potentials. This can be used to map modified gravity and dark energy models onto the effective fluid modifications in Eqs. (8) and (9) using the evolution of their perturbations in the super- and subhorizon limits.[152, 202, 203]

2.2.3. *Effective field theory of dark energy and modified gravity*

A more systematic approach to covering the range of possible dark energy and modified gravity models than by introducing two free effective functions of time and scale is through the effective field theory of cosmic acceleration (EFT)[111, 157–164] (also see Refs. 209–211). The gravitational action is written here in unitary gauge, where the time coordinate absorbs a scalar field perturbation in the metric $g_{\mu\nu}$, and is built from the combination of geometric quantities that are invariant under time-dependent spatial diffeomorphisms. Those quantities are then assigned free time-dependent coefficients. To quadratic order, describing the cosmological background evolution and linear perturbations of the model space, one obtains in the low-energy limit[159, 160]

$$
\begin{aligned}
S = \frac{1}{2\kappa^2} \int d^4x \sqrt{-g} \Big\{ & \Omega(t)R - 2\Lambda(t) - \Gamma(t)\delta g^{00} + M_2^4(t)(\delta g^{00})^2 \\
& - \bar{M}_1^3(t)\delta g^{00}\delta K^\mu{}_\mu - \bar{M}_2^2(t)(\delta K^\mu{}_\mu)^2 - \bar{M}_3^2(t)\delta K^\mu{}_\nu \delta K^\nu{}_\mu \\
& + \hat{M}^2(t)\delta g^{00}\delta R^{(3)} + m_2^2(t)(g^{\mu\nu} + n^\mu n^\nu)\partial_\mu g^{00}\partial_\nu g^{00} \Big\} + S_{\rm m} \left[\psi_{\rm m}; g_{\mu\nu}\right],
\end{aligned}
\tag{15}
$$

adopting the notation of Ref. 196. R and $R^{(3)}$ are the four-dimensional and spatial Ricci scalars, $K_{\mu\nu}$ denotes the extrinsic curvature tensor, n^μ is the normal to constant-time surfaces, and δ indicates perturbations around the background. ΛCDM is recovered for $\Omega = 1$, constant Λ, and all remaining coefficients vanishing. For quintessence models, $\Omega = 1$ with Λ and Γ describing the scalar field potential and kinetic terms. $M_2^2 \neq 0$ is introduced in k-essence and $\bar{M}_1 \neq 0$ in the cubic Galileon and Kinetic Gravity Braiding models. All the coefficients except for m_2 are used to embed Horndeski theories, however, with the restriction that $2\hat{M}^2 = -\bar{M}_3^2 = \bar{M}_2^2$.[161, 212] The second condition ensures the restriction to second-order spatial derivatives in the equations of motion with $2\hat{M}^2 \neq \bar{M}_2^2$ in beyond-Horndeski theories.[161] Finally, $m_2 \neq 0$ is introduced in Lorentz covariance violating Hořava-Lifshitz gravity.[213]

There is a total of nine coefficients in the action (15), where the scale factor $a(t)$ or the Hubble parameter H of the spatially homogeneous and isotropic background adds a tenth function. For simplicity spatial flatness and a matter-only universe with pressureless dust will be assumed here. The Friedmann equations,

$$
H^2\left(1 + \frac{\Omega'}{\Omega}\right) = \frac{\kappa^2\rho_{\rm m} + \Lambda + \Gamma}{3\Omega}, \quad (H^2)'\left(1 + \frac{1}{2}\frac{\Omega'}{\Omega}\right) + H^2\left(3 + \frac{\Omega''}{\Omega} + 2\frac{\Omega'}{\Omega}\right) = \frac{\Lambda}{\Omega},
\tag{16}
$$

following from variation of the action with respect to the metric, then provide two constraints between the first three background coefficients in Eq. (15). Hence, for specified matter content and spatial curvature, the space of dark energy and modified gravity models embodied by Eq. (15) can be characterized by eight free functions of time. In particular, the cosmological background evolution and linear perturbations of Horndeski theories is described by five coefficients only.

Similarly to Eq. (15), the effective action can also be built from the geometric quantities that can be introduced in an Arnowitt-Deser-Misner (ADM) 3+1 decompostion of spacetime with the uniform scalar field hypersurfaces as the constant time hypersurfaces.[111, 161] Variation of the action then defines an equivalent set of functions describing the cosmological background as well as the scalar and tensor modes of the perturbations. The formalism separates out the expansion history H as the free function determining the cosmological background evolution, which relates to Ω, Γ, and Λ through Eqs. (16). The linear fluctuations of Horndeski scalar-tensor theories are then characterized by four time-dependent functions: the kineticity $\alpha_K \equiv (\Gamma + 4M_2^4)/(H^2\kappa^2 M^2)$ that parametrizes the contribution of a kinetic energy of the scalar field; the evolution rate of the gravitational coupling $\alpha_M \equiv (M^2)'/M^2$ with effective Planck mass $M^2 \equiv \kappa^{-2}(\Omega + \bar{M}_2^2)$; the braiding parameter $\alpha_B \equiv (H\Omega' + \bar{M}_1^3)/(2H\kappa^2 M^2)$ describing the braiding or mixing of the kinetic contributions of the scalar and metric fields; and the alteration in the speed of gravity $\alpha_T \equiv -(\bar{M}_2^2)/(\kappa^2 M^2)$ with $c_T^2 = 1 + \alpha_T$. Additional terms are introduced if generalizing the formalism to include further higher-derivative scalar-tensor terms, for instance the beyond-Horndeski function $\alpha_H \equiv (2\hat{M}^2 - \bar{M}_2^2)/(\kappa^2 M^2)$,[161] or encompass vector-tensor and tensor-tensor theories.[164] The formalism has also been extended to encompass more general dark sector interaction models.[195] ΛCDM is recovered when $\alpha_i = 0$ $\forall i$.

The perturbed modified Einstein and scalar field equations as well as a reduced system of differential equations in this formalism can be found in Refs. 111 or 203. In particular, in Horndeski gravity ($\alpha_H = 0$) at subhorizon scales, time derivatives of the metric potentials and large-scale velocity flows can be neglected with respect to spatial derivatives and matter density fluctuations. At leading order in k, one obtains for the effective modifications

$$\mu_{QS} = \frac{2\left[\alpha_B(1+\alpha_T) - \alpha_M + \alpha_T\right]^2 + \alpha(1+\alpha_T)c_s^2}{\alpha c_s^2 \kappa^2 M^2}, \tag{17}$$

$$\gamma_{QS} = \frac{2\alpha_B\left[\alpha_B(1+\alpha_T) - \alpha_M + \alpha_T\right] + \alpha c_s^2}{2\left[\alpha_B(1+\alpha_T) - \alpha_M + \alpha_T\right]^2 + \alpha(1+\alpha_T)c_s^2}, \tag{18}$$

where $\alpha \equiv 6\alpha_B^2 + \alpha_K$ and the sound speed of the scalar mode is

$$c_s^2 = -\frac{2}{\alpha}\left[\alpha_B' + (1+\alpha_T)(1+\alpha_B)^2 - \left(1 + \alpha_M - \frac{H'}{H}\right)(1+\alpha_B) + \frac{\rho_m}{2H^2 M^2}\right]. \tag{19}$$

One finds that α_K, and hence M_2^4, does not contribute in the subhorizon limit, but it can give rise to a clustering effect on very large scales.[196] Importantly, when $\alpha_H \neq 0$, the velocity field and time derivatives contribute at leading order in the subhorizon limit of the field equations such that a quasistatic approximation of μ and γ becomes inaccurate without additional information on the growth rate of matter density fluctuation f or time derivatives of Φ and ζ.[203]

Stability of the background cosmology to the scalar and tensor modes requires[111]

$$\frac{M^2\alpha}{(1+\alpha_{\rm B})^2} > 0, \quad c_{\rm s}^2 > 0, \quad M^2 > 0, \quad c_{\rm T}^2 > 0, \qquad (20)$$

which also implies $\alpha > 0$.

Finally, it should be noted that there are more free functions in the theory space spanned by the coefficients in Eq. (15), or H and α_i, than what can be measured by geometric probes and large-scale structure observations. In particular, this allows for a model space that is degenerate with ΛCDM,[196] producing the same expansion history and linear scalar fluctuations as standard cosmology despite $\alpha_i \neq 0$, which also contains genuinely self-accelerated models (Sec. 2.1.2).[38] For Horndeski theories one of the degeneracy conditions is[38, 196] (cf. Eq. (6) in Sec. 2.1.2)

$$\alpha_{\rm T} = \frac{\kappa^2 M^2 - 1}{(1+\alpha_{\rm B})\kappa^2 M^2 - 1}\alpha_{\rm M}, \qquad (21)$$

where self-acceleration is produced at late times for[38] (Sec. 2.1.2)

$$\frac{\Omega'}{\Omega} = \alpha_{\rm M} + \frac{\alpha_{\rm T}'}{1+\alpha_{\rm T}} \lesssim -\mathcal{O}(1). \qquad (22)$$

As we shall see next, testing the modifications in the propagation of gravitational waves breaks the degeneracy in Eqs. (21) and (22).

2.2.4. *Cosmological propagation of gravitational waves*

With the direct detection of gravitational waves, further constraints on linear cosmological modifications of gravity can be obtained from their effects on the propagation of the wave. A parametrization for these modifications was introduced in Ref. 185 with the wave equation

$$h_{ij}'' + \left(3 + \frac{H'}{H} + \nu\right) h_{ij}' + \left(c_{\rm T}^2 k_H^2 + \frac{\tilde{\mu}^2}{H^2}\right) h_{ij} = \frac{\Gamma}{H^2}\gamma_{ij}, \qquad (23)$$

where $h_{ij} \equiv g_{ij}/g_{ii}$ is the linear traceless spatial tensor perturbation. In addition to a possible impact on the expansion history H, the gravitational modifications are characterized by the Planck mass evolution rate ν, the speed of the wave $c_{\rm T}$, the mass of the graviton $\tilde{\mu}$, and the source term $\Gamma\gamma_{ij}$. The modifications can generally be time and scale dependent. GR is recovered in the limit of $\nu = \tilde{\mu} = \Gamma = 0$ and $c_{\rm T} = 1$. The effects of these modifications on the gravitational waveform have been studied in detail in Ref. 186, where also the connection to parametrizations in the strong-field regime have been considered (see Sec. 3.4). Vector-tensor theories, for instance, can introduce $c_{\rm T} \neq 1$ with $\nu = \tilde{\mu} = \Gamma = 0$, and bimetric massive gravity produces $\tilde{\mu} \neq 0$, $\Gamma\gamma_{ij} \neq 0$, $\nu = 0$, and $c_{\rm T} = 1$.[185] Here, we shall focus on scalar-tensor theories, which are described by $\nu = \alpha_{\rm M}$, $c_{\rm T}^2 = 1 + \alpha_{\rm T}$, and $\tilde{\mu} = \Gamma = 0$, directly relating the modifications in the propagation of the wave to the modifications in the large-scale structure (Sec. 2.2.3) as well as the modifications required for a genuine self-acceleration effect in Eq. (22).

Particular attention has been given to constraints on c_T. For instance, the observation of ultra high energy cosmic rays implies a strong constraint on gravitational Cherenkov radiation from a subluminal propagation of the waves as otherwise the radiation would decay at a rate proportional to the square of their energy $\mathcal{O}(10^{11}\text{ GeV})$ and not reach Earth.[37, 214] For galactic $\mathcal{O}(10\text{ kpc})$ or cosmological $\mathcal{O}(1\text{ Gpc})$ origin, the relative deviation in c_T is constrained to be smaller than $\mathcal{O}(10^{-15})$ or $\mathcal{O}(10^{-19})$, respectively. The constraint is, however, only applicable to subluminal deviations. Conservatively, the effect could furthermore be screened in case of galactic origin and the constraints inferred from the short wavelengths of the gravitational waves may generally not be applicable to the low-energy effective modifications considered in the cosmological background and large-scale structure.[38, 132, 215] The first caveat is evaded by bounds on the energy loss in pulsars, which constrains both subluminal and superluminal deviations in c_T at the subpercent level for Horndeski theories that rely on Vainshtein screening.[132, 216] In particular this implies the inviability of Galileon theories[38] if combined with deficiencies in the large-scale structure[217] such as an observational incompatibility of galaxy-ISW cross correlations,[114, 217, 218] provided the applicability of the cosmological time variation of the scalar field at the binary system. However, binary pulsars may conservatively still be considered screened by other shielding mechanisms for more general theories. Both the first and second caveat can be avoided by a direct cosmological measurement of c_T over linear, unscreened distance scales.

It was first anticipated in Refs. 197 and 38 that a constraint on the relative deviation between the speeds of gravity and light of $\mathcal{O}(10^{-15})$ should be obtained from the comparison of the arrival times between a gravitational wave measured in LIGO & Virgo and the electromagnetic counterparts emitted by a neutron star merger or a merger between a neutron star and a black hole. This is a direct consequence of resolution of the detectors limited to $\mathcal{O}(100\text{ Mpc})$ for such an event and from higher likelihood of an event at larger volume with larger distances as well as emission time uncertainties of $\mathcal{O}(1\ s)$. It was also estimated that a few such events with simultaneous signals should be expected per year of operation of the detectors. Such a measurement has recently been realized with GW170817[23] and the wealth of counterpart observations,[2] providing a constraint in agreement with the predictions of Refs. 197 and 38. In particular, for Horndeski theory the nearly simultaneous arrival implies a breaking of the degeneracy in the large-scale structure with $\alpha_T \simeq 0$ and that a genuine cosmic self-acceleration effect is incompatible with the observed cosmic structure[38, 39] (Secs. 2.1.2 and 2.2.3). Specifically, tensions arise in the observed galaxy-ISW cross correlations that directly test the evolving gravitational coupling α_M yielding self-acceleration through its impact on the evolving gravitational potentials. Furthermore, $c_T = 1$ implies $G_{4X} = G_5 = 0$ in the Horndeski Lagrangian, first pointed out in the context of the gravito-Cherenkov constraints[114] and discussed for the arrival time measurements in Ref. 219.[c] Finally,

---

[c]Note that while a constant G_5 does not modify c_T, its contribution to the action can be removed

it is also worth noting that the GW170817 constraint arises from a merger in the NGC 4993 galaxy of the Hydra constellation at $z = 0.01$. While applicable for tests of cosmic self-acceleration, operating in the same redshift regime, the speeds of gravity and light may not necessarily be equal at higher redshifts, also unleashing the possibility of a dark degeneracy in the high-redshift cosmic structure.

Further constraints on c_T have been discussed as forecasts for arrival time comparisons with supernovae emissions,[38, 185, 197, 216] which however are limited to galactic events that may be screened and are limited to a low rate of a few events per century. LIGO black hole mergers[220] or the comparison between the arrival times of the waves in the detectors[221] only provide weak constraints on c_T. Forecasts have also been discussed for eclipsing binary systems observable with the LISA detectors.[40] Finally, order unity bounds on c_T can also be placed at early times from the B-mode power spectrum of the CMB.[222]

Besides the speed of the gravitational waves, further constraints can be inferred for the modified damping term in Eq. (23) with the running Planck mass adding to the Hubble friction. For early-time modifications, effects on the CMB B-modes have been employed for constraints in Ref. 223. For bounds on low-redshift modifications relevant to cosmic acceleration (Sec. 2.1.2), Ref. 185 suggested the use of Standard Sirens[224, 225] and first forecasts have been inferred in Refs. 38, 226, and 227. More specifically, gravitational waves provide a distance measurement by the decay of the wave amplitude with the luminosity distance. Combined with an identification of the redshift of the source this therefore provides a luminosity distance-redshift relation $d_L^{GW}(z)$, or a *Standard Siren*. Due to the additional damping of the wave with the running Planck mass on top of the Hubble friction, a comparison of $d_L^{GW}(z)$ inferred from Standard Sirens and $d_L(z)$ inferred, for instance, from electromagnetic standard candles such as Type Ia supernovae yields a constraint on ν (or α_M and M).[38, 185] More specifically, $d_L^{GW}(z) = M(0)d_L(z)/M(z)$.[186, 226, 227] GW170817 has provided the first Standard Siren,[13] yielding a constraint of $\sim 10\%$ on H_0. It was shown in Ref. 38 that percent-level constraints on the expansion history at low redshifts from Standard Sirens can yield a conclusive 5σ tension for the minimal modification in scalar-tensor theories necessary for a genuine cosmic self-acceleration (also see Ref. 39). This can be achieved either with LISA[228, 229] or possibly with LIGO, Virgo, and KAGRA before. Importantly, this constraint will also be applicable to theories beyond Horndeski gravity with $\alpha_H \neq 0$, where the dark degeneracy is reintroduced in observations of the large-scale structure,[196] preventing fully exhaustive and general conclusions without the use of gravitational wave measurements.

---

by integration by parts and can therefore be considered vanishing. It can furthermore be combined with a constant G_3 to form the unique divergence-free tensor that can be contracted with $\partial^\mu \phi \partial^\nu \phi$, i.e., the Einstein tensor $G_{\mu\nu}$ from $\mathcal{L}_5$ with the contribution of $\Lambda g_{\mu\nu}$ from $\mathcal{L}_3$, and that can hence be removed by integration by parts.

2.2.5. *Parametrizing linear modifications*

In Secs. 2.2.2–2.2.4, we have seen how a small number of effective modifications introduced with generalized frameworks for linear cosmological perturbation theory can enter as free functions of both time and scale in the computations of the formation of large-scale structure and the propagation of gravitational waves. Here, we shall briefly discuss different choices that are frequently adopted for the time and scale dependence of the two closure relations in Sec. 2.2.2 as well as for the time dependent functions in the effective field theory formalism (Sec. 2.2.3).

For the closure functions $\mu(a, k)$ and $\gamma(a, k)$ and equivalent expressions a range of simple expansions in the scale factor a and the wavenumber k, smoothly interpolated bins in a (or z) and k, or PCA tests have been adopted (see Ref. 20 for a review). A motivation for the segregation of the scale-dependence in μ and γ was presented in Ref. 230 and shall briefly be summarized here. With adiabatic initial conditions and the weak equivalence principle, Ref. 230 points out that the linear gravitational potentials and density fluctuations are directly related to the initial conditions through the transfer functions T_Φ, T_Ψ, T_Δ such that $\mu \sim k_H^2 T_\Psi/T_\Delta$ and $\gamma = T_\Phi/T_\Psi$, where $T_\chi^2(k)$ encapsulates the k-dependent shape of the power spectrum in $\chi = \Phi, \Psi, \Delta$. For local four-dimensional theories of gravity, where transfer functions depend on k^2 only, and restricting to at most second spatial derivatives in the equations of motion, the quasistatic μ and γ must then be described by five functions of time $p_i(t)$ only with

$$\mu(k, t) = \frac{1 + p_3(t)k^2}{p_4(t) + p_5(t)k^2}, \quad \gamma(k, t) = \frac{p_1(t) + p_2(t)k^2}{1 + p_3(t)k^2}. \tag{24}$$

Generally, these functions are applicable at leading order as further, inverse powers of k can appear by accounting for time derivatives and reducing the modified Einstein and field equations to four equations of motion.[203] Note that a similar form to Eq. (24) for μ and $\frac{1}{2}(1+\gamma)$ was first proposed in Ref. 231 with p_4 and the analog of p_1 set to unity and power laws of $a(t)$ adopted for the remaining p_i. One can also notice the possibility of a large-scale modification of gravity with a small-scale ΛCDM limit of $\mu(k \to \infty) = \gamma(k \to \infty) = 1$, which can for instance be tested on ultra large scales.[196, 202] Often, however, the small-scale limit is adopted, where the mass scale involved is assumed to be small, for instance for a gravitational modification driving cosmic acceleration, with the typical observational probes lying well in the subhorizon limit. Hence, in this case only time dependent modifications of μ and γ are considered but this does, for instance, not capture viable chameleon models.

Focusing on time-dependent modification only, similar parametrizations have been considered for $\mu(a)$ and $\gamma(a)$ and the EFT functions in Sec. 2.2.3, for instance, power laws in a or a proportionality of the deviations p_i to $\Omega_{\mathrm{DE}}(a)/\Omega_{\mathrm{DE}}(a = 1)$,[232] where $\Omega_{\mathrm{DE}}(a) \equiv \kappa^2 \rho_{\mathrm{DE}}/(3H^2)$, in which case tests of μ and γ restrict to only two constants, $p_3(a = 1)/p_5(a = 1)$ and $p_2(a = 1)/p_3(a = 1)$.

Typically these modifications have been studied with vanishing modifications at early times (however, see, e.g., Refs. 201, 222, 223 and 233). Dropping the motivation of modified gravity as genuine alternative to dark energy, plausibly given the implications of $c_T = 1$ for cosmic acceleration discussed in Secs. 2.1.2 and 2.2.4, one may argue that parametrizations should be generalized to include early-time effects. It is worth noting, however, that modifications of gravity may still be indirectly related to cosmic acceleration, for instance, as a non-minimally coupled dark energy field or as a remnant of an effect that tunes the cosmological constant, preventing a vacuum catastrophe, in which case effects may still be expected at late times only. Contrariwise if focusing on aspects of direct cosmic self-acceleration, one may instead integrate out condition (4) to find the minimal running of the Planck mass or the speed of gravity required for a genuine self-acceleration from modifying gravity, using Eq. (22). With $c_T = 1$, one finds that[39] $M_{\min}^2 = \kappa^{-2}(a_{\mathrm{acc}}/a)^2 e^{C(\chi_{\mathrm{acc}} - \chi)}$ for $a \geq a_{\mathrm{acc}}$, where $a_{\mathrm{acc}} \equiv [\Omega_m/(1 - \Omega_m)/2]$ is the scale factor at which the expansion of the late-time universe becomes positively accelerated, $C \equiv 2H_0 a_{\mathrm{acc}}\sqrt{3(1 - \Omega_m)}$, and χ denotes the comoving distance. $M_{\min}^2$ is then fully specified by a given expansion history, e.g., matching ΛCDM, and does not require an additional choice of parametrization for the time dependence of α_M. To minimize the impact of the modification on the large-scale structure, it furthermore follows that $\alpha_B = \alpha_M$ from Eq. (17).[39] To quantify the amount of self-acceleration allowed by an observational dataset, one may then introduce the parametrization $\alpha_M = \lambda\,\alpha_{M,\min}$ with constant λ, where $\lambda > 1$ allows sufficient gravitational modification for self-acceleration, $\lambda = 1$ is the minimal scenario, $\lambda < 1$ needs an additional contribution from dark energy or a cosmological constant, $\lambda = 0$ corresponds to ΛCDM or GR, and $\lambda < 0$ corresponds to a scenario where the modification of gravity acts to decelerate the expansion and is counteracted by the introduction of dark energy or a cosmological constant.

Finally, it is worth noting that sampling general EFT or α_i coefficients in comparison to observations can be an inefficient process if additionally imposing the stability of the theory with the criteria in Eq. (20) as not each sample is guaranteed to yield a stable model. Moreover the contours of the viable parameter space can as a consequence produce edges and leave ΛCDM in a narrow corner that may only be sparsely sampled and make a statistical interpretation of evidence against or in favor of an extended theory of gravity more difficult. Reference 234 therefore argued that the functions to be sampled for the linear perturbations in EFT adopting $\alpha_T = 0$ should be M^2, c_s^2, $\alpha > 0$ in addition to a constant value for α_B today or at an initial epoch. The α_i functions can be reconstructed from these expressions, where it should be noted that c_s^2 yields a homogeneous linear second-order differential equation for B with $B'/B \equiv (1 + \alpha_B)$ through Eq. (19) such that a real and unique solution is guaranteed for a real boundary condition on α_B.

2.3. *Nonlinear cosmology*

The simple and generalized treatment of the cosmological background (Sec. 2.1) and linear perturbations (Sec. 2.2) for modified gravity and dark energy models enables consistent and efficient computations of the evolution of the background universe, the growth of large-scale structure from the Hubble scale to a few tens of Mpc, and the cosmological propagation of gravitational waves. However, the linear framework fails at describing structures at increasingly smaller scales, where there is a great wealth of observational data already available and becoming available with future galaxy surveys like Euclid[235] or LSST.[236] However, cosmological tests are difficult in the nonlinear regime and tests of gravity are additionally complicated in the presence of screening mechanisms. Nevertheless, a number of frameworks have been developed to extend the parametrized treatment of the effects of modified gravity to quasilinear scales with perturbation theory and interpolation functions (Sec. 2.3.1) as well as to perform a parametrization in the deeply nonlinear regime of cosmological structure formation using spherical collapse and halo model approaches (Sec. 2.3.2).

2.3.1. *Weakly nonlinear regime*

To describe the matter density fluctuations in a universe governed by a modified theory of gravity, Ref. 152 proposed a phenomenological extension of the linear parametrized post-Friedmann formalism (Sec. 2.2.2) for the modeling of the modified nonlinear matter power spectrum by

$$P(k, z) = \frac{P_{\text{non-GR}}(k, z) + c_{\text{nl}}\Sigma^2(k, z)P_{\text{GR}}}{1 + c_{\text{nl}}\Sigma^2(k, z)}, \tag{25}$$

where $P_{\text{GR}}(k, z)$ and $P_{\text{non-GR}}(k, z)$ denote the nonlinear matter power spectra with background expansion of the modified model adopting either GR or the gravitational modification in the absence of screening. The weighting function $\Sigma^2(k, z)$ governs the efficiency of the screening, where c_{nl} controls the scale of the effect and could also be time dependent. The weight

$$\Sigma^2(k, z) = \left[\frac{k^3}{2\pi^2}P_{\text{lin}}(k, z)\right]^n \tag{26}$$

was adopted in Refs. 152 and 165 with $P_{\text{lin}}(k, z)$ denoting the linear power spectrum of the modified gravity model. The one-loop perturbations of Dvali-Gabadadze-Porrati[31] (DGP) and $f(R)$ gravity[237] are well described by $n = 1$ and $n = 1/3$, respectively, with a $c_{\text{nl}}(z)$ that can be fitted to the computations. While the combination of Eq. (25) with the weighting function in Eq. (26) recovers $P(k, z)$ at quasilinear scales to good accuracy, increasingly more complicated $\Sigma^2(k, z)$ need to be devised to reproduce power spectra from N-body simulations at increasingly nonlinear scales.[173]

Alternatively, the combination of the spherical collapse model and the halo model (see Sec. 2.3.2) with linear perturbation theory and one-loop computations or a simple quasilinear interpolation motivated by $c_{nl}\Sigma^2(k,z)$ provides an accurate description for $P\left(k \lesssim 10\,h\,\mathrm{Mpc}^{-1}\right)$ for a range of scalar-tensor theories.[191, 238, 239] The combination of generalized perturbative computations to one-loop order with a generalized modified spherical collapse model promises to be a good approach to designing a nonlinear extension to the linear parametrized post-Friedmannian framework.[174] The particulars are, however, still being developed. It is worth noting that for chameleon, symmetron, and dilaton models a generalized parametrization covering the linear and nonlinear scales has been developed in Ref. 63 (Sec. 4) and applied to the modeling of the nonlinear matter power spectrum.[238]

Recently, there has been much progress on the further generalization of the computation of higher-order perturbations for modified gravity and dark energy theories (see, e.g., Refs. 166–172). While for general modifications of gravity the next-order perturbations introduce new EFT coefficients to the ones discussed in Sec. 2.2.3,[170] it should be noted that for Horndeski theory these terms depend on G_{4X} and G_5,[166, 169] which are vanishing due to the $\alpha_T \simeq 0$ constraint from GW170817 (Sec. 2.2.4) and hence do not expand the parameter space at second-order in the quasistatic perturbations.

2.3.2. *Deeply nonlinear regime*

The cosmological structure formation in the nonlinear regime can be studied with the spherical collapse model, where a dark matter halo is approximated by a spherically symmmetric top-hat overdensity with its evolution described by the nonlinear continuity and Euler equations from an initial condition to the time of its collapse. For a metric theory of gravity with a pressureless non-relativistic matter fluid, these equations become[155, 240, 241]

$$\dot{\delta} + \frac{1}{a}\nabla \cdot (1+\delta)\mathbf{v} = 0, \tag{27}$$

$$\dot{\mathbf{v}} + \frac{1}{a}(\mathbf{v}\cdot\nabla)\mathbf{v} + H\mathbf{v} = -\frac{1}{a}\nabla\Psi, \tag{28}$$

where $\delta \equiv \delta\rho_m/\bar{\rho}_m$, dots indicate derivatives with respect to physical time, and comoving spatial coordinates have been adopted. Combining these equations, one obtains

$$\ddot{\delta} + 2H\dot{\delta} - \frac{1}{a^2}\nabla_i\nabla_j(1+\delta)v^i v^j = \frac{1}{a^2}\nabla_i(1+\delta)\nabla^i\Psi. \tag{29}$$

For a spherical top-hat density with $\mathbf{v} = A(t)\mathbf{r}$ of amplitude A and from the continuity equation, one finds

$$\frac{1}{a^2}\nabla_i\nabla_j v^i v^j = \frac{4}{3}\frac{\dot{\delta}^2}{(1+\delta)^2}. \tag{30}$$

This yields the spherical collapse equation

$$\ddot{\delta} + 2H\dot{\delta} - \frac{4}{3}\frac{\dot{\delta}^2}{(1+\delta)} = \frac{1+\delta}{a^2}\nabla^2\Psi, \tag{31}$$

from which one infers the evolution of a spherical shell at the edge of the top hat

$$\frac{\ddot{\zeta}}{\zeta} = H^2 + \dot{H} - \frac{1}{3a^2}\nabla^2\Psi, \tag{32}$$

where $\zeta(a)$ denotes the physical top-hat radius at a and we have used that mass conservation implies a constant $M = (4\pi/3)\bar{\rho}_{\rm m}(1+\delta)\zeta^3$.

The particular choice of metric gravitational theory enters through the background evolution and the Poisson equation for Ψ. One can generally parametrize the modification of the Poisson equation as

$$\nabla^2\Psi \equiv \frac{a^2}{2}\left(1 + \frac{\Delta G_{\rm eff}}{G}\right)\kappa^2\delta\rho_{\rm m}, \tag{33}$$

which yields the spherical collapse equation

$$\frac{\ddot{\zeta}}{\zeta} = H^2 + \dot{H} - \frac{\kappa^2}{6}\left(1 + \frac{\Delta G_{\rm eff}}{G}\right)\delta\rho_{\rm m}. \tag{34}$$

Interpreting the gravitational modifications as an effective fluid with energy-momentum tensor $T^{\mu\nu}_{\rm eff}$ (Sec. 2.1.1) the first two terms on the right-hand side of Eq. (34) can be rewritten as $H^2 + \dot{H} = -\kappa^2\left[\bar{\rho}_{\rm m} + (1 + 3w_{\rm eff})\bar{\rho}_{\rm eff}\right]/6$ using the Friedmann equations.

One can further define the comoving top-hat radius $r_{\rm th}$ with $\zeta(a_i) = a_i r_{\rm th}$ at an initial scale factor $a_i \ll 1$ and $y \equiv \zeta/(a r_{\rm th})$, where mass conservation, $\bar{\rho}_{\rm m}a^3 r_{\rm th}^3 = \rho_{\rm m}\zeta^3$, implies $\rho_{\rm m}/\bar{\rho}_{\rm m} = y^{-3}$. This yields

$$y'' + \left(2 + \frac{H'}{H}\right)y' + \frac{1}{2}\Omega_{\rm m}(a)\left(1 + \frac{\Delta G_{\rm eff}}{G}\right)(y^{-3} - 1)y = 0, \tag{35}$$

which is typically solved by setting initial conditions in the matter-dominated regime with $y_i \equiv y(a_i) = 1 - \delta_i/3$ and $y'_i = -\delta_i/3$.

It is worth noting that in modified gravity models, Birkhoff's theorem can be violated, causing a shell crossing and departure of the overdensity from its initial top-hat profile over time.[242, 243] Nevertheless, the top hat still provides a good approximation if additionally accounting for the evolution of the surrounding environmental density with an analogous relation to Eq. (35) and its impact on $G_{\rm eff}$.[191, 244–246]

A useful quantity to define for the description of nonlinear structure formation is the spherical collapse density $\delta_{\rm c}(z)$, the extrapolation of the initial overdensity δ_i leading to collapse in Eq. (35) at redshift z with the linear growth factor $D/D_i \equiv \delta_{\rm lin}/\delta_i$. D is obtained from solving Eq. (34) in the linearized limit that is time dependent only in GR but can be both time and scale dependent in modified gravity models. To avoid a scale dependence entering through the extrapolation, one may adopt the GR linear growth factor. Importantly, it is really the initial overdensity δ_i that is the relevant quantity for structure formation, which allows one to define this effective extrapolation in this computationally more convenient manner as long as the same extrapolation is also adopted for comparable quantities such as the variance of the linear matter power spectrum. With $\delta_{\rm c}$ one can then model modified cluster properties such as cluster profiles[247–250] and concentrations,[191, 248] halo bias,[191, 240] or the halo mass function,[191, 240, 244, 246, 251–255] and similar computations can also be performed for modified void properties.[256, 257] Those quantities can then be combined in the halo model[258–260] to compose the modified halo model power spectrum[191, 240, 251, 253, 261] in the deeply nonlinear regime (also see Ref. 262 for a related approach that can be mapped[191, 246] and Ref. 239 for a review). The one-halo term can also be combined with higher-order perturbations to improve accuracy on quasilinear scales (Sec. 2.3.1). For higher efficiency in the modeling of cluster properties and the halo model power spectrum, one may also consider the direct parametrization of δ_c instead of using G_{eff} in Eq. (33). Some fitting functions are, for instance, available for $f(R)$ gravity.[243, 261, 263]

It should be noted that at deeply nonlinear cosmological scales baryonic effects become important and need to be accounted for in the comparison of model predictions to observations. Usually, fitting functions are adopted for this that have been matched to observations and hence may conservatively also be used to model the gaseous and stellar components in modified gravity. With improved physical description,[263, 264] however, the baryonic effects may themselves be used as test of gravity[265, 266] or to discriminate between universal and matter-specific couplings. Alternatively, statistical techniques can be applied such as a density weighting in the matter power spectrum that break degeneracies between baryonic effects, the variation of cosmological parameters, and modified gravity signatures that may even be unscreened by the statistic.[267]

A general parametrization of G_{eff} in the spherical collapse equation (35), embedding the variety of screening mechanisms encountered in modified gravity theories, has been proposed in Ref. 174. The parametrization is modularly built from transitions in the effective gravitational coupling of the form

$$\frac{\Delta G_{\text{eff}}}{G} \sim b \left(\frac{r}{r_0}\right)^a \left\{\left[1 + \left(\frac{r_0}{r}\right)^a\right]^{1/b} - 1\right\}, \tag{36}$$

which has the limits

$$\frac{\Delta G_{\text{eff}}}{G} \sim \begin{cases} b\left(\frac{r}{r_0}\right)^{a(b-1)/b}, & \text{for } (b > 0) \bigwedge [(r \ll r_0, a > 0) \bigvee (r \gg r_0, a < 0)], \\ -b\left(\frac{r}{r_0}\right)^a, & \text{for } (b < 0) \bigwedge [(r \ll r_0, a > 0) \bigvee (r \gg r_0, a < 0)], \\ 1, & \text{for } (r \ll r_0, a < 0) \bigvee (r \gg r_0, a > 0). \end{cases} \tag{37}$$

Here, r_0 denotes the screening scale, a determines the radial dependence of the gravitational coupling in the screened regime[d] together with the interpolation rate b between the screened and unscreened G_{eff}. The form of the transition in Eq. (36) is motivated by the screening profile of the Vainshtein mechanism in DGP gravity, where the expression becomes exact. While the range of screening and suppression mechanisms in literature can be mapped onto Eq. (36),[174] one may alternatively wish to adopt other transition functions instead.

---

[d]Note that this is not the scale factor of the FLRW metric.

For some models there are multiple transitions in G_{eff} such as the screening regime on small scales and the Yukawa suppression on large scales encountered in chameleon models. For general modifications of gravity therefore, one may consider the combination[174]

$$\frac{G_{\text{eff}}}{G} = A + \sum_i^{N_0} B_i \prod_j^{N_i} b_{ij} \left(\frac{r}{r_{0ij}}\right)^{a_{ij}} \left\{\left[1 + \left(\frac{r_{0ij}}{r}\right)^{a_{ij}}\right]^{1/b_{ij}} - 1\right\}, \quad (38)$$

with integers $i, j > 0$. The parameter A describes the relative deviation to the gravitational constant in the fully screened regime, which can be different from unity for modified gravity models, whereas B_i denotes the enhancement in the fully unscreened limit of a particular transition.

The effective modification can be implemented in the spherical collapse model with the replacement $r \to \zeta = a\, r_{\text{th}}\, y$ and $y_0 \equiv r_0/(a\, r_{\text{th}})$ in Eq. (38), defining the G_{eff} of Eq. (35), which yields

$$\frac{\Delta G_{\text{eff}}}{G} = B\, b \left(\frac{y_h}{y_0}\right)^a \left\{\left[1 + \left(\frac{y_0}{y_h}\right)^a\right]^{1/b} - 1\right\} \quad (39)$$

for one element in Eq. (38). A single general element $N_0 = N_1 = 1$ can then be modeled by seven parameters p_{1-7} in addition to $p_0 = A$, where

$$a = \frac{p_1}{p_1 - 1} p_3, \qquad b = p_1, \qquad B = p_2 \quad (40)$$

and the dimensionless screening scale is given by

$$y_0 = p_4 a^{p_5} (2G\, H_0 M_{\text{vir}})^{p_6} \left(\frac{y_{\text{env}}}{y_h}\right)^{p_7}. \quad (41)$$

For $p_7 \neq 0$ an environmental dependence enters such that y is solved for the collapsing halo y_h and for its environment y_{env}. The parametrization is general enough to allow a mapping of chameleon, symmetron, Vainshtein, and k-mouflage screening as well as Yukawa suppressions or linear shielding with simple analytic expressions for p_{1-7} that are determined by the model parameters (see Ref. 174). For instance, p_2 is given by a Brans-Dicke coupling, or more generally the unscreened modification that can be matched to the linear μ of PPF or EFT in Sec. 2.2.2. Furthermore, p_4 includes a dependence on cosmological parameters, usually $p_5 = -1$, $p_6 = 0$ for DGP and $p_7 \neq 0$ for chameleon models. Using the scaling method of Ref. 219 one further finds that the radial dependence of G_{eff} in the screened limit is determined by a particular combination of the powers of the second and first derivative terms, s and t, and of the derivative-free terms u dominating the scalar field equation in this regime as well as the dimensionality of the matter distribution v, $\Delta G_{\text{eff}}/G \sim r^{(3s+2t+u-v)/(s+t+u)}$.[174] Hence, these powers determine $p_3 = (3s+2t+u-v)/(s+t+u)$ and can in principle directly be read off from the action of a theory.[219]

The parametrization in Eq. (38) enables a generalized computation of the spherical collapse density δ_c in modified gravity or dark sector models from which cluster properties and the halo model matter power spectrum may be computed. It is worth noting that one can for instance use Eq. (41) to map δ_c from one set of parameters to another, which may be used to build a direct parametrization of δ_c. One may also use Eq. (38) in N-body simulations, employing techniques such as developed in Refs. 268 or 269.

3. Parametrized Post-Newtonian Formalism

Model-independent tests of gravity have very successfully been conducted in the low-energy static regime, where a post-Newtonian expansion can be performed and parametrized for generalizations of the gravitational interactions. Many stringent constraints on departures from GR have been inferred in this limit, particularly from observations in the Solar System. The formalism is, however, not suitable for cosmology, where the evolution of the background needs to be accounted for, a problem that has been addressed with the frameworks discussed in Sec. 2 (also see Sec. 4). Moreover, the screening mechanisms motivated by cosmological applications also introduce further complications in this low-energy static expansion. Section 3.1 briefly reviews the post-Newtonian series and Sec. 3.2 presents its parametrization for generalized tests of gravity. Section 3.3 discusses how screening mechanisms can be incorporated in the formalism. Finally, a parametrization of gravitational waveforms inspired by PPN is briefly discussed in Sec. 3.4.

3.1. *Post-Newtonian expansion*

Slowly evolving weak-field gravitational phenomena are well described by the low-energy static limit of GR, a regime particularly applicable to Solar-System tests of gravity.[34, 145] The metric of such a system can be expanded in orders[e] of $(v/c)^i$ with $h_{\mu\nu}^{(i)}$, neglecting cosmological evolution and assuming an asymptotic Minkowskian limit such that

$$g_{\mu\nu} = \eta_{\mu\nu} + h_{\mu\nu} = \eta_{\mu\nu} + h_{00}^{(2)} + h_{ij}^{(2)} + h_{0j}^{(3)} + h_{00}^{(4)}. \tag{42}$$

The virial relation determines the order of the Newtonian potential $U \approx \mathcal{O}(2)$. The matter density for a perfect non-viscous fluid ρ is of the same order, which follows from the Poisson equation. The pressure in the Solar System is of the order of the gravitational energy ρU and the specific energy density Π is of that of U. Hence, for the energy-momentum tensor, one finds to fourth order,

$$T_{00} = \rho[1 + \Pi + v^2 - h_{00}^{(2)}], \tag{43}$$

$$T_0{}^i = -\rho v^i, \tag{44}$$

$$T^{ij} = \rho v^i v^j + p\delta^{ij}. \tag{45}$$

With the system slowly evolving in time, it holds that $d/dt = \partial_t + \mathbf{v} \cdot \nabla \approx 0$, which indicates that spatial derivatives are an order lower than time derivatives. Adopting

---

[e]For clarity the speed of light c is kept stated explicitly here to emphasize the dimensionless counting of orders in the velocity although c has generally been set to unity (Sec. 1).

the standard post-Newtonian gauge with diagonal and isotropic metric, one finds for GR the conditions

$$h^\mu_{i,\mu} - \frac{1}{2}h^\mu_{\mu,i} = 0\,, \tag{46}$$

$$h^\mu_{0,\mu} - \frac{1}{2}h^\mu_{\mu,0} = -\frac{1}{2}h_{00,0}\,, \tag{47}$$

where the relations can change for modified theories of gravity.

The components of the Ricci tensor $R_{\mu\nu}$ are to fourth order

$$2R_{00} = -\nabla^2 h_{00} - (h_{jj,00} - 2h_{j0,j0}) + h_{00,j}\left(h_{jk,k} - \frac{1}{2}h_{kk,j}\right) - \frac{1}{2}|\nabla h_{00}|^2 + h_{jk}h_{00,jk} \tag{48}$$

and to second order

$$R_{0j} = -\frac{1}{2}\left(\nabla^2 h_{0j} - h_{k0,jk} + h_{kk,0j} - h_{kj,0k}\right)\,, \tag{49}$$

$$R_{ij} = -\frac{1}{2}\left(\nabla^2 h_{ij} - h_{00,ij} + h_{kk,ij} - h_{ki,kj} - h_{kj,ki}\right)\,. \tag{50}$$

With the Einstein equations $R_{\mu\nu} = 8\pi G(T_{\mu\nu} - \frac{1}{2}g_{\mu\nu}T)$ one finds at second order from R_{00} that $-\frac{1}{2}\nabla^2 h^{(2)}_{00} = 4\pi G\rho$, where the Newtonian potential can be defined as $U \equiv h^{(2)}_{00}/(2G)$. Employing the gauge condition (46) and using that $T_{ij} = 0$ at second order, yields $4\pi G\rho\delta_{ij} = -\frac{1}{2}\nabla^2 h^{(2)}_{ij}$ and, hence, $h^{(2)}_{ij} = 2GU\delta_{ij}$. For $h^{(3)}_{0j}$ one obtains

$$8\pi G\rho v_j = -\frac{1}{2}\nabla^2 h^{(3)}_{0j} - \frac{1}{2}GU_{,0j} \tag{51}$$

with the gauge conditions (46) and (47). Equation (51) is solved with the Green's function for the Poisson equation and by defining the post-Newtonian potentials

$$V_i \equiv \int \frac{\rho' v'_i}{|\mathbf{x} - \mathbf{x}'|}d^3x'\,, \tag{52}$$

$$W_i \equiv \int \frac{\rho'[\mathbf{v}' \cdot (\mathbf{x} - \mathbf{x}')](x - x')_i}{|\mathbf{x} - \mathbf{x}'|^3}d^3x' \tag{53}$$

such that $h^{(3)}_{0j} = -\frac{7}{2}GV_j - \frac{1}{2}GW_j$. This follows from employing the identity

$$\frac{\partial}{\partial t}\int \rho' f(\mathbf{x}, \mathbf{x}')d^3x = \int \rho'\mathbf{v}' \cdot \nabla' f(\mathbf{x}, \mathbf{x}')d^3x\,[1 + \mathcal{O}(2)]\,, \tag{54}$$

obtained from the vanishing of the total time derivative with $\rho' \equiv \rho(\mathbf{x}', t)$ and $v'_i = \partial x'_i/\partial t$. Furthermore, one has

$$R^{(4)}_{00} = -\frac{1}{2}\nabla^2(h^{(4)}_{00} + 2U^2 - 8\Phi_2) \tag{55}$$

with the definition

$$\Phi_2 \equiv \int \frac{\rho' U'}{|\mathbf{x} - \mathbf{x}'|}d^3x'\,. \tag{56}$$

The 00 matter component is

$$T_{00} - \frac{1}{2}g_{00}T = \rho\left(v^2 - \frac{1}{2}h_{00}^{(2)} + \frac{1}{2}\Pi + \frac{3}{2}\frac{p}{\rho}\right) \tag{57}$$

and one can define new post-Newtonian potentials[145] Φ_3 and Φ_4, employing the Green's function for the Laplacian, such that

$$h_{00}^{(4)} = -2G^2U^2 + 4G\Phi_1 + 4G^2\Phi_2 + 2G\Phi_3 + 6G\Phi_4. \tag{58}$$

3.2. Parametrizing the post-Newtonian expansion

The coefficients of the potentials found for the expansion of the metric in Eq. (42) depend on the particular gravitational theory assumed. Theory-independent tests of gravity can be performed by adopting a parametrized post-Newtonian (PPN) formalism for the expansion of the metric,[144, 145]

$$g_{00} = -1 + 2GU - 2\beta G^2U^2 - 2\xi G\Phi_W + (2\gamma + 2 + \alpha_3 + \zeta_1 - 2\xi)G\Phi_1$$
$$+2(3\gamma - 2\beta + 1 + \zeta_2 + \xi)G^2\Phi_2 + 2(1 + \zeta_2)G\Phi_3 + 2(\gamma + 3\zeta_4 - 2\xi)G\Phi_4$$
$$-(\zeta_1 - 2\xi)G\mathcal{A} - (\alpha_1 - \alpha_2 - \alpha_3)Gw^2U - \alpha_2 w_i w_j GU_{ij} + (2\alpha_3 - \alpha_1)w^i GV_i,$$
$$\tag{59}$$

$$g_{0i} = -\frac{1}{2}(4\gamma + 3 + \alpha_1 - \alpha_2 + \zeta_1 - 2\xi)GV_i - \frac{1}{2}(1 + \alpha_2 - \zeta_1 + 2\xi)GW_i$$
$$-\frac{1}{2}(\alpha_1 - 2\alpha_2)Gw_iU - \alpha_2 w_j GU_{ij}, \tag{60}$$

$$g_{ij} = (1 + 2\gamma GU)\delta_{ij} \tag{61}$$

with the PPN parameters γ, β, ξ, α_n, and ζ_n. Physically, γ parametrizes the amount of curvature caused by a unit rest mass and is the analog to the gravitational slip parameter in Eq. (9) on cosmological scales. The parameter β quantifies the amount of nonlinearity in the gravitational superposition, ξ captures preferred location effects, ζ_n and α_3 describe violations in the conservation of energy, momentum, or angular momentum, and the α_n parametrize preferred frame effects. Here, w^i is the system velocity in a universal rest frame and new potentials have been introduced that appear, for instance, in vector-tensor or bimetric theories of gravity.[145] GR is recovered when $\gamma = \beta = 1$ and all other parameters vanish. The parameter values for a range of other gravitational theories can be found in Ref. 34, where also a summary of constraints on the PPN parameters is presented. In particular, a bound of $|\gamma - 1| \lesssim 10^{-5}$ in the Solar System was inferred with the Cassini mission from the Shapiro time delay of a radio echo passing the Sun and orbital dynamics.[270] This can be compared to the cosmological parameter γ in Sec. 2.2.2, which implies that for larger cosmological effects modifications of gravity should be scale or environment dependent, for instance, due to a screening mechanism, or that the new field does not couple to baryons.

3.3. *Incorporating screening mechanisms*

Because of the linearization of the field equations in the post-Newtonian expansion, the nonlinear interactions that cause the screening effects are removed, which precludes a simple direct mapping of the screened models to the PPN formalism and the straightforward comparison to observational parameter bounds. Screening effects can also depend on ambient density, giving rise to both low-energy limits where screening operates and where it does not. Different approaches to performing a PPN expansion in the presence of screening mechanisms have been examined in Refs. 146, 147 and 271. Reference 271 derived expressions for the parameters γ and β and the effective gravitational coupling for a variety of scalar-tensor theories that screen at large values of the Newtonian potential by allowing an environmentally dependent mass for the scalar field, where the relations were found in terms of Newtonian and Yukawa potentials restricting to static spherically-symmetric systems. A Lagrange multiplier method[272, 273] was employed in Ref. 146 to perform the post-Newtonian expansion for the cubic Galileon model with Vainshtein screening to order $(v/c)^2$. Expansions based on such transformations can, however, be mathematically involved for more complex gravitational theories and the method has not been extended to large-field screening like the chameleon effect. A unified, more systematic, and efficient method for determining the effective field equations dominating in regimes of derivative or large-field screening or no screening was developed through a scaling approach in Ref. 219, where the post-Newtonian expansion to order $(v/c)^4$ for scalar-tensor theories with Vainshtein and chameleon screening implementing this method was presented in Ref. 147. Further applications to k-mouflage, linear shielding, or a Yukawa suppression can be found in Ref. 174.

Screening mechanisms may be incorporated in the PPN formalism following two different approaches:[219] either by an extension of the formalism through introducing new potentials; or by promoting the PPN parameters to functions of time and space. In the second approach, the parameters in Eqs. (59)–(61) become functions of the coordinates with G being replaced by an *effective gravitational coupling* G_{eff} and the parameter ξ introduced by preferred location effects with the spatially dependent expansion being promoted to a matrix ξ_{ij} to accommodate the screened models.[147] More precisely, for chameleon and cubic Galileon models, Ref. 147 finds

$$G_{\text{eff}} = G^{(0)} + \alpha^q \frac{\psi^{(q,p)}}{2U}, \tag{62}$$

$$\gamma = 1 - \frac{2\alpha^q \psi^{(q,p)}}{2G^{(0)}U + \alpha^q \psi^{(q,p)}}, \tag{63}$$

$$\beta = 1 + \alpha^q \left(\beta_{\text{BD}}^{(q)} + \beta_{\text{Scr}}^{(q)} \right), \tag{64}$$

$$\xi_{ij} = \alpha^q (1 - \delta_{ij}) \frac{(\epsilon_{jkl} W_k V_l)_{\text{eff}}}{\epsilon_{ikl} V_k W_l}, \tag{65}$$

where α is the scaling parameter,[147, 219] relating the mass scale introduced with the new field to the Planck mass. The parameters q and p denote the orders of the two simultaneous expansions in α and in (v/c), respectively, with the leading orders $q = -1/2$ and $p = 1$ for the cubic Galileon model and $q = 1/(1-n)$ and $p = 2/(n-1)$ for chameleon models with Jordan-frame potential $\alpha(\phi - \phi_{\min})^n$. Furthermore, $G^{(0)} = G\phi_0^{-1}$, where ϕ_0 is the background scalar field and ψ is its perturbation, solved by the corresponding perturbation equations. In particular, we note the recovery of the GR result $\gamma = 1$ in the screened limit $\alpha \to \infty$ for the cubic Galileon and $\alpha \to 0$ for the chameleon model (as well as the GR results $G_{\text{eff}} = G^{(0)}$, $\beta = 1$, and $\xi_{ij} = 0$). This is because of $\alpha^q \to 0$ in Eqs. (62)–(65) for the corresponding values of q in the two models. The parameter value for β in Brans-Dicke gravity is represented by β_{BD}. Furthermore, $\beta_{\text{Scr}} = \beta_{\text{Cubic}}$ or $\beta_{\text{Scr}} = \beta_{\text{Cham}}$ are additional terms that contribute for the cubic Galileon and chameleon models, respectively, where

$$\beta_{\text{BD}}^{(-\frac{1}{2})} = \frac{-\Phi_{\text{BD}}^{(-\frac{1}{2},3)} + G^{(-\frac{1}{2})}\frac{\delta h_{00}^{(0,4)}}{\delta G^{(0)}} + \gamma^{(-\frac{1}{2})}\frac{\delta h_{00}^{(0,4)}}{\delta\gamma^{(0)}}}{2G^{(0)2}U^2 + 4G^{(0)2}\Phi_2},$$

(66)

$$\beta_{\text{Cubic}}^{(-\frac{1}{2})} = \frac{\Phi_{\text{Cubic}}^{(-\frac{1}{2},3)}}{2G^{(0)2}U^2 + 4G^{(0)2}\Phi_2},$$

(67)

$$\beta_{\text{Cham}}^{(\frac{1}{1-n})} = \frac{-\Phi_{\text{Cham}}^{(\frac{1}{1-n},\frac{2n}{n-1})}}{2G^{(0)2}U^2 + 4G^{(0)2}\Phi_2}.$$

(68)

The new potentials introduced in Eqs. (65)–(68) can be found in Ref. 147.

Alternatively to this approach, attributing new potentials to the new corrections of the metric and assigning them their own parameter in the spirit of the PPN formalism, Ref. 219 finds for the cubic Galileon,

$$g_{00} = g_{00}^{\text{PPN}} + \sigma_1 \left(\psi^{(-\frac{1}{2},1)} + \tilde{\Phi}_1^{(-\frac{1}{2},3)} - 3\mathcal{A}_\psi^{(-\frac{1}{2},3)} - \mathcal{B}_\psi^{(-\frac{1}{2},3)} + 6G^{(0)}\tilde{\Phi}_2^{(-\frac{1}{2},3)} \right)$$
$$+ \sigma_2\psi^{(-\frac{1}{2},3)} + \sigma_{\text{Cubic}}\Phi_{\text{Cubic}}^{(-\frac{1}{2},3)},$$

(69)

$$g_{0i} = g_{0i}^{\text{PPN}} + \sigma_1 \left(\frac{1}{2}\mathcal{V}_i + \frac{3}{2}\mathcal{W}_i \right),$$

(70)

$$g_{ij} = g_{ij}^{\text{PPN}} - \sigma_1\psi^{(-\frac{1}{2},1)},$$

(71)

where $g_{\mu\nu}^{\text{PPN}}$ is defined by Eqs. (59)–(61). Here all potentials arising from $\psi^{(-\frac{1}{2},1)}$ are parametrized by the coefficient σ_1, those from $\psi^{(-\frac{1}{2},3)}$ by σ_2, and from $\Phi_{\text{Cubic}}^{(-\frac{1}{2},3)}$ by σ_{cubic}, where for the cubic Galileon $\sigma_1 = \sigma_2 = \sigma_{\text{Cubic}} = \alpha^{-\frac{1}{2}}$, which thus vanish in the screened limit $\alpha \to \infty$.

3.4. *Parametrizing gravitational waveforms*

From the measured change in the orbital period of binary pulsar systems due to the energy loss through gravitational wave emission one can infer an upper bound on the self-acceleration of the center of mass from violation of momentum conservation, which constrains the PPN parameter ζ_2.[34] Binary pulsars, however, only constrain the lowest two orders of a parametrized expansion of the gravitational waveform.[183]

In the same spirit as the PPN formalism, a parametrized post-Einsteinian (ppE) framework was introduced in Ref. 179 to parametrize the effects of departures from GR in the dynamical strong-field regime on the gravitational waveforms from the binary coalescence of compact objects with

$$h(f) = \left(1 + \sum_j \bar{\alpha}_j u^j\right) e^{i \sum_k \bar{\beta}_k u^k} h_{\mathrm{GR}}(f), \tag{72}$$

where $u \equiv (\pi \mathcal{M} f)^{1/3}$ with chirp mass $\mathcal{M}$ and frequency f. The GR waveform is reproduced when the parameters $\bar{\alpha}_j$ and $\bar{\beta}_k$ vanish. A particular subclass of the ppE framework is the generalized inspiral-merger-ringdown waveform model[180] recently used by the strong-field tests of GR by LIGO,[183, 184] where parametrizations of departures from GR are restricted to fractional changes in the parameters that determine the gravitational wave phase,

$$h(f) = e^{i\delta\Phi_{\mathrm{gIMR}}} h_{\mathrm{GR}}(f), \tag{73}$$

$$\delta\Phi_{\mathrm{gIMR}} = \frac{3}{128\eta} \sum_{i=0}^{7} \phi_i \delta\chi_i (\pi \tilde{M} f)^{(i-5)/3} \tag{74}$$

with total mass $\tilde{M}$, symmetric mass ratio $\eta = m_1 m_2/(m_1 + m_2)^2$, and i-th order post-Newtonian GR phase ϕ_i. The connection of the ppE framework and the generalized inspiral-merger-ringdown waveform model with the parametrization of modified gravitational wave propagation in Sec. 2.2.4 (and therefore with EFT in Sec. 2.2.3) is discussed in Ref. 186. For instance, one finds that $\sum_j \bar{\alpha}_j u^j = \ln[M(k, z)/M(k, 0)]$ for $\nu \equiv d \ln M^2/d \ln a$ and that $\sum_k \bar{\beta}_k u^k$ is determined by an integration over c_{T} and $\tilde{\mu}$ (see Ref. 186). Particularly, for only time-dependent modifications, it follows that $\bar{\alpha}_j = \bar{\beta}_k = 0$ except for non-vanishing $\bar{\alpha}_0$ and $\bar{\beta}_{\pm 3}$. This implies that the deviation in c_{T} is of fourth post-Newtonian order, $\tilde{\mu}$ of $\mathcal{O}(1)$, and ν of $\mathcal{O}(0)$. Furthermore, the modification $\delta\Phi_{\mathrm{gIMR}}$ is equivalent to that found for $\sum_k \bar{\beta}_k u^k$ with ν being irrelevant due to the absence of a modification of amplitude in the parametrization.

An interesting open problem is the incorporation of screening effects in the modifications of the gravitational waveform. One approach was recently proposed in Ref. 147, based on the post-Newtonian expansion of Refs. 181 and 182, which compute the gravitational waveform for a compact binary system in scalar-tensor gravity to post-Newtonian order $(v/c)^4$. After casting the Einstein equations into a *relaxed form* together with harmonic gauge conditions, those are solved as a retarded integral over the past null cone, whereby the integral is split into a near-zone and radiation-zone part.[274–276] To first approximation the effects of modified gravity are screened in the near zone and only enter in the radiation zone and through the boundary conditions.[147]

4. Further Parametrizations

Besides the PPF, EFT, PPN, or ppE formalisms, there are a range of further parametrization frameworks that have been developed for astrophysical tests of gravity. A few of those alternative approaches shall be briefly discussed here.

The PPN formalism neglects the evolution of the cosmic background and assumes asymptotic flatness, hence, it needs to be adapted for cosmological applications. Such an extension is provided by the parametrized post-Newtonian cosmology[176] (PPNC) framework that is based on four free functions of time. Thereby the post-Newtonian expansion is adopted for small regions of space and then patched together to determine the cosmological large scales. This is accomplished by performing coordinate transformations with local scale factors that are associated with a global scale factor employing an appropriate set of junction conditions. Four free functions of time $\{\alpha, \gamma, \alpha_c, \gamma_c\}$ are then introduced to describe modifications of the Poisson equation of the gravitational potentials Φ and Ψ up to $(v/c)^3$. Hereby, γ and α can be linked to the PPF parameters in Sec. 2.2.2 but also to the PPN parameters in Sec. 3.2. A number of dark energy, scalar-tensor, and vector-tensor theories can be described by the formalism. However, the framework so far does not provide a relativistic completion nor a description for Yukawa interactions or screening mechanisms.

A more direct analogy to the post-Newtonian formalism for relativistic cosmology than the frameworks discussed in Sec. 2 is given by the post-Friedmannian formalism of Ref. 177, where the expansion of the metric is performed in inverse powers of the light speed in a cosmological setting in Poisson gauge (also see Ref. 277 for an application). After some field redefinition and applying the linearized Einstein equations, the formalism reproduces linear cosmological perturbation theory, thus providing a link between the post-Newtonian limit on small scales and the large-scale structure. A parametrization of the gravitational modifications in the formalism has yet to be developed.

An analogous approach to the effective field theory of dark energy and modified gravity (Sec. 2.2.3) was conducted for the perturbations of a static spherically symmetric system in Ref. 278. Thereby the ADM spacetime decomposition was employed for a $2 + 1 + 1$ canonical formalism that separates out the time and radial coordinates. The effective action was then built from scalar quantities of the canonical variables of this spacetime foliation and the lapse, which encompasses scalar-tensor theories of gravity. Reference 278 derived three background equations that can be used for the generalized study of screening mechanisms as well as the linear perturbations around this background for stability analyses.

Finally, a unified parametrization covering chameleon, dilaton, and symmetron models on large and small cosmological scales as well as in Solar-System and laboratory tests was introduced in Ref. 63. It uses the cosmological time variation of the mass $m(t)$ of the scalar field ϕ and its coupling $\beta(t)$ at the minimum of its effective potential $V_{\text{eff}}(\phi)$. The starting point is the scalar-tensor action in Einstein frame with $\mathcal{L}_\phi = -(\nabla\phi)^2/2 - V(\phi)$ and a conformal factor $A(\phi)$ defined by $g_{\mu\nu} = A^2(\phi)\tilde{g}_{\mu\nu}$, where $\tilde{g}_{\mu\nu}$ denotes the Einstein frame metric. The scalar-field equation in the presence of pressureless matter is $\Box\phi = \beta\rho_m + dV/d\phi$ with

$$\beta(\phi) \equiv M_{\rm Pl}\frac{d\ln A}{d\phi}, \tag{75}$$

where $M_{\rm Pl}$ denotes the bare Planck mass. This defines an effective scalar field potential $V_{\rm eff} = V(\phi) + [A(\phi) - 1]\rho_m$ with

$$m^2(\phi) = \frac{d^2 V_{\rm eff}}{d\phi^2}\bigg|_{\phi_{\rm min}}, \tag{76}$$

where $\nabla^2\phi = V_{\rm eff,\phi}$ and $\phi_{\rm min}$ is the minimum of the potential $V_{\rm eff}$. Appropriate choices of $m(\phi)$ and $\beta(\phi)$ recover the chameleon, symmetron, and dilaton models. Screening occurs in the regime $|\phi_\infty - \phi_c| \ll 2\beta(\phi_\infty)M_{\rm Pl}\Psi$, where ϕ_c and ϕ_∞ are the scalar field values in the minima inside and outside of a body. For $m^2 \gg H^2$, the scalar field also remains at the minimum in the cosmological background such that its cosmological evolution is given by $d\phi/dt = 3H\beta A\bar{\rho}_m/(m^2 M_{\rm Pl})$.

This allows one to reconstruct the scalar-tensor Lagrangian from $m(t)$ and $\beta(t)$. Using

$$\phi(a) - \phi_{\rm i} = \frac{3}{M_{\rm Pl}}\int_{a_{\rm i}}^a da\,\frac{\beta(a)\bar{\rho}_m(a)}{a\,m^2(a)}, \tag{77}$$

where $\phi_{\rm i}$ is the initial scalar field value at $a_{\rm i}$, one finds

$$\int_{\phi_{\rm i}}^\phi \frac{d\phi}{\beta(\phi)} = \frac{3}{M_{\rm Pl}^2}\int_{a_{\rm i}}^a da\,\frac{\bar{\rho}_m(a)}{a\,m^2(a)}, \tag{78}$$

$$V = V_{\rm ini} - \frac{3}{M_{\rm Pl}^2}\int_{a_{\rm i}}^a da\,\frac{\beta^2(a)\bar{\rho}_m^2(a)}{a\,m^2(a)}. \tag{79}$$

Spherical collapse, halo model, and cosmological N-body simulations in this unified approach have been studied in Refs. 279 and 238. While the approach covers screening effects that operate at large Ψ, it does not cover k-mouflage or Vainshtein screening operating through derivatives of Ψ.

5. Summary and Outlook

Intensive efforts have been devoted to the development of generalized frameworks that enable the systematic exploration and testing of the astronomical and cosmological implications of the wealth of proposed modified gravity, dark energy, and dark sector interaction models. This chapter reviewed a number of different formalisms devised for this purpose. In the low-energy static limit of gravity, the PPN expansion has been highly successful in conducting model-independent tests of gravity and inferring stringent constraints on departures from GR, mainly from observations in the Solar System. An overview was presented of the post-Newtonian expansion, its parametrization, and extensions developed to incorporate the screening mechanisms motivated by cosmological modifications of gravity.

In cosmological settings the PPN formalism is not suitable as the expansion of the background needs to be accounted for. With the objective of imitating the success of PPN, a number of cosmological counterpart frameworks have been developed, generally commonly referred to as PPF formalisms. These aim at consistently unifying the effects of modified gravity, dark energy, and dark sector interactions on the cosmological background evolution and the formation of linear and nonlinear large-scale structure. This includes parametrizations of the equation of state, the growth rate of structure, closure relations for cosmological perturbation theory, EFT, generalized perturbation theory, interpolation functions between linear and nonlinear regimes as well as phenomenologically parametrized screening mechanisms.

The new era of gravitational wave astronomy facilitates further powerful tests of gravity. Modified gravity and novel interactions can manifest both through cosmological propagation effects or in the emitted gravitational waveforms. A parametrization formalism for the propagation effects has been reviewed and formalisms inspired by PPN for the parametrization of modified waveforms has also briefly been discussed. It has also been described how these formalisms can be connected to EFT and PPF.

The PPN and PPF formalisms are generally separated frameworks. It, however, seems feasible to undertake more general steps towards a *unified parametrization for gravity and dark entities* suitable to all scales and types of observations. More specifically, if restricting to scalar-tensor theories of the chameleon, dilaton, or symmetron types a unification can be realized through a parametrization of the cosmological time variation of the mass of the scalar field and the coupling at the minimum of its effective potential.[268] This reconstructs a Lagrangian that can be used to connect the different parametrization frameworks as it encapsulates the full nonlinear freedom of these models, enabling applications to cosmology, the low-energy static limit as well as gravitational waves and laboratory tests. More generally, focusing on a new scalar degree of freedom a unification can in principle be achieved[174] by performing a reconstruction[234, 280] of general scalar-tensor theories from the linear EFT functions, which can then be connected to nonlinear PPF frameworks as well as the PPN and waveform formalisms. Further approaches to unifying PPN and PPF such as with PPNC or a post-Friedmannian expansion have also briefly been inspected. Completing the development of a global framework for tests of gravity and the dark sector is subject to current research. Applications of such a framework promise to remain a very interesting and active field of research over the next decades with the wealth of high-quality observational data becoming available for tests of gravity, spanning a wide range of scales and encompassing a great variety of observational probes.

Acknowledgments

This work was supported by a Swiss National Science Foundation (SNSF) Professorship grant (No. 170547), a SNSF Advanced Postdoc.Mobility Fellowship (No. 161058), and the Science and Technology Facilities Council Consolidated Grant for Astronomy and Astrophysics at the University of Edinburgh.

Chapter 3

Simulation Techniques

Claudio Llinares

Institute for Computational Cosmology, Department of Physics,
Durham University, Durham DH1 3LE, UK

Institute of Cosmology and Gravitation, University of Portsmouth,
Dennis Sciama Building, Portsmouth PO1 3FX, UK

claudio.llinares@port.ac.uk

The standard paradigm of cosmology assumes General Relativity (GR) is a valid theory for gravity at scales in which it has not been properly tested. Developing novel tests of GR and its alternatives is crucial if we want to give strength to the model or find departures from GR in the data. Since alternatives to GR are usually defined through non-linear equations, designing new tests for these theories implies a jump in complexity and thus, a need for refining the simulation techniques. We summarize existing techniques for dealing with modified gravity (MG) in the context of cosmological simulations. N-body codes for MG are usually based on standard gravity codes. We describe the required extensions, classifying the models, not according to their original motivation, but by the numerical challenges that must be faced by numericists. MG models usually give rise to elliptic equations, for which multigrid techniques are well suited. Thus, we devote a large fraction of this review to describing this particular technique. Contrary to other reviews on multigrid methods, we focus on the specific techniques that are required to solve MG equations and describe useful tricks. Finally we describe extensions for going beyond the static approximation and dealing with baryons.

1. Introduction

Tracking the evolution of the matter distribution in the late Universe is a multi-scale problem in both space and time. For instance, the evolution of the largest voids in the Universe have associated scales of tens of Mpc and gigayears. On the contrary, typical scales associated with merger events in high density regions are kpc and megayears or event shorter. Thanks to this, studying the evolution of the Universe as a whole is a highly non-trivial task. This chapter is about how to calculate the evolution of matter over the largest possible dynamical range (i.e. maximizing the difference between the largest and smallest simulated scales). While these are the most precise calculations that can be done, they are also the most expensive and require the implementation of complex numerical techniques.

According to the scales involved, there are four ways in which we can study the evolution of the distribution of matter in the Universe:

- *Background cosmology:* the matter distribution is assumed to be uniform. In this case there is no gravitational collapse and the only dynamics is associated with the expansion rate. The resulting equations are ordinary differential equations which can be solved analytically or with Runge-Kutta methods.
- *Linear cosmology:* the matter distribution is assumed to have perturbations which are small when compared to the background density. This means that it is possible to linearise the equations with respect to the density perturbations. In the case of General Relativity (GR), this approximation gives rise to ordinary differential equations for the density perturbations. By using this approach, it is possible to obtain high accuracy predictions for the redshift $z = 0$ power spectrum for scales that correspond to frequencies $k \lesssim 0.1 \, h/\mathrm{Mpc}$.
- *Non-linear cosmology:* when the matter perturbations become of order one, it is not possible to linearise the equations with respect to the density any more. Under these conditions, it is necessary to keep all the terms in an expansion with respect to the density. However, as cosmological evolution is not affected by the presence of black holes, non-linear terms can still be neglected in the velocities and the metric perturbations. The resulting equations are typically partial differential equations in space and time, and complex simulation techniques are required.
- *Relativistic or post-Newtonian cosmology:* including relativistic species or super-horizon scales requires the addition of higher order terms in the Einstein's equations. The solution of these equations can be obtained with specialised codes which take into account post-Newtonian corrections.[281, 282]

This chapter will deal with the third class of calculations in a modified gravity (MG) context. With few exceptions, existing MG codes can simulate only scalar tensor theories, which means that they follow only one additional degree of freedom. This requires solving the equations of motion for this field in addition to the standard N-body equations. The chapter is structured according to the class of equations that describe this field. We will describe the methods step by step increasing the complexity of the equations and thus of the required solvers. A summary of this classification, as well as the methods that are commonly used and existing implementations are given in Table 1. An in-depth discussion of the definition of the models and, in particular, of the recurrent concept of screening mechanism can be found in Chapter 1. For a more detailed coverage of N-body simulations in modified gravity models, the techniques developed, and their potential applications to other fields, the interested readers are referred to the book Ref. 283.

Dark energy models share similarities with models that are intended to explain the dark matter component of our Universe (a.k.a. MOND). For this reason, there is also a lot to learn from the implementation of these particular models. We summarize in Table 2 existing methods and implementations for this family of theories.

The robustness of the methods and their implementation can be quantified by comparing results with analytic or linear solutions. However, due to the lack of analytic solutions in the fully non-linear regime, different implementations of the models are validated also by comparing codes that were written independently by different authors. Such a comparison was presented for instance in Refs. 284 and 285. Furthermore, a methodical analysis of the accuracy of these codes was presented in Ref. 286. Several quantities were compared: power spectrum at different redshifts, velocity power spectrum, mass function, density and velocity dispersion profile of halos and field distributions. As the base codes used for each MG implementation were different, the comparison was made on the prediction of the deviations from GR rather that the absolute quantities. It was found that all the MG codes that took part in the experiment agree in these predictions up to an accuracy of 1%. In particular, this accuracy was found for the power spectrum of density perturbations in a range of scales that goes from below $k = 0.1$ to $k \sim 7 \, h/\mathrm{Mpc}$, which is consistent with the scales that will be included in the next generation of galaxy surveys.

I close this introduction by giving a warning note. Perceptive readers may notice the absence of diagrams in the text. This is actually deliberate and intended to force the readers to do the diagrams by themselves. Different people will make the diagrams in different ways. This will increase the variety of interpretations of the methods described here, which in turn, will increase the chances of new ideas to be triggered by this modest contribution.

2. The N-body Equations for Cold Dark Matter Simulations

A change of the gravitational theory constitutes a change of one of the fundamental pillars on which cosmological simulations are built. Thus, before searching for simulation methods for MG, it is important to define in a precise way what cosmological simulations are and which equations are involved. This is not a trivial task in the most general case. So in order to keep the discussion focused on the changes that are required to simulate alternative gravitational models, and not to get lost in technicalities of more advanced simulations, we will restrict the discussion to the simplest type of cosmological simulations. These are simulations that deal with non-relativistic collisionless fluid (i.e. cold dark matter). Some details of what is necessay to simulate other kinds of fluids will be given in the last section of this chapter.

The simplest definition we could imagine is the following:

Definition 2.1. *A cold dark matter cosmological simulation is the solution of Newtonian equations of motion in an expanding universe for a continuous density field of non-relativistic collisionless matter.*

While this definition might be good enough when applied to standard gravity simulations, it is important to be more precise in the MG case. This is because there is no guarantee that the approximations that were made to derive the Newtonian

Table 1: Classification of modified gravity models and their implementations. In all the cases, L represents a differential operator and f is a scalar function.

Type of equation	Model	Method	References
$L(\phi, \dot{\phi}, \ddot{\phi}) = f(\rho, \phi)$ L, f linear	Dynamical DE	Modified expansion	ART[287]
	Coupled DE	Modified expansion + G_{eff}	Refs. 288, 289
	Non-local	Modified expansion + G_{eff}	RAMSES[290]
	Vector DE	Modified expansion, power spectrum and growth history	Ref. 291
$L(\phi, \partial_x\phi, \partial_x^2\phi) = f(\rho, \phi)$ L linear in $\partial_x^2\phi$.	$f(R)$	Multigrid (Jordan frame)	Ref. 292 MLAPM[173] ECOSMOG[293] MG-GADGET[284]
		Multigrid (Einstein frame)	Isis[285]
	Generalized chameleon	Multigrid (Einstein frame) Multigrid (Jordan frame)	MLAPM[294] ECOSMOG[279]
	Symmetron	Multigrid	MLAPM[295] Isis[285] ECOSMOG[279]
	Dilaton	Multigrid ECOSMOG[279]	MLAPM[296]
$L(\phi, \partial_x\phi, \partial_x^2\phi) = f(\rho, \phi)$ L non-linear in $\partial_x^2\phi$	DGP	Multigrid with smoothing	DGPM[101]
		Fourier with smoothing	Refs. 297, 298
		FFT-relaxation with operator splitting	Ref. 299
		Multigrid with operator splitting	ECOSMOG[78]
	Cubic Galileon	Multigrid with operator splitting	ECOSMOG[300]
	Quartic Galileon	Multigrid with operator splitting	ECOSMOG[301]
$L(\ddot{\phi}, \dot{\phi}, \partial_x\phi, \partial_x^2\phi) =$ $f(\rho, \dot{\phi}, \phi)$	Symmetron	Leap-frog	Solve[302] Isis[303]
	$f(R)$	Implicit	ECOSMOG[304]
	Disformal	Leap-frog	Isis[305]

limit of GR are valid in MG. We need to keep track of all these approximations, for which it is necessary to derive the equations starting from basic principles. For instance, for some specific MG models, the gravitational slip (i.e. the ratio between time and spatial scalar perturbations of the metric) is not zero. In these cases, new equations will have to be solved to take into account the new degrees of freedom that may arise. The form of these new equations can be only derived in a fully relativistic set up (i.e. starting from the basic principles defined by the Einstein-Hilbert action and making necessary approximations). Thus, we re-define a cosmological simulation as follows:

Definition 2.2. *A cold dark matter cosmological simulation is the solution of Einstein's equation (or their modified versions) up to first order in the metric perturbations and velocities. Furthermore, the energy-momentum tensor of the matter component is assumed to be isotropic and to represent non-relativistic collisionless matter.*

With the exception of the code presented in Ref. 306, the integration of the equations is performed in a cubic region of space assuming periodic boundary conditions. Given this definition, it is now easy to derive the equations that need to be solved in the cosmological codes. We describe the derivation for the standard gravity case first and then describe the modifications that should be made to this derivation when dealing with extended models.

With few exceptions, these equations are solved with N-body methods, which rely on the fundamental concept of N-body particle. These techniques are based on the idea of discretizing the continuous density fields and describing them with a large set of particles. Thus, the relatively small set of equations that define the physical model (in our case Einstein's equations) is substituted by a much larger set of equations that describe the trajectory of these particles. Since there is not a one-to-one correspondence between the N-body particles and the physical particles that are associated to the continuous fields, the interactions between the N-body particles must be smoothed out. See Ref. 307 for a detailed discussion on the concept of N-body particles and methods. During the rest of this chapter we will use the terms *cosmological simulation* and *N-body simulation* as synonyms. However, the distinction between the concept of simulation and the method that we use to run them should always be kept in mind.

2.1. *General relativity*

The aim is to solve Einstein's equation for a cosmological non-relativistic collisionless fluid. In N-body simulations, the density of such a fluid is described by a set of particles. So we need to re-write Einstein's equation in a way that its solution can be described by particles. We do this by taking its divergence:

$$\nabla^a T_{ab} = 0, \tag{1}$$

Table 2: Classification of the MOND family of modified gravity models and their implementations.

Model	Method	References
AQUAL	Algebraic	Ref. 308
		AMIGA[309]
	Multigrid	Ref. 310
		Ref. 311
		AMIGA[312]
		Solve[313]
	Multigrid (on spherical grid)	NMODY[314, 315]
QUMOND	Poisson based	Solve[313]
		Ref. 316
		RAyMOND[317]
		POR[318]
Bi-metric	Multigrid	Solve[313]

which describes the conservation of energy. The energy momentum tensor associated with a set of N-body particles is then assumed to be the addition of Dirac's δ functions associated with each particle. This gives a set of geodesics equations for each particle:

$$\frac{d^2 x_n^a}{d\tau^2} + \Gamma_{bc}^a \frac{dx_n^b}{d\tau} \frac{dx_n^c}{d\tau} = 0, \tag{2}$$

where Γ_{bc}^a are the Christoffel symbols, the index n runs over all the particles and τ is proper time (see for instance exercise 1.9 in Ref. 319). In order to get the equations for the N-body codes, we need to fix the reference frame, which can be defined by fixing the metric. The most general perturbation of the Friedmann-Lemaitre-Robertson-Walker metric in Newtonian gauge is given by

$$ds^2 = a^2(t) \left[-(1 + 2\Psi) dt^2 - 2B_i dx^i dt + (1 - 2\Phi) \delta_{ij} dx^i dx^j + h_{ij} dx^i dx^j \right], \tag{3}$$

where t is conformal time and Φ, Ψ, B_i, and h_{ij} are scalar, vector and tensor perturbations. Different codes use a different independent variable for the integration. Common choices are supercomoving time,[320] expansion factor $a(t)$ or conformal time. The supercomoving time has the advantage of making the hydro-dynamical equations independent of a, which allows the implementation of standard algorithms defined on Minkowski space-time. With the exception of the works presented in Refs. 281 and 282, vector and tensor perturbations are neglected. Substituting this metric in Eq. (2) and changing the parameter of the geodesics from proper time to conformal time gives the following equation for the spatial coordinates of each particle

$$\ddot{x}_n^i + H(a)\dot{x}_n^i + \left(\frac{\partial \Phi}{\partial x^i} \right)_n = 0, \tag{4}$$

where the dots represent partial derivatives with respect to conformal time and $H(a) = \dot{a}/a$ and second order terms in the velocities were neglected. The treatment of the damping term in this equation (second term in the left hand side) can be simplified with the following definition

$$p_n^i \equiv a\dot{x}_n^i, \tag{5}$$

which gives

$$\dot{x}_n^i = \frac{p_n^i}{a(t)} \tag{6}$$

$$\dot{p}_n^i = -a(t)\left(\frac{\partial \Phi}{\partial x^i}\right)_n \tag{7}$$

which are the Hamilton equations that can be found in N-body codes. Note the presence of expansion factors in these equations. As the Universe expands, objects become farther and farther away from each other, with a consequent weakening of the gravitational force between them. These factors take into account this effect. It may be surprising that the effect appears in the so called comoving coordinates defined by Eq. (3), in which the background position of the objects does not change with time. Even if the objects do not move, their interdistance (which is a covariant quantity and thus, independent of the reference frame), still increases with time, which results is smaller and smaller forces as time passes.[a]

In order to complete the set of equations that will be solved with the N-body code, we need to find an equation for the force that appears in Eq. (7). This will be given by the Einstein's equations themselves[b]:

$$G_{ab} + g_{ab}\Lambda = T_{ab}. \tag{8}$$

Definition 2.2 says that we need to linearise these equations with respect to the metric perturbations and velocities (but not with respect to the density, which is treated non-perturbatively).[c] So we need to write down the perturbed Einstein's equations for the metric (3). In the GR case, the ij component of Einstein's equations says that the two scalar perturbations are equal. This is also valid in MG

---

[a]It is important not to confuse the *comoving coordinates* with the *comoving distance*. The second is an alternative definition of distance used by cosmologists which is defined as the distance with respect to the distance that is calculated using the background metric. Under this alternative definition, galaxies do not depart from each other because of the expansion, however, the effect of the expansion is still present in the Poisson and Hamilton equations.

[b]Note that I define the Einstein's equation as containint a cosmological constant. This may open up the discussion about this equation representing GR or the ΛCDM cosmological model within GR. Different authors have different opinions about these definitions. In order to fix the terminology and to avoid getting lost in philosophical discussions about what is GR and what is modified gravity, I define GR as in Eq. (8). Everything else will be referred to as modified gravity.

[c]The linearisation of the velocities mentioned here should not be confused with the linearisation that is usually made when solving the equations in the context of linear cosmology. Once the equations are determined, no restrictions exist in the amplitude of the velocities when solving for them with N-body codes.

in the Einstein frame, but it is not necessary valid in the Jordan frame. However, as the conformal factor that relates both frames is very close to one, it is usually assumed that both perturbations are equal (i.e. the present generation of simulation codes do not include gravitational slip).

Neglecting vector and tensor perturbations gives rise to equations that depend only on one degree of freedom. This means that all the information provided by Einstein's equations can be condensed into one equation. It is costumary to obtain this information from the 00 component of Einstein's equations. By substituting the metric given by Eq. (3) in Eq. (8), neglecting second order terms in Φ, and substracting the order zero equation (i.e. Friedmann equation) we obtain a generalization of Poisson's equation in an expanding universe:

$$\nabla^2 \Phi = 4\pi G a^2(t)\delta\rho, \tag{9}$$

where we have assumed the following energy-momentum tensor:

$$T_{ab} = (\rho_0 + \delta\rho)u_a u_b + (P_0 + \delta P)(g_{ab} + u_a u_b), \tag{10}$$

where u^a is the 4-velocity of the fluid, ρ_0 and P_0 are background density and pressure and $\delta\rho$ and δP are perturbations on these quantities (which are not assumed to be small). Furthermore, appropriate gauge choices have to be made to remove time derivatives of the metric perturbation. See Refs. 321–323 and 324 for a discussion on this issue. The function $a(t)$ in Eqs. (6), (7) and (9) is obtained as a solution of the zeroth order Einstein's equations (i.e. the Friedmann equations):

$$\frac{\dot{a}(t)}{a(t)} = H_0^2 \left[\frac{\Omega_m}{a(t)^3} + \Omega_\Lambda \right]. \tag{11}$$

The routines that solve Eq. (9) are some of the more complex routines in N-body codes and will be the main focus of the sections that follow. Several algorithms exist at present. For instance, particle mesh (PM) codes solve explicitly Poisson's equation on a grid (through multigrid or Fourier based methods) and estimate the force with discretization formulae for the derivatives. The forces are then interpolated back to the particle positions using specific smoothing kernels. The density that is required to solve the Poisson's equation is calculated from the particle distribution using the same type of interpolation.[d]

Alternatively, in the case of direct summation codes, Poisson's equation is integrated once analytically and thus, the gravitational force can be obtained by adding up individual contributions from each particle:

$$(\nabla\Phi)_n = 4\pi G \sum_m \sum_n \frac{M_n M_m}{|x_n^i - x_m^i|^2} \mathbf{x}_{mn}, \tag{12}$$

---

[d]We remind the reader that in order to ensure momentum conservation, the same kernel must be used to estimate the density on the grid and to interpolate the force to the particles' position. The definitions of commonly used interpolation kernels can be found in Ref. 307.

where the indexes m and n run over all the particles and $\mathbf{x}_{mn}$ is a unitary vector between the particles m and n. In practice, these very expensive summations can be substituted by approximated summations in which the geometry of the long range interactions is simplified. Furthermore, it is possible to combine direct summation and PM algorithms to construct even faster codes. Finally, there are the so called multipole methods, which rely on a multipolar expansion of Poisson's equation.

Note that direct summation and multipole algorithms can be implemented only because Poisson's equation is linear (i.e. superposition principle holds). In general, this property does not apply to equations that arise in MG models and thus, direct summation, multipole or Fourier techniques are not used for simulating MG. The exact implementation of these methods in the standard gravity case is beyond the scope of this chapter and can be found for instance in Refs. 325–327 or 328.[e]

To summarize, in the standard gravity case, it is possible to translate the definition of N-body simulation given at the beginning of this section into the following more practical definition:

Definition 2.3. *A standard gravity cold dark matter cosmological simulation is the solution of Eqs. (6) and (7) for a large set of particles. The force that appears in these equations can be obtained by either solving Eq. (9) on a grid (in the case of particle-mesh codes) or using Eq. (12) (in the case of tree codes) together with Eq. (11).*

The next section reformulates this definition for the MG models that were already implemented in codes and for which already exist non-linear simulations.

2.2. *Modified gravity*

Following Definitions 2.2 and 2.3, the implementation of a MG model in an N-body code has two distinctive stages. First, the N-body equations have to be determined. This means, that it is necessary to find a generalization of Eqs. (6), (7), (9) and (11). The choice of these generalized equations is not unique. For instance, very complex theories can be mapped into simple ones. Furthermore, in several cases it is possible to approximate complex equations by simple ones for which simpler methods can be applied without loss of accuracy. The decisions taken at this stage can make the difference between being able to simulate a model with minimal effort and not simulate it at all. Once the equations have been fixed, it is possible to implement them in N-body codes. In many cases the resulting equations largely differ from the type of equations that can be found in GR. However, these equations may exist

---

[e]Further information on how simulations work can be found in the presentation of individual codes listed in Tables 1 and 2 and references therein. The ultimate way of understanding how an N-body code works is to write one from scratch. This is not such as big task as it seems. State-of-the-art codes typically contain more that 100,000 lines and require several years of development. However, minimalistic implementations for training purposes, can consist only in 2,000 to 3,000 lines which can be easily written.

in different fields of physics. Multidisciplinary collaborations and communication between different communities are crucial when it comes to the implementation stage.

2.2.1. *Scalar field theories*

Almost all the theories that have been simulated can be defined by adding a scalar field to the Einstein-Hilbert action. There is almost infinite freedom in the way this can be done. However, all this diversity can be condensed into four different properties of the action. So the most general action that has been simulated is:

$$S = \int \sqrt{-g} R d^4 x + \int \sqrt{-g} \left[f(\partial \phi, \partial^2 \phi) + V(\phi) \right] d^4 x +$$
$$\int \sqrt{-\tilde{g}} \mathcal{L}_M (m(\phi), \tilde{g}_{ab}(g_{ab}, \phi)) d^4 x = S_G + S_{KG} + S_M, \quad (13)$$

where S_G is the action that describes the geometry, S_{KG} the one that gives the Klein-Gordon equation for an uncoupled field and S_M is the matter action. The four characteristics that define different classes of theories are the following:

- $f(\partial \phi, \partial^2 \phi)$: a free function of first and second covariant derivatives of the field ϕ (i.e. the kinetic term), which includes contractions between derivatives.
- $V(\phi)$: a potential in which scalar field oscillates or roles.
- $m(\phi)$: a coupling between the scalar field and the matter fields.
- $\tilde{g}_{ab}(g_{ab}, \phi)$: a relation between the Jordan and Einstein frame metrics.

Different choices of these four ingredients give different families of theories. The classification that is relevant for numericists is the one that highlights the numerical complexity of the resulting equations. In this sense, the exact form of the free functions does not matter, but only its functional dependence. The complexity varies from very simple models in which the scalar field is uncoupled and only affects the background evolution of the Universe up to theories that introduce non-linear couplings with the matter fields which can lead to fully non-linear partial differential equations in space and time. Table 3 summarizes this classification with increasing complexity of the equations and methods. Note that this classification is inspired in both the original definition of the models as well as the numerical techniques that have been applied. However, the classification has some flexibility: theories from a given class may be mapped into theories that belong to different classes (see for instance Ref. 329 for a detailed discussion on the relation between classes 4 and 5 in Table 3). Implementation techniques that are applied to each class follow.

Class 1: Unperturbed fields: The simplest imaginable scalar fields are the ones that have a standard kinetic term (i.e. $f = (\partial \phi)^2$) and are uncoupled. As the field is not affected by metric nor matter perturbations, it is expected that it will not have spatial perturbations and thus, it will only evolve in the background. The only effects that this will produce in the matter distribution are through the energy of

Table 3: Classification of scalar field models that has been simulated.

Class	Kinetic term	Potential	Coupling	Jordan metric	Model
1	$(\partial\phi)^2$	$V(\phi)$	1	g_{ab}	Ratta & Peebels SUGRA
2	$(\partial\phi)^2$	$V(\phi)$	$m(\phi)$ (non-universal)	g_{ab}	Coupled DE
3	$(\partial\phi)^2$	$V(\phi)$	1	$A(\phi)g_{ab}$	Chameleon Symmetron Dilaton
4	$(\partial\phi)^2$	$V(\phi)$	1	$A(\phi)g_{ab} + B(\phi)\phi_{,a}\phi_{,b}$	Disformal
5	$(\partial\phi)^2$ + higher order	$V(\phi)$	1	g_{ab}	Galileon

the field, which will affect the expansion of the Universe. Two modifications have to be made to the codes to include this class of models: background expansion tables (which are needed because of the dependence of Eqs. (6), (7) and (9) on a) and the power spectrum that is required for generating the initial conditions (i.e. the growth factor). Implementation of these models has no overhead with respect to GR simulations and thus it is possible to simulate these models with the same resolution that can be reached in state-of-the-art standard gravity simulations.

Class 2: Fields with explicit coupling to dark matter: the simplest way to add a coupling to matter is to couple only the dark matter component and leave the baryons uncoupled. This will allow the model to pass Solar System tests and at the same time allow the use of very simple equations which do not include screening mechanisms. The consequence that such coupling has in the N-body equations is the addition of an effective gravitational constant which is a function of the field plus an additional friction term.

Class 3: Fields with conformal coupling: Giving a field dependence to the Jordan frame metric provides an extra term in the conservation of energy which acts as a fifth force in the geodesics equation (Eqs. (6) and (7)). In addition to this, the equation of motion for ϕ becomes non-linear. Different choices of $V(\phi)$ and $A(\phi)$ give rise to different non-linear terms which in turn give rise to different screening mechanisms. Thanks to these screening mechanisms, these fields can be coupled to the baryons and at the same time pass Solar System tests. The standard kinetic term gives rise to a hyperbolic Klein-Gordon equation which can be solved with the methods described in Sec. 4. However, it is customary to assume the quasi-static approximation (i.e. to neglect the time derivatives of the field). This will give rise to elliptic equations and enable us to implement the multigrid methods described in Sec. 3.

Class 4: Fields with disformal coupling: The addition of derivatives of the field in the definition of the Jordan frame metric introduces a coupling with the time derivatives of the field in both the Klein-Gordon equation as well as in the geodesics. This means that the equations contain terms such as $\dot{\phi}^2\rho$ and $\dot{\phi}^2\nabla\phi$. These terms can only be treated with the non-static solvers described in Sec. 4 (however, see Ref. 896).

Class 5: Fields with non-standard kinetic terms: The addition of higher order terms in the kinetic part of the scalar field action can give rise to fully non-linear equations of motion for the field (i.e. the equations are non-linear in the second derivatives of the field). These equations were included in the codes in the quasi-static limit using the multigrid techniques presented in Sec. 3. The full non-linearity makes the additional techniques presented in Sec. 3.5.3 mandatory. Note that these models do not include coupling to matter (both explicitly or through the definition of the Jordan metric). This means that the geodesics equations are unchanged and the effects of the scalar field on the matter distribution are through the energy of the scalar field which will appear as an additional term in the right hand side of the Poisson's equation (Eq. (9)).

Theories such as $f(R)$ are not included in Table 3 because, even if it is possible to map them into scalar-tensor theories, their definition have different motivation (i.e. they are not defined in the Einstein frame by adding a scalar field, but by adding a function of the Ricci scalar to the Einstein-Hilbert action in the Jordan frame). Details on these other theories, are given below in independent sections.

2.2.2. *Non-local gravity*

A particular representation of the model can be defined by adding a term $m^2 R\Box^{-2}R$ to the Einstein-Hilbert action[52, 330, 331] and was already implemented in the RAM-SES code.[290] It can be shown that this non-trivial non-local action, can be recast into a local action which includes two scalar fields. The resulting equations of motion for these fields are linear and thus, it is possible to use the simple methods that are applicable to class 2 described in Sec. 2.2.1.

2.2.3. *The $f(R)$ family of theories*

The action that defines this family of theories is usually given in the Jordan frame:

$$S_{f(R)} = \int \frac{\sqrt{-\tilde{g}}}{16\pi G} [R + f(R)] \, d^4x + S_m(\tilde{g}_{ab}, \psi), \tag{14}$$

where f is a free function and ψ are matter fields. The only model that has been simulated to date was proposed in Ref. 72 and it is defined with the following function

$$f(R) = -m^2 \frac{c_1(R/m^2)^n}{c_2(R/m^2)^n + 1}. \tag{15}$$

The linearized Einstein's equation includes two scalar degrees of freedom, which means that in addition to a modified Poisson's equation, the model requires an extra equation to track this extra field. These two degrees of freedom can be parametrized as the usual gravitational potential Φ (i.e. the scalar perturbation of the metric defined by Eq. (3)) and the derivative of the free function itself $f_R \equiv \frac{df}{dR}$. After minimizing the action (14) and assuming the quasi-static limit, we obtain the following linearized equations for these two degrees of freedom:

$$\nabla^2 \Phi = \frac{16\pi G}{3}\delta\rho - \frac{1}{6}\delta R(f_R) \tag{16}$$

$$\nabla^2 f_R = -\frac{8\pi G}{3c^2}\delta\rho + \frac{1}{3c^2}\delta R(f_R), \tag{17}$$

where $\delta R(f_R)$ is the perturbation of the Ricci scalar with respect to the background cosmological value. Its functional dependence with the field f_R is non-linear and given by:

$$\delta R(f_R) = \bar{R}(a)\left(\sqrt{\frac{\bar{f}_R(a)}{f_R}} - 1\right), \tag{18}$$

where $\bar{R}(a)$ and $\bar{f}_R(a)$ are background quantities, the free parameters of the model were fixed such that the expansion is consistent with observations (i.e. such that it gives an effective cosmological constant) and $n = 1$. Under these conditions, the model has only one free parameter f_{R0}, which sets the normalization of the background field $\bar{f}_R(a)$. The set of equations that are necessary to describe non-linear cosmology is completed with the geodesics equations. As these theories do not include additional fields besides the metric, the geodesics are unchanged and thus, the trajectory of the particles is determined by $\nabla\Phi$. By combining Eqs. (16) and (17) it is possible to see that the MG contribution to the acceleration is given by $c^2\nabla f_R/2$.

Equations (16) and (17) substitute Eq. (9) in Definition 2.3. These two equations must be solved together, however, the second equation is independent of the metric perturbation Φ and thus, can be solved independently to obtain f_R in every time step of the simulation. After this is done, the field f_R can be introduced in Eq. (16) to obtain the metric perturbation and thus, the force that is required to move the particles.

A different implementation can be constructed by taking into account that the theory can be mapped into a theory that includes a chameleon field:[87]

$$S_{f(R),E} = \int \frac{\sqrt{-g}}{16\pi G}\left[R + (\delta\phi)^2 + V(\phi)\right]d^4x + S_m(\tilde{g}_{ab}, \psi), \tag{19}$$

where g_{ab} and $\tilde{g}_{ab}$ are Einstein and Jordan frame metrics, respectively. This mapping can be done by defining a scalar field ϕ such that

$$\exp\left(-\frac{2\beta\phi}{M_P}\right) = f'(R) \equiv f_R, \tag{20}$$

where M_P is the Planck mass and the coupling constant is $\beta = 1/\sqrt{6}$. The relation between the Einstein and Jordan frame metrics is

$$\tilde{g}_{ab} = \exp\left(-\frac{2\beta\phi}{M_{PL}}\right) g_{ab}. \tag{21}$$

In the new frame (i.e. the Einstein frame where $f(R) = 0$), the equations for the two degrees of freedom (Eqs. (16) and (17)) are substituted with the following:

$$\nabla^2 \Phi = 4\pi G \delta \rho - \rho_{f_R} \tag{22}$$

$$\nabla^2 f_R = -\frac{1}{a}\Omega_m H_0^2 (\eta - 1) + a^2 \Omega_m H_0^2 \left[\left(1 + 4\frac{\Omega_\Lambda}{\Omega_m}\right)\left(\frac{f_{R0}}{f_R}\right)^{\frac{1}{n+1}} - \left(a^{-3} + 4\frac{\Omega_\Lambda}{\Omega_m}\right)\right], \tag{23}$$

where $\rho_{f_R} = \dot{f}_R^2 + |\nabla f_R|^2 + V_{\text{eff}}(f_R)$ is the energy of the scalar field (which is assumed to be small) and η is the matter density in terms of the background density.

As the part of the action (19) that defines the geometry is the same as in GR, the MG effects on the mater distribution appear as a fifth force in the geodesics equations:

$$\frac{d^2\mathbf{x}}{d\tau^2} + \nabla\tilde{\Phi} + \frac{1}{2}\nabla f_R = 0, \tag{24}$$

where τ and $\tilde{\Phi}$ are super-comoving time and potential, which are the variables that are used in the only code that includes this representation of the model.[285]

Once a frame is chosen, we need to find a way of solving the field equations. In both cases, the field f_R satisfies the equation of a chameleon field. This is a quasi-linear equation, which can be solved with the non-linear multigrid methods described in Sec. 3. In order to ensure that the field is positive and to increase the stability of the solvers, it is customary to solve the equations for a transformed field u, which is defined by

$$f_R = -\frac{A}{a(t)^b}e^u, \tag{25}$$

where the normalization A and the dependence b with the expansion factor was chosen differently by different authors. While this change of variables provides robustness to the solvers, it forces them to evaluate an exponential function in every point of the grid and every iteration step. This is a very expensive function to calculate and thus the implementation of this change reduces the performance of the solvers. A workaround to this problem, which is based on a different change of variables that enables the implementation of an explicit solver, is described in Ref. 332.

The two formalisms described above are mathematically equivalent. However, the numerical errors associated with the solution of these two set of equations may be different and thus, there is no guarantee that the end result of simulations ran in the Jordan or Einstein frame will be the same. This has been tested in Ref. 286, were three different implementations of the model were compared (two of them written in the Jordan frame and one in the Einstein frame). The differences between the predictions provided by these codes are below 1%, which is enough for the accuracy required by present and near future surveys.

2.2.4. *Models with higher number of dimensions (DGP)*

Models defined in higher dimensional space-times, such as the DGP model,[31] can be mapped into the scalar tensor theories described in Sec. 2.2.1, which include the Vainshtein screening mechanism (class 5 in Table 3).[78, 101, 333] Once this is done, it is possible to apply the same techniques applied to fully non-linear equations (see Sec. 3 and in particular Sec. 3.5.3).

2.2.5. *Vector fields*

The only vector-tensor theory that has been simulated is defined by the following action:[291]

$$S = \int \sqrt{-g} \left[-\frac{R}{16\pi G} - \frac{1}{4} F_{ab} F^{ab} - \frac{1}{2} \left(\nabla_a A^a \right)^2 + R_{ab} A^a A^b \right] d^4x, \qquad (26)$$

where[f]

$$F_{ab} = \partial_a A_b - \partial_b A_a. \qquad (27)$$

The equations of motion for the four extra degrees of freedom are

$$\Box A_a + R_{ab} A^b = 0. \qquad (28)$$

As the part of the action that defines the geometry is the same as in GR, Einstein's equations are unchanged, with the exception of the addition of the energy-momentum tensor of the vector field in the source. This extra term does not add numerical complexity. However, the equation of motion for the vector (Eq. (28)) may be extremely difficult to solve in the most general case.

In the only existing simulations,[291] the perturbations of the vector field were not tracked and only the effects on the background expansion and the linear cosmology were taken into account. This means that Eqs. (6), (7) and Eq. (9) were kept as described in Definition 2.3. Only the solution of Eq. (11) was changed while running the simulation. In addition to this, the initial conditions were modified to take into account the presence of the vector field at high redshift. These modifications apply to both the initial power spectrum as well as the growth rate. In other words, these authors made appropriate approximations which map the theory into the class 1 defined in Table 3 of Sec. 2.2.1.

2.2.6. *The MOND case*

From the point of view of late time cosmology, it is possible to classify MG theories into two large families which are motivated as alternative solutions to the two main open problems in cosmology: dark matter and dark energy. While this book focuses in the second class of theories, the equations associated with both families of theories share similarities. Thus, there is a lot to learn from numerical implementations of

---

[f]Note that a different sign convention was used by these authors.

solutions of the MOND equations, which are representative of alternatives to dark matter.

While there are several relativistic extensions, the original MOND idea is defined in a non-relativistic set up.[334] The theory substitutes the usual Newtonian potential with a modified potential, which is solution of

$$\nabla \cdot \left[\mu \left(\frac{|\nabla \Phi_M|}{a_0} \right) \nabla \Phi_M \right] = 4\pi G \delta \rho, \tag{29}$$

where μ is a free function and a_0 a free parameter which can be fixed, for instance, by requiring the theory to explain rotation curves of galaxies in the absence of dark matter. A classical action for this equation can be found in Ref. 335. The equation is quasi-linear and thus can be solved with the same multigrid methods that will be described in Sec. 3.

An alternative method commonly used to solve the MOND equation consists in equating the left hand side of Eq. (29) with the left hand side of the standard Poisson's equation. Integrating once gives

$$\mu \left(\frac{|\nabla \Phi_M|}{a_0} \right) \nabla \Phi_M + \nabla \times \mathbf{k} = \nabla \Phi, \tag{30}$$

where the curl term $\mathbf{k}$ is an integration constant (in the sense that its divergence is equal to zero). It can be shown that this term is exactly zero for specific symmetries and that behaves at least as r^{-3} for non-symmetric configurations.[335] By assuming that the term is second order, it is possible to obtain an algebraic equation for the spatial gradient of the modified potential, which is straightforward to solve once the Newtonian potential is calculated with the usual methods that are implemented in GR codes. The method is exact for 1D or 2D codes respecting certain symmetries.

Note that while this method is approximate, it conserves the two main features of the MOND theories: forces decaying as $1/r$ at large distances and a screening mechanism based on the gradient of the field.[g] The validity of the approximation for the highly non-symmetric configurations associated with the cosmic web was tested in Refs. 312 (unpublished results associated with this paper can be found also in Ref. 313). Among other things, it has been found that the curl term has the effect of boosting structure formation at intermediate scales. Since this result was obtained in cosmological models that do not include dark matter (the dark matter effect is given by MOND), extrapolation of these results to the dark energy case must be made with care. See Sec. 3.3 for a description of how this approximate solution can be used as an initial guess for iterative solvers of Eq. (29).

---

[g]In fact, the MOND community does not use the expression "screening mechanism". Instead, they refer to the "Newtonian limit" of the theory. However, these two terms make reference to the same thing: the fact that these theories include mechanisms that can hide the MG effects when needed. In the MOND case, the Newtonian limit (where the MG effects disappear) is obtained when the gradient of the MOND potential is large.

An alternative method is based on a different realization of the MOND theory (QU-MOND).[336] First simulations with this model were presented in Ref. 313. See also other implementations presented in Refs. 316, 317 and 318. The gravitational potential of this theory is described by the following set of equations:

$$\nabla^2 \Phi_N = 4\pi G \delta\rho \tag{31}$$

$$\nabla^2 \Phi_M = \nabla \cdot \left[\nu \left(\frac{|\nabla\Phi_N|}{a_0} \right) \nabla\Phi_N \right], \tag{32}$$

where Φ_N is the Newtonian potential which is solution of the standard Poisson's equation, Φ_M is the MOND potential which will affect the trajectory of the particles and the free function ν is not to be confused with the μ interpolating function required in the AQUAL model. Note that this representation of the MOND idea is straightforward to implement: once the Newtonian potential was obtained by solving Eq. (31) with usual methods, it can be discretized on a grid and used as source of the Poisson's equation associated to the MOND potential (32), which is also linear and can be solved with the same methods used for solving Eq. (31).

A different representation of the MOND idea can be derived from a bi-metric Lagrangian and was presented in Ref. 337. This particular theory can be simulated by combining solutions of standard Poisson's solvers and non-linear solvers of Eq. (29). A particular implementation can be found in Ref. 313.

2.2.7. *Code units*

The reference frame adopted in simulations is usually defined by the metric given by Eq. (3) (i.e. comoving spatial coordinates with some particular choice of time). On top of this, it is still necessary to choose units. It is customary to describe the space and time coordinates in terms of the size of the box and the age of the Universe $(1/H_0)$. The unit of mass is usually such that the mass of the particles is order one or that the mean density of the Universe is exactly one.

In the MG case, it is also necessary to chose units for the extra degrees of freedom. These are usually defined in terms of the background value of the field, its vacuum value or the Planck mass. Rescaling the fields with a given power of the expansion factor may also help in removing some terms from the Klein-Gordon or geodesics equations.

Note that these units are *not* natural units and thus, the speed of light and Planck constant are not equal to one in these simulations. They must be included explicitly with the correct units or in the normalization of the fields. This happens for instance when taking into account time derivatives in the equations for the extra degree of freedom.

2.2.8. *A word on performance*

Modified gravity simulations usually need to track additional degrees of freedom, so more calculations are likely to be required. This will result in an overhead with respect to GR cosmological simulations, which in some cases may make MG simulations infeasible. This overhead will naturally depend on the methods that are implemented, which in turn depends on the class of models we are dealing with. The simplest models that can be simulated (i.e. those that only modify the background expansion or are condensed in a rescaling of the gravitational constant) have a straightforward implementation, which consists in substituting the tables that have the information about the background cosmology or the units associated to the force. This kind of simulations has zero overhead and thus, can be run with the same resolution used for standard gravity simulations. The same applies to non-local gravity (Sec. 2.2.2) when appropriate approximations are made.

The situation is different for models which give rise to non-linear elliptic equations (second and third classes of models presented in Table 1). This kind of equations are typically solved with iterative methods. The convergence rate strongly depends on the specifics of both the models and the solvers. See for instance Ref. 332 for an example of how different integration variables can have a dramatic change in the convergence rate of the solvers. The overhead associated to these kinds of simulations typically lies between a factor of 2 to 10 or 20.

Finally, the most complex models we can deal with give rise to hyperbolic equations (fourth class of models in Table 1). The only exact solvers that exist at present are conditionally stable, which means that they require a Courant-Friedrich-Levy condition on the time step. The number of time steps associated to the MG solver increases by typically two to three orders of magnitude with respect to the standard gravity part of the code. Implementation of implicit solvers for this family of models will largely increase the competitiveness of these simulations in terms of speed.

3. Multigrid Methods for Non-Linear Partial Differential Equations

Most of the models summarized in the previous section are described by non-linear hyperbolic equations (i.e. non-linear wave equations). This is different to what happens in the GR case, where an appropriate choice of the gauge can hide the time derivatives of the metric perturbations. In order to keep the spirit of standard GR simulations (in which the metric perturbations are not evolved in time starting from some initial conditions, but calculated at every time step) it is usually assumed that the terms that include time derivatives in the MG equations are sub-dominant or important only in the background. This is the so-called quasi-static limit of the models, in which the additional degrees of freedom are described by elliptic equations. The non-linearity of these equations prevents us from applying techniques that rely on the superposition principle (such as Fourier based solvers or direct summation of forces). Instead, codes take advantage of multigrid techniques, which are the subject of this section.

To fix the notation, let us assume that we are interested in solving an equation that has the following form:

$$L(\partial^2\phi, \partial\phi, \phi) = S(\rho, \phi) \quad \text{in } \Omega \qquad (33)$$

$$\phi(x) = f(x) \qquad \text{in } \partial\Omega \qquad (34)$$

where L is a differential operator (not necessarily linear), S is a non-linear scalar function of the density and the scalar field itself, f is a function that defines the boundary conditions, Ω is the domain where the equation is defined and $\partial\Omega$ the boundary of Ω. We are interested in finding a numerical solution of this equation on a grid that covers the simulation domain. The method consists in discretizing the equation on the grid and using an iterative method to obtain improved solutions starting from an initial guess. In this section we will summarize different aspects of the iteration method as well as multi-level techniques for accelerating the convergence rate.

The multigrid techniques are described in great detail in Refs. 338–340 and 341. Furthermore, a lot of information and tricks needed to solve MG equations can be found in the papers associated with specific implementations of the models summarized in Tables 1 and 2. Finally, as these methods are widely used in several other contexts, such as fluid dynamics, geophysics, medicine, etc., information may be found also in non-astrophysical journals.

3.1. *Gauss-Seidel iterations*

Before setting up an iteration scheme, it is necessary to discretize equation (33) on the grid. Such discretization takes the following form:

$$L^l[\phi^l] = S^l[\rho^l, \phi^l]. \qquad (35)$$

The index l represents the grid in which the problem is described. The fields ρ^l and ϕ^l are the density and the required solution on the grid. Finally, L^l is a discretization of the differential operator L. Equation (35) represents a very large set of non-linear algebraic equations, whose number is given by the number of nodes in the grid. Our aim is to approximate the solution of the differential equation (33) by the solution of the algebraic equation (35).

The discretization of the equation can be obtained by discretizing each derivative of the differential operator, writing them as linear combinations of values of the field at the neighbouring nodes. For instance, second order formulas for the second derivatives can be obtained using a 19-points stencil:

$$\partial_x^2 \varphi_{i,j,k} = \frac{1}{h^2}\left(\varphi_{i+1,j,k} + \varphi_{i-1,j,k} - 2\varphi_{i,j,k}\right) \qquad (36)$$

$$\partial_x\partial_y \varphi_{i,j,k} = \frac{1}{4h^2}\left(\varphi_{i+1,j+1,k} - \varphi_{i+1,j-1,k} - \varphi_{i-1,j+1,k} + \varphi_{i-1,j-1,k}\right), \qquad (37)$$

where the indexes i, j, k corresponds to the indexes of each grid in each Cartesian direction.

Equation (35) is solved iteratively with a Gauss-Seidel scheme. In every iteration step of this scheme, the solution is updated in every element of the grid by finding the root of the following function

$$T^l[\phi^l] \equiv L^l[\varphi^l] - S^l. \tag{38}$$

In the linear case, it is possible to find an analytic expression for the updated value $\tilde{\phi}^l$, which is a linear combination of the values of the field in the neighbouring nodes. This gives rise to the so called *explicit* solvers. For instance, the 1D standard Poisson's equation, can be discretized as follows:

$$\frac{\phi_{i+1} - 2\phi_i + \phi_{i-1}}{h^2} = \rho_i. \tag{39}$$

In an explicit solver the solution in a given iteration step can be updated for instance in the following manner:

$$\tilde{\phi}_i = \frac{1}{2}\left[\phi_{i-1} + \phi_{i+1} - h^2\rho_i\right]. \tag{40}$$

Exact details will depend on the type of discretization and exact algorithm used, but in general, each iteration step consists in updating the solution in every grid point i with a linear combination of the values that surround i and that correspond to the previous step.

In the non-linear case, it is customary to use *implicit* solvers, which consist in approximating the root of T^l by applying one step of a Newton-Raphson algorithm:

$$\bar{\phi}^l = \phi^l - \frac{T^l}{\partial T^l / \partial \phi^l}. \tag{41}$$

Reference 332 has shown that implicit solvers are not the only route to deal with non-linear problems. It is in fact possible to obtain explicit algorithms by making appropriate changes of variables.

In order to fully specify a Gauss-Seidel scheme, we need to fix the order in which the iterations are made in each sweep through the grid. For instance, in the lexicographic ordering, the update of the field in the cell (i, j, k) is followed by a calculation in the cell $(i + 1, j, k)$ and so on. A better convergence rate and paralelization properties can be obtained by altering this scheme. In the most popular scheme, each complete iteration step consists in two sweeps in which the update of a given cell (i, j, k) is followed by an update on the cell $(i + 2, j, k)$. The two sweeps start in different cells: $(0, 0, 0)$ and $(1, 0, 0)$ and thus, the scheme resembles a two colour chess board. Similar schemes exists which consist in four or even eight colours.

3.2. *Multigrid acceleration*

The quality of a given solution ϕ^l is usually determined by studying the amplitude of the residuals, which are defined as:

$$\epsilon^l = L^l[\phi^l] - S^l[\rho^l, \phi^l]. \tag{42}$$

The smaller the residual, the better the solution will be. An associated quantity, which can be used to gauge the quality of the given solver, is the so called convergence rate, which can be defined as the ratio of residuals at two consecutive iteration steps. It can be shown through analysis of the convergence rate of individual Fourier modes[338–341] that the Gauss-Seidel algorithm is very efficient in smoothing out the residuals, but has a very slow convergence rate when the solution includes modes whose wavelength is much larger than the spatial resolution of the grid. This is due to the local nature of the method: in each iteration step, information travels from each cell to its closest neighbours only and thus a very large number of iterations is required to transfer information across the domain.

It is possible to accelerate the convergence rate for these long modes by combining solutions obtained on coarser grids and coarser grids.[h] The typical way of arranging these grids is to define a set of grids whose resolutions are one half, one quarter and so on of the target resolution. The coarsest grid of this set usually contains two nodes per dimension, but this is not a strict rule. The coarsest grid can even be changed while the iterations progress. In order to prove convergence properties of the method, it is enough to work with two grids. The algorithm is the following: a number of iterations is performed on the finest grid l (i.e. the grid in which we want to obtain the solution). When the convergence rate slows down (or a fixed number of iterations has been completed), iterations are made to approximate the error of this intermediate solution. As most of the error will come from the large modes that did not converge yet, this is done on the next coarse grid $l-1$. So the new equation for which iterations are made on the coarse grid is

$$L^{l-1}[\delta\varphi^{l-1}] = R(\epsilon^l),\qquad(43)$$

where R is a *restriction operator* which transfers fields from the fine to the coarse grid. Once a given number of iterations is made on the coarse grid, the additional information that was collected related to the large modes is added to the solution on the fine grid:

$$\bar{\phi}^l = \phi^l + P(\delta\phi^{l-1}).\qquad(44)$$

The transfer between the coarse and fine grids is made with a *prolongation operator* P. See for instance Refs. 338–340 or 341 for a detailed description of convergence properties of this method.

Application of the method to real problems requires repeating this process on more than one grid level. By doing this, it is possible to reconstruct the solution on the finest grid by combining solutions of coarser and coarser grids which treat each set of Fourier modes separately. The decision about changing the grid level can be made by taking into account the behaviour of the residuals and convergence rate

---

[h]These different grid levels should not be confused with the different grid levels that exist in codes that include adaptive mesh refinements (AMR). In that case, different grids are used to increase the resolution locally, while in the multigrid case described here, resolution is actually downgraded below the target resolution where we want to obtain the solution.

or by using fixed schemes. Popular schemes are the V cycle (in which the iterations are made starting from the finest grid, all the way down to the coarsest grid and then back to the finest grid) or the W cycle in which jumps between coarse levels are more often that between fine levels. Well behaved solvers typically reduce the residual by one order of magnitude after each V cycle.

In the non-linear case, it is not possible to add up solutions obtained in different grids and thus, the iterations in the coarse levels are not made to approximate the error $\delta\phi^{l-1}$, but the solution ϕ^{l-1} itself. So Eq. (43) is substituted by

$$L^{l-1}[\varphi^{l-1}] = -R(\epsilon^l(\varphi^l, S^l)) + \epsilon^{l-1}(R(\varphi^l), R(S^l)). \tag{45}$$

The correction then is made in the following way:

$$\bar{\varphi}^l = \varphi^l + P(\varphi^{l-1} - R(\varphi^l)). \tag{46}$$

This extension to the method is known with the name of full approximation storage (FAS).

3.3. *Initial guess for the iterative solvers*

Making appropriate choices for the initial guess with which the iterations start can be crucial for reaching convergence in a reasonable number of iterations. In general, this choice depends on the particulars of the equation to be solved. The decision is usually made by trial and error. The initial guess for obtaining solutions on the domain grid (i.e. the grid that covers the entire simulation box) can be obtained in the following ways:

- Background value of the field.
- Minimum of the effective potential.
- Extrapolation in time from previous time step.
- Solution of auxiliary PDEs. This method has been applied to the MOND equation (29) in Ref. 313 (see Appendix A in this reference). An auxiliary PDE can be constructed by taking the divergence of the field $\nabla\Phi_M$, which can be obtained after neglecting the curl term $\nabla\times\mathbf{k}$ in Eq. (30). This is a linear Poisson's equation and thus, can be solved with Fourier based algorithms which are both efficient and easy to implement. When using this solution as the initial guess, the non-linear multigrid solver that is required to solve Eq. (29) does not need to make iterations to obtain the main features of the solution, but only needs to recover the effects associated with the curl field, which are small.

The initial guess for obtaining solutions in the refined levels is usually determined by interpolating values from the closest coarse grid. This naturally applies only to AMR codes.

3.4. *Convergence criteria*

An important decision to be made when developing a multigrid solver is related to the criterion that is used to stop the iterations. Criteria that are too permissive may give inaccurate solutions, while criteria that are too strict may force the solver to make many more iterations than required to understand the underlying physics and may become extremely inefficient (especially in the context of N-body simulations, where the equations must be solved several hundreds or even thousands of times in a single simulation). When designing a convergence criterion, there are three different solutions to take into consideration:

- ϕ_n^l = approximate solution of the discretized equation (35) on the grid l which was obtained after n iterations.
- ϕ^l = exact solution of the discretized equation (35) on the grid l.
- ϕ = exact solution of the PDE (33).

The following are important facts associated with these solutions:

- The multigrid method converges after infinite number of iterations to the correct solution of the discretized equation:

$$\phi_\infty^l = \phi^l. \tag{47}$$

- The fact that the convergence rate η in the grid l is close to one (which means that the solution does not evolve when making iterations) does *not* imply that the iterations have converged (it only implies that the iterations for the modes associated with the grid l have converged):

$$\left(\eta \equiv \frac{\phi_n^l}{\phi_{n+1}^l} \sim 1\right) \nRightarrow \left(\frac{\phi_n^l}{\phi_\infty^l} \sim 1\right). \tag{48}$$

- The solution of the discretized equation is *not* the solution of the differential equation evaluated at the nodes of the grid:

$$\phi_\infty^l \neq \phi. \tag{49}$$

This is because the discretization of the equation introduces an error, which depends on the order of the discretization formulas used and on the resolution reached by the grid. A commonly used estimation of the truncation error τ^l is[i]

$$\tau^l = L^{l-1}\left[R(\phi^l)\right] - R(L^l\left[\phi^l\right]). \tag{50}$$

---

[i]The truncation error can be estimated by comparing the result of applying the discretized differential operator in two consecutive grids l and $l-1$ to the analytic solution of the equation: $\tau^l = L^{l-1}[\phi] - L^l[\phi]$. In practice, the analytic solution is substituted by the only solution that we have, which is the fine grid solution ϕ^l. In order to be able to apply the coarse grid differential operator L^{l-1} to the fine solution ϕ^l, we need to restrict it by applying the restriction operator R. This gives rise to the first term of the right hand side of Eq. (50). Similarly, once we applied the differential operator in both grids, we need to restrict the fine one to be able to make the comparison (which gives rise to the second term of the right hand side of this equation).

- The residual is not a dimensionless quantity, which means that it is possible to make it arbitrarily small by changing the units:

$$[\epsilon^l] \neq 1 \tag{51}$$

- The residual is not necessarily uniform on the grid and may indeed have large variations across the domain:

$$\epsilon^l(\phi_n^l) = \epsilon^l(\phi_n^l, x) \tag{52}$$

This is because the convergence rate is not uniform. More complex features in the solution in specific regions will require more iterations.

Convergence criteria are usually based on amplitude of the residual, so the first thing we need to do is to fix is a norm for the residual. Common choices are norm 2 or ∞:

$$||.||_2^l = \Sigma_{i,j,k} \left(\epsilon_{i,j,k}^l\right)^2 \tag{53}$$
$$||.||_\infty^l = \max |\epsilon_{i,j,k}^l|, \tag{54}$$

where the indices (i, j, k) run over all the grid points in the 3D box.

A few commonly adopted criteria follow:

- Comparison with initial guess:

$$\frac{||\epsilon_n||}{||\epsilon_0||} < a. \tag{55}$$

This criterion has the problem that if the initial solution is already good, then the solver will make more iterations than necessary. It is possible to also have the opposite case, in which the initial solution is too bad. In this case, reducing the residual by a given fraction of the initial residual, may not be good enough. The technique is implemented in several solvers with standard or MG. However it must be used with care.

- Comparison with a fixed constant:

$$||\epsilon_n|| < a. \tag{56}$$

This criterion has the problem that a must be chosen according to the units used in the code.

- Comparison with truncation error:

$$\frac{||\epsilon_n||}{||\tau_n||} < a. \tag{57}$$

In this case, the iterations will stop once the error associated with the discretization of the equations dominates.

In all these criteria, a is a small free parameter.

3.5. *Extended multigrid techniques*

Details of the solver strongly depend on the properties of the equation that we are dealing with. Convergence properties depend on both the differential operator and the source of the equation. In some cases it may be necessary to apply additional tricks to reach convergence in a reasonable number of iterations. We describe a few examples.

3.5.1. *Selective iterations*

In situations in which the complexity of the solution strongly depends on the position in the domain (such as for instance systems in which the density is almost uniform in a large part of the domain and has strong variations in a small region) the solvers do not usually converge uniformly in the whole domain. In some regions the solution will quickly approach the desired solution, while in others, it may require a much larger number of iterations to reduce the local residual. In this cases, it is possible to implement *selective iterations*, which means that some local criterion is established to stop the iterations. Thus, the iterations will be made only in the region in which the solution has not converged and thus the overall computational effort will be reduced.

3.5.2. *Continuation method*

The continuation method can be applied to non-linear equations for which a particular parameter defines a linear limit.[338] In these cases, it may be convenient to solve the linear equation in the first place and obtain intermediate solutions changing the values of the parameter such that the solution evolves from the linear case towards the fully non-linear situation in a continuous way.

3.5.3. *Dealing with fully non-linear equations: Operator splitting*

Models that include Vainstein screening mechanism such as DGP or Galileon give rise to fully non-linear partial differential equations for the additional degrees of freedom. The equations are still second order in space and time, but there is a non-linear dependence with the second spatial derivatives, which represent a major challenge for multigrid solvers. Naive multigrid implementations such as those described in Secs. 3.1 and 3.2 have poor convergence properties in high density regions, which make simulations impractical or simply impossible to run. We describe here two different strategies that are commonly used to deal with this problem. A comparison between different Vainstein solvers can be found in Ref. 286.

The first attempts to solve these equations implemented workarounds to this problem which consisted in using a spherically symmetric approximation (which gives equations that can be solved with a Fourier based algorithm)[298, 333] or by smoothing out the density distribution before sending it to the multigrid solver.[101, 297, 333] The second approach can improve the convergence rate and make the fully non-linear simulations feasible. However, it comes with a price in accuracy.

An alternative method, the operator splitting technique, was proposed in Ref. 299 and implemented afterwards in the code ECOSMOG.[78, 301] Let us fix notation by assuming that we are interested in finding solutions of the following equation:

$$\nabla^2\phi - A\left[\left(\nabla^2\phi\right)^2 - (\partial_i\partial_j\phi)(\partial^i\partial^j\phi)\right] = B\delta\rho, \tag{58}$$

which has the same form of the DGP equation of motion and where A and B are model dependent constants and ∂_i are partial derivatives with respect to spatial coordinates. In order to define the Gauss-Seidel iterations, we will need to solve a discretized version of the equation for the field $\phi_{l,m,n}$ in each node (l, m, n). If we can ensure that the discretization of the term $(\partial_i\partial_j\phi)(\partial^i\partial^j\phi)$ in the node (l, m, n) is independent of $\phi_{l,m,n}$, then we can see the equation as an equation for $(\nabla^2\phi)_{l,m,n}$ and solve it analytically for $\phi_{l,m,n}$ once we obtained $(\nabla^2\phi)_{l,m,n}$. This can be done by decomposing the term $\partial_i\partial_j\phi$ into trace and traceless parts:

$$\partial_i\partial_j\phi = T_{ij} + S_{ij}, \tag{59}$$

where

$$T_{ij} = \frac{\nabla^2\phi}{3}I \tag{60}$$

$$S_{ij} = \partial_i\partial_j\phi - \frac{\nabla^2\phi}{3}I, \tag{61}$$

where I is the identity operator. Substitution of these definitions in $(\partial_i\partial_j\phi)(\partial^i\partial^j\phi)$ gives

$$(\partial_i\partial_j\phi)(\partial^i\partial^j\phi) = \left(\frac{\nabla^2\phi}{3}\right)^2 + \left[2S_{ij}T^{ij} + S_{ij}S^{ij}\right], \tag{62}$$

which in turn gives the following alternative version of the equation of motion (58)

$$\nabla^2\phi - A\left\{\frac{2}{3}\left(\nabla^2\phi\right)^2 - \left[2S_{ij}T^{ij} + S_{ij}S^{ij}\right]\right\} = B\delta\rho. \tag{63}$$

The term inside the square brackets in this expression fulfils the property that we are interested in: the usual discretization that can be obtained in the cell (l, m, n) by applying Eq. (37) is independent of the value of the solution $\phi_{l,m,n}$ in that cell.

It has been found that this way of arranging the terms largely improves the convergence rate of multigrid solvers. However, it does not solve all the issues with this model since the solution of Eq. (63) for $\nabla^2\phi$ may give rise to square roots of negative numbers in under-dense regions, where density perturbations are negative. This does not occur in the DGP case, but in the quartic Galileon[301, 342] model. See Ref. 343 for a detailed discussion on this issue.

3.6. *Summary*

This section summarizes the decisions that need to be taken when developing multi-grid codes. This list of decisions should not be considered as a simple summary, but as a check list of properties of the solver that can be checked when things do not go as expected. So if the solver that the reader is developing has poor convergence rate, it may be worth checking the following properties of his or her implementation:

- Parametrization of the theory: Can we find an alternative representation of the theory which simplifies the equations?
- Integration variables (independent variables as well as fields).
- Stencil and discretization formula for the PDE inside the domain.
- Stencil and discretization formula for the PDE at the boundaries.
- Initial guess for the iterations.
- Type of iterations (explicit, implicit, over-relaxation).
- Sweeping strategy (2, 4, 8 colours?).
- Number of multigrid levels.
- Number of iterations per multigrid level (which may be different from level to level).
- Multigrid scheme (V, W, adaptive, etc.).
- Restriction and prolongation operators.
- FAS vs. no FAS scheme.
- Convergence criterion.
- Extended techniques: selective iterations, continuation methods, operator splitting, etc.

Taking these decisions constitutes a form of art, whose outcome depends on the particular details of the equations to be solved. What works for one equation may not work for others.

Once a newly implemented solver can pass several tests and is accepted to be accurate enough, it may be a good idea to optimize it. A few ways in which this can be done are discussed in Refs. 344 and 332. The first of these optimization methods consists in reducing the spatial resolution when solving the Klein-Gordon equation for the additional degree of freedom. The impact of resolution in these solutions was also discussed in Ref. 303.

In cases in which the code needs to solve both, the Poisson's equation for the metric perturbation and the Klein-Gordon equation for the scalar field, it is possible to increase the speed of the simulations by solving both equations in parallel. As the GR and MG solvers work with very similar routines, it is possible to shear the loops that have to be made through the grid to calculate different quantities. This method was implemented in the code Isis[285] for the symmetron model and it was found to give an improvement in speed of about 20%.

4. Non-Static Solvers for Non-Linear Partial Differential Equations

The dynamics of MG degrees of freedom is usually described by non-linear hyperbolic equations (please, see Refs. 345, 346 or your favourite book on partial differential equations or the notes of your undergrad courses if in need of refreshing classification of equations). A common practice is to adopt the so called quasi-static approximation, which consists of assuming that the time derivatives of these fields are small or important only in the background. This approximation usually gives rise to non-linear elliptic equations, which can be solved with the methods described in Sec. 3. This section is about going beyond this approximation.

4.1. *The quasi-static approximation*

Terms that include time derivatives of the fields are not exclusive to the field equations associated with MG. For instance, in the GR case, it is necessary to deal with time derivatives when deriving the usual Poisson's equation from basic principles. The effects associated with these terms were analysed in Refs. 321–323 and 324.

The validity of the quasi-static approximation in a MG context was discussed for first time in Ref. 292. In that early work, the time derivatives were not take into account when running the simulations, but estimated afterwards by post-processing the simulation data (i.e. by applying appropriate discretization formulas which combine static solutions that were obtained at different time steps). The results show that the terms that include time derivatives in the $f(R)$ equations are subdominant when compared with the spatial derivatives. A similar comparison was made by myself using symmetron simulations that were presented in Ref. 302. I found that, while it is true that globally the time derivatives are sub-dominant, they may be larger than the spatial derivatives when restricting the analysis to under-dense regions (i.e. voids), where the Laplacian of the field is very small.

These results can only be confirmed by running simulations beyond the static limit and comparing the data with the output of static codes, which do not rely of the quasi-static approximation. This has been done in the linear regime for the $f(R)$ model[347] and in the non-linear regime for two different models: symmetron and $f(R)$.[286, 302–304] The approximation was found to be good enough for predicting the matter and velocity power spectrum. For instance, in the case of the symmetron model, a difference of 0.2% was found between static and non-static power spectra in a range of scales from $k = 0.05$ to $k = 10$ h/Mpc.[303] Differences in the mass function were studied in Ref. 286 and also found to be well below 1% for masses between 3×10^{12} and 10^{14}. In the case of $f(R)$ model, differences between static and non-static results are within the same ranges.[304] However, it is worth keeping in mind that these non-linear codes contain simplifications. Furthermore, the impact of these additional terms on different observables is still unknown.

It is worth noticing that there are models, such as disformal gravity, in which the time derivatives of the field appear not only in the kinetic part of the equations of motion, but also in the coupling to matter.[305] This provides a very rich phenomenology which does not exist in models that are conformally coupled. Neglecting time derivatives when running simulations with this kind of model may prevent us from predicting the unique signature of new physics that we are looking for. So non-static simulations are crucial for testing these models.

A problem that arises when running non-static simulations is related to the determination of initial conditions for the additional degrees of freedom. Models that include screening mechanisms are expected to give rise to small perturbations at the start of the simulations. However, the exact shape and amplitude of the spectrum of these perturbations may still be important. This issue was discussed in Ref. 305, where it was shown that the end result of the simulations is independent to the amplitude that is chosen for the initial perturations of the field. In this particular model (disformally coupled symmetron), this is not only a consequence of the screening, but also of the fact that there is a phase transition at a redshift of about one. The scalar field receives a large kick when this happens, which forces it to lose memory of any small perturbations that may have existed at early times. Note that this result may not be valid for models that do not include screening of the field at high redshift. In this case, initial conditions will have to be generated more carefully, for instance, as a realization of a power spectrum generated with linear theory.

4.2. *Solving hyperbolic equations in N-body simulations*

The first solvers for hyperbolic equations in the context of MG N-body simulations were applied to the symmetron model and are based on the leap-frog algorithm.[302, 303] The method consists in converting the second order Klein-Gordon equation into a first order system of equations with the following change

$$q = a^b \dot{\phi}, \tag{64}$$

where b is chosen such that commonly existing terms cancel. The resulting equation has the following form:

$$\dot{\phi} = \frac{q}{a^b} \tag{65}$$

$$\dot{q} = S(\rho, \phi), \tag{66}$$

where S is a model dependent function. Note that these equations are schematic and that details may depend on the model. These equations are similar to Hamilton equations for a set of particles whith the difference that instead of evolving the position of the particles, they evolve the amplitude of the field at every point of space. So by discretizing the field on a grid, it is possible to apply the same algorithm that is commonly used to integrate Hamilton's equations in N-body codes. The method was applied to uncoupled scalar fields within GR for instance in Refs. 348 and 349. Implementation to coupled scalar fields can be found in Refs. 302, 303 and 305. As in the model studied in these papers the scalar field undergoes very fast oscillations, the simulation has to deal with two different time scales (one associated with normal gravity and a much shorter one associated with the scalar field). This problem has been solved by having two different time steps in the simulation: a main time step as in a usual GR code and a much smaller one for modified gravity. As the time step that is employed to move the particles is the one associated with GR, these simulations still have the limitation that the particles can not see the oscillations of the scalar field and so the impact of the scalar waves in the matter distribution is still an open question. The same method has been implemented in 1D in spherical coordinates[343, 350] or 2D cartesian grid.[67]

The leap-frog algorithm has the disadvantage that it is conditionally stable, which means that it requires a Courant-Friedrich-Levy condition on the time step: For the method to work, the time step has to be such that no information is propagated in more than one spatial cell element within the step.[351, 352] While this condition does not affect the quality of the results, it forces the simulation to work with extremely small time steps, which makes them very expensive. A work around to this problem consists of introducing numerical viscosity which suppress small scale oscillations of the solution and can be done by means of implicit methods.

A different implementation of non-static solvers was presented in Ref. 304 and applied to the $f(R)$ model. In this case, the time derivatives of the field were estimated with a down-wind discretization:

$$\left(\frac{\partial \phi}{\partial t}\right)_n = \frac{\phi_n - \phi_{n-1}}{t_n - t_{n-1}}, \tag{67}$$

$$\left(\frac{\partial^2 \phi}{\partial t^2}\right)_n = \frac{(d\phi/dt)_n - (d\phi/dt)_{n-1}}{t_n - t_{n-1}}, \tag{68}$$

which involves information from a given time step n and the previous one $n-1$. Once this is done, it is possible to translate the terms that depend on these derivatives to the right hand side of the Klein-Gordon equation:

$$\nabla^2 \phi_{n+1} = f(\phi_{n+1}, \rho_{n+1}) + g(\dot{\phi}_n, \ddot{\phi}_n), \tag{69}$$

where f and g are model dependent functions. This way of discretizing the Klein-Gordon equation gives rise to an elliptic equation for the scalar field at the time step $n+1$ and, which can be solved with static solvers described in Sec. 3. Detailed discussion of up-wind and down-wind discretizations can be found for instance in Refs. 351 and 352.

Additional information about integration of non-static equations in gravity can be found in the GR literature. References 281, 282 and 306 present examples of implicit and explicit solvers that can be used for going beyond the static limit in GR. Furthermore, as discussed in Sec. 3, the equations associated with MG are not unique to gravity, but are used also in different contexts such as geophysics, engineering (especially in hydrodynamical applications) or medicine. Communication with these communities is essential for making progress in this field.

5. Baryonic Simulations

The only difference that exists between simulating dark matter and simulating baryons is that in the baryonic case, it is not possible to neglect non-gravitational interactions between the particles. This gives rise to a plethora of thermodinamic and quantum processes between different elements of the periodic table, transitions between them, transfer of energy between these elements and photons, etc. The complexity of the solutions when including all these effects increases disproportionally: A simulation that includes these interactions and which has infinite resolution may end up producing humans that may run simulations which may generate other humans that may run other simulations and so on. Naturally, we do not have the computational resources that are required to run such a simulation, so we need to make simplifications. These simplifications are based on the idea of dividing the solutions of the equations into different sets of scales. For the large scales, the equations are solved self-consistently to follow the evolution of dark matter and baryonic perturbations. The effects that occur at scales that are smaller than the resolution of the simulation are treated analytically following different processes with approximate models. As we are not interested in studying the formation of humans in a cosmological context but only galaxies, these sub-grid recipes have also a cutoff in resolution.

The complexity of baryonic simulations increases as the field progresses. The first baryonic simulations included only adiabatic gas and no sub-grid physics. Later on, cooling of the gas was included, followed by recipes for stellar formation, stellar evolution and different kinds of feedback (from supernovae, black holes, etc). The state-of-the-art today includes magnetic fields. Information about these kind of simulations within GR can be found in the presentation of some of the latest simulations and references therein.[353–356] The following sections describe how to include MG in these simulations.

5.1. *Including MG in hydrodynamical solvers*

In order to understand the evolution of baryons in the non-linear regime, we need to reformulate the definition of cosmological simulation provided in Sec. 2 (Definition 2.2). In particular, we need to relax the assumption of collisionless matter that is encoded in the energy-momentum tensor. Doing this will prevent us from describing the fluid by following geodesics (Eq. (2)). So to derive alternatives to these equations, we need to repeat the analysis that was presented in Secs. 2.1 and 2.2, which resulted in Definition 2.3. I provide a schematic of how this can be done below. More details on these derivations can be found in the works that describe specific implementations.[284, 357–359]

Let us assume, for instance, a gravitational model described by the Lagrangian (13). Minimizing the action with respect to the metric will give rise to the following equation:

$$G_{ab} = \frac{\delta \tilde{g}_{ab}}{\delta g_{ab}} \tilde{T}_{ab} + T_{ab}^{\phi}, \tag{70}$$

where the two energy-momentum tensors are the usual Jordan frame matter tensor and the Einstein frame energy momentum tensor of the scalar field:

$$\tilde{T}_{ab} \propto \frac{1}{\sqrt{-g}} \frac{\delta S_M}{\delta \tilde{g}_{ab}}, \tag{71}$$

$$T_{ab}^{\phi} \propto \frac{1}{\sqrt{-g}} \frac{\delta S_{KG}}{\delta g_{ab}}. \tag{72}$$

Taking the divergence of Eq. (70) gives a generalized equation of conservation of energy. After projecting the result in the time and space directions, we will end up with the usual continuity and Euler equations that can be found in hydrodynamical codes plus MG terms that involve derivatives of T_{ab}^{ϕ}. In the Einstein frame, these new terms take the form of an additional energy in the continuity equation and a fifth force in the Euler equation and depend on the scalar field itself and its derivatives in space and time. Thus, they can be obtained by solving the Klein-Gordon equation for the field with the methods described in Sec. 3 or 4.

According to this analysis, implementing MG hydrodynamical solvers may be straightforward: a MG hydrodynamical solver can be obtained simply by plugging the solution of the Klein-Gordon equation that already exists in collisionless MG codes into a standard hydrodynamical solver. However, additional complexity may arise, which could lead to the need for more complex implementations. For instance, in the case where the free function f in Eq. (13) corresponds to a standard kinetic term, the energy momentum associated to the scalar field is:

$$T_{ab}^{\phi} = \nabla_a \phi \nabla_b \phi - \frac{1}{2} g^{cd} \nabla_c \phi \nabla_d \phi g_{ab} - V(\phi) g_{ab}, \tag{73}$$

which is diagonal for a diagonal metric (i.e. when neglecting vector and tensor modes). However, in the more general case in which f includes cross derivatives, this assumption may be broken, which may force us to take into account non-diagonal components of the equations. Also the Klein-Gordon equation for these models is sourced by the trace of the energy momentum tensor. In the case of baryons, there will be a pressure term sourcing the equation, which is not included in existing simulations.

5.2. *Including MG in sub-grid models and semi-analytics*

The sub-grid models contain information about baryonic processes that occur at the very small scales that lie below the resolution of the self-consistent part of the simulation. These processes are certainly affected by the presence of gravity and thus, depend on the assumed gravitational theory. For instance, the formation of stars depends on how gas clouds fragment and collapse under their own gravity. Different gravitational theories will give rise to different evolution of the clouds. This in turn should give rise to a different initial mass function, which is one of the ingredients in these sub-grid models. In present simulations, these extremely complex gravitational effects are included implicitly in the models (i.e. in most of the cases, the gravitational constant does not appear explicitly in the equations). These highly non-linear phenomena are described by simple equations and condense the complexity into a set of few free parameters. In the GR case, these parameters are choosen by trial and error: Large number of test runs is performed varying model parameters until a few observables such as the stellar formation rate density, the luminosity function of galaxies, etc. agree with the observed quantities. This process of tuning the parameters of the simulations is very costly and done by large collaborations.

The generalization of the sub-grid models to MG is a two step process. First it is necessary to re-derive the analytic expressions that describe different aspects of the models. Second, the model parameters have to be recalculated. In practice, constructing these models and tuning the simulations is an extremely complex task, which can not be done by the small MG community. The state-of-the-art simulations with MG do not include specifics of MG in the definition of the models and instead use sub-grid models and parameters given by the GR simulations that are usually run for comparison. Thus, existing simulations are not yet self-consistent.

For models that include screening mechanisms, the approximation that is made when not redefining the sub-grid models might be justified by the fact that the impact of sub-grid physics is dominant in high density regions, where the additional degrees of freedom should be screened. Examples of these kind of simulations can be found in Refs. 360 and 359. The impact of MG in some of the small scale problems that the ΛCDM model historically faces, and whose proposed solutions within GR are usually related to the impact of baryons, was presented in Refs. 361, 362 and 363.

A different approach usually applied in a GR context to the determine statistical properties of galaxies consists in taking into account baryonic effects only in the post-processing phase of the simulations. In this case, the simulations follow only the dark matter component. A set of recipes is applied *a posteriori* on the resulting merger trees of dark matter halos to determine the properties of the galaxies that form inside them. A change in the underlying gravitational model, will affect the outcome of these calculations through differences in both the merger trees and details on the semi-analytic model itself. As in the case of sub-grid physics used in full hydro simulations, applying GR semi-analytic codes to MG merger trees may not give realistic results, specially in the MG models that include coupling. Several quantities have to be redefined. A simple example is given by the hydrostatic equilibrium, which is obtained by opposing pressure forces to gravity forces and which is usually part of the semi-analytic models. In the MG case, the fifth force should be included in the analysis. Another example is the distribution of spin parameter of the dark matter halos. We know from simulations that the distribution in MG is log-normal as in the GR case, but the parameters (which are not free, but obtained from simulations) are different. So a correct implementation of a semi-analytic model in MG will require changing these and other similar quantities in the codes. Furthermore, retuning of the free parameters of the models will be required. See Refs. 364 and 365 for examples of MG gravity implementations.

The question of whether the differences found between GR and MG galaxy formation simulations are real or a consequence of poorly tuned baryonic models is still open. For instance, differences were found in rotation curves of galaxies in Ref. 360. Are these differences based on reality or simply on the fact that the simulations need re-tuning? Can we recover the GR rotation curves in a MG context by retuning parameters? Only by re-running the simulations we will be able to give answers to these questions. This is extremely challenging, but by doing it we may open the door for establishing new tests which may give us the definitive answer about gravity. The risk is high, but certainly worth taking.

Acknowledgments

CLL acknowledges support from STFC consolidated grant ST/L00075X/1 & ST/P000541/1 and ERC grant ERC-StG-716532-PUNCA. CLL thanks Shaun Cole for carefully reading the manuscript and anonymous referees and editor whose comments greatly improve the quality of the chapter.

Chapter 4

Approximation Methods in Modified Gravity Models

Baojiu Li

Institute for Computational Cosmology, Department of Physics,
Durham University, Durham DH1 3LE, UK
baojiu.li@durham.ac.uk

We review some of the commonly used approximation methods to predict large-scale structure formation in modified gravity models for the cosmic acceleration. These methods are developed to speed up the often slow N-body simulations in these models, or directly make approximate predictions of relevant physical quantities. In both cases, they are orders of magnitude more efficient than full simulations, making it possible to explore and delineate the large cosmological parameter space. On the other hand, there is a wide variation of their accuracies and ranges of validity, and these are usually not known *a priori* and must be validated against simulations. Therefore, a combination of full simulations and approximation methods will offer both efficiency and reliability. The approximation methods are also important from a theoretical point of view, since they can often offer useful insight into the nonlinear physics in modified gravity models and inspire new algorithms for simulations.

1. Overview

Up until today, N-body simulations are the only tool to be able to predict the large-scale structure formation in the nonlinear regime with any desired accuracy. The need for full simulations in modified gravity models is even stronger because of the difficulty to model the effects of nonlinear physics using other approaches, which is why a great deal of effort has been devoted to the development of simulation algorithms and codes in the past decade or so. Since the first few N-body code[292, 299, 333] that solves the nonlinear equation in $f(R)$ gravity appeared over a decade ago, ever more advanced codes, e.g., those with adaptive mesh refinement (AMR) nature, have been developed.[173, 294–296, 366] With the development of massively parallelised versions of the earlier codes, our ability of simulating modified gravity models has improved by orders of magnitude during this period (e.g., Refs. 284–286, 293, 300, 301, 332, 344, 367–369; see Chapter 3 for review). It is with little doubt that advancements in supercomputing infrastructure and paradigm shift in programming design will allow further improvements in the coming years.

However, numerical simulations have their own limitations in practice. A major drawback, from this point of view, is their lack of efficiency, as is evident in the effort

poured into making existing codes faster to cope with the ever increasing demand
for numbers of independent realisations, volume and resolution, in order to match
those of current and future observational surveys. Unlike ΛCDM, which is unique
in being widely accepted as a standard paradigm, there are many modified gravity
models, which makes it more difficult to allocate much effort to simulating individual
models or performing a continuous parameter search. Furthermore, simulations are
often used as a black box, making the underlying physics intractable, which does not
help in developing reliable theoretical templates used in model constraints. Finally,
although a simulation can predict structure formation down to very small scales,
it may be that information from such small scales is not strictly necessary and
trustable (e.g., due to the uncertainties in modelling baryonic physics).

This is certainly not to downplay the importance of simulations, which are not
replaceable in many aspects, e.g., when we need accurate model predictions deep
in the nonlinear regime and when the physics governing galaxy formation and evolution needs to be taken into account. However, if approximation methods can be
developed for the modified gravity models, and demonstrated to be valid in certain
regimes, then it would be realistic to use observations in those regimes to make
model constraints much more efficiently.

In this chapter we review some of the main approximation methods that have
been developed in recent years to make predictions of the modified gravity effect
on large-scale structure formation. We will follow a logic of introducing methods
that are closer to full simulations and then describing ones that can directly predict
observables. The former cover fast, approximate simulations and rescaled simulations, which produce particle snapshots like full simulations, while the latter predict quantities based on simplified treatments of the underlying physics, such as
perturbation theory and spherical collapse. It is worthwhile to point out, before
we get into the details, that these approaches (especially the latter) usually have
unknown validity regimes *a priori* or free parameters describing physical effects that
are not part of the modelling, and these must be validated or calibrated using full
simulations.

2. Approximate Simulation Techniques

In this section we briefly review some recent algorithm developments to allow
approximate particle snapshots to be obtained in much more efficient ways than
full modified gravity simulations. We will describe three classes of methods in the
order of moving further away from the full simulation technique and thus with
increasing efficiency (and less well-controlled accuracy).

2.1. *Truncated refinement-level simulations*

This class of method is built upon the observation that many of the modified gravity
models of interest in cosmology have the screening property, which means that their
deviation from GR is small in high-density regions. A major source of cost of modern

adaptive-mesh-refinement (AMR) simulations is the use of refinements where the Poisson equation (in GR) and the modified gravity partial differential equations (PDE)s have to be solved. Since screening happens exactly in those regions where AMR is needed, one may ask whether a high resolution for the modified gravity PDE is necessary: after all, if the fifth force is weak in high-density regions, then a small inaccuracy in it will probably have little impact on the simulation as a whole.

In Ref. 344 the authors propose a truncated simulations, where the standard Newtonian gravity is solved on all levels of the simulation mesh, while the scalar field that determines the fifth force is only solved on the domain grid (the lowest-resolution mesh that covers the whole simulation domain). The fifth force from the domain grid is then interpolated to the finer refinement levels. In this manner, both the fifth and the Newtonian forces are solved 'accurately', but to different accuracies as dictated by the fineness of the domain and most refined mesh, respectively. More explicitly, let G_{eff} be Newton's constant that includes the effects of both the Newtonian (F_N) and the fifth (F_5) forces, then in the full and truncated simulations one has, respectively,

$$G_{\text{eff}}^{\text{full}} = 1 + \frac{F_5^{\text{full}}}{F_N^{\text{full}}}, \quad G_{\text{eff}}^{\text{trunc}} = 1 + \frac{F_5^{\text{trunc}}}{F_N^{\text{full}}}.$$

In dense regions, where the truncated fifth force has a fractional error ϵ and $F_5^{\text{full}} = \eta F_N^{\text{full}}$, with $\epsilon, \eta \ll 1$, one has $F_5^{\text{trunc}} = F_5^{\text{full}}(1 + \epsilon) = F_N^{\text{full}}\eta(1 + \epsilon) \approx \eta F_N^{\text{full}} = F_5^{\text{full}}$, and so $G_{\text{eff}}^{\text{trunc}} \approx G_{\text{eff}}^{\text{full}}$. The left panels of Fig. 1 show how the truncations at different refinement levels result in $G_{\text{eff}}^{\text{trunc}} \approx G_{\text{eff}}^{\text{full}}$ even for the most aggressive truncation (i.e., only solving the scalar field on the domain level). The right panel shows that the truncation has little impact on the matter power spectrum ($< 1\%$ at $k < 5h\text{Mpc}^{-1}$). The same has been found in most other observables (e.g., $< 3\%$ for halo abundance, mass and density profiles, $< 0.05\%$ for halo positions and velocities). The gain in simulation speed is over an order of magnitude, marking a significant improvement.

Although the idea of allowing lower accuracy in the fifth-force calculation in high-density regions sounds physical, and is shown to work well for Vainshtein-type screened scenarios such as the DGP model, one should be cautious in generalising it to other types of screened theories. Indeed, it was found[332] that this approach works less well for chameleon models, e.g., $f(R)$ gravity, possibly because the Vainshtein mechanism is very efficient at screening the fifth force inside a matter clump, while the chameleon screening depends also on environments and works less well for matter clumps in underdense or intermediate-density regions. The symmetron mechanism has an explicit dependence on the local matter density and thus this approach may work well for it, but to date this has not been checked explicitly.

Another approach that follows a similar logic is that proposed earlier in Ref. 268. There, again the standard Newtonian force is solved accurately using AMR, but the fifth force is obtained in an approximate manner on all refinement levels including the domain one. More explicitly, because for chameleon and Vainshtein models the scalar field value can be calculated analytically for spherical tophat density profiles,

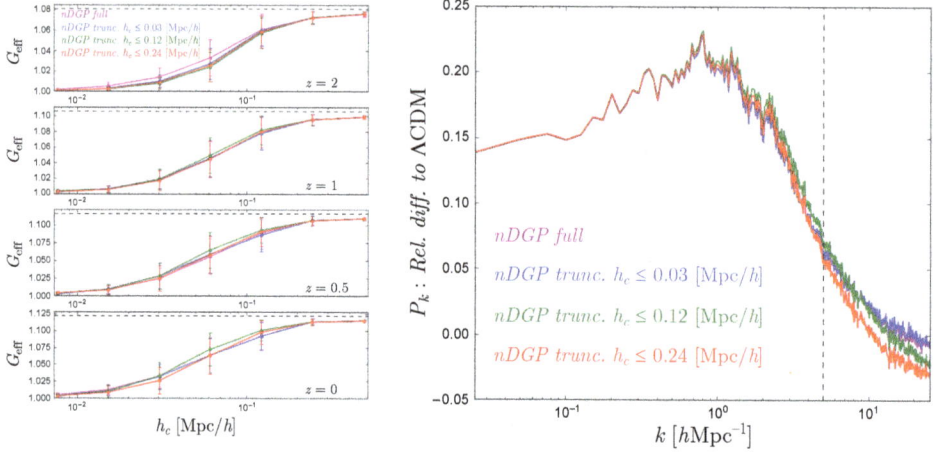

Fig. 1: *Left*: Total-force-to-normal-gravity ratio G_{eff} at particle positions as a function of the cell size of the finest refinement level the particles are in, for four redshifts ($z = 2$, 1, 0.5, 0) and for nDGP simulations truncated at four different refinement levels, as shown in the legends (e.g., $h_c \leq 0.03$ means that the scalar field is not directly solved but is interpolated from coarser levels on all refinement levels with cell size $\leq 0.03h^{-1}\mathrm{Mpc}$). The full curve (magenta) means there is no truncation. *Right*: The enhancements of the matter power spectrum wrt ΛCDM in the 4 truncated simulations as in the left panels. The vertical dashed line indicates $k = 5h\mathrm{Mpc}^{-1}$, which is the k value below which the different N-body simulations codes compared by Winther *et al.* (2015)[286] have agreement better than 1%. The figure is adapted from Barreira *et al.* (2015).[344]

they use this analytical approximation (with matter density estimated locally) to calculate the fifth force during the simulation. The agreements with full simulations are good at $k \leq 1h\mathrm{Mpc}^{-1}$, with details depending on model and redshift.

2.2. *COLA*

The truncated approach above boils down to sacrificing the accuracy of the fifth force calculation to gain improvement in efficiency. One can certainly take this logic one step further, and ask whether it is possible to give up the high accuracy of the Newtonian force calculation in return of further increases in performance.

Indeed, a majority of computer time in any AMR simulation is spent on the gravity solver and particle movements on refinements, which is particularly true for high-resolution simulations. Refinements are crucial for accurately resolving structures and substructures on small, highly-nonlinear, scales. But if one is only interested in mildly-nonlinear scales, e.g., the matter power spectrum down to $k \sim 0.3\ h\mathrm{Mpc}^{-1}$, then not using refinements at all may not be too bad an approximation.

In ΛCDM AMR simulations, the relaxation method is often used to solve the standard Poisson equation on refinements, partially because it is hard to make fast Fourier transform (FFT) work for the irregular-shaped refinement domains. If there is no refinement, the Poisson equation can be solved straightforwardly using FFT, which is efficient as the simulation becomes a pure particle-mesh (PM) one.

The situation is slightly worse for modified gravity, because even for a cubic periodic box the nonlinear PDEs of these equations cannot be solved using FFT (or at least the simple version of it), but instead has to be solved using relaxation. Nevertheless, by completely abandoning refinements one basically follows a more drastic approach than the truncated simulations described above, by calculating both the Newtonian and fifth forces accurately but with a lower resolution.

An alternative method in addition to pure PM simulations is the so-called Comoving Lagrangian Acceleration (COLA) approach,[370] which is also an approximation scheme. This method takes advantage of the fact that the second-order Lagrangian perturbation theory (2LPT) prediction of the the evolution of particle trajectories, $\vec{x}_{2\mathrm{LPT}}$, can be obtained with little computational effort. As a result, one can write the full particle position as $\vec{x} \equiv \vec{x}_{2\mathrm{LPT}} + \delta\vec{x}$, where $\delta\vec{x}$ denotes the difference. The standard acceleration equation $\ddot{\vec{x}} = -\vec{\nabla}\Phi$ then can be written as

$$\delta\ddot{\vec{x}} = -\vec{\nabla}\Phi - \ddot{\vec{x}}_{2\mathrm{LPT}}, \tag{1}$$

where now the known $\ddot{\vec{x}}_{2\mathrm{LPT}}$ serves as an effective friction. This method is flexible and can be very efficient — 2LPT accuracy is guaranteed even if the simulation consists of a single time step — if one is interested in large scales only the simulation can afford to have much fewer time steps, but with increasing number of time steps small-scale structures can be recovered as well. If one is interested in mildly nonlinear scales and requires a large number of simulations, say, to estimate the covariance matrices of observables, then COLA can be an efficient substitute for full N-body simulations.

The COLA approach is extended to modified gravity and massive neutrino scenarios in Refs. 371–373. The basic framework requires two major generalisations: (i) the 2LPT solution $\vec{x}_{2\mathrm{LPT}}$ needs to be generalised to modified gravity models, some of which involve scale-dependent linear growth factors (or scale-dependent Newton's constant) and additional source terms in the Poisson equation; (ii) at nonlinear scales the screening effects should be included, which is achieved using spherical approximations to the fifth force as in Ref. 268.

Figure 2 shows that the MG-COLA approach is able to reproduce the modified gravity effects in $f(R)$ and nDGP models to a few percent on length scales down to $k \sim \mathcal{O}(1)h\mathrm{Mpc}^{-1}$ but with much few time steps and less computational time. Due to the approximate modelling of screening it is not straightforward to see how an increase of time steps improves the agreement with full simulations on all scales — one will perhaps need to use both more steps and accurate computation of the screening.

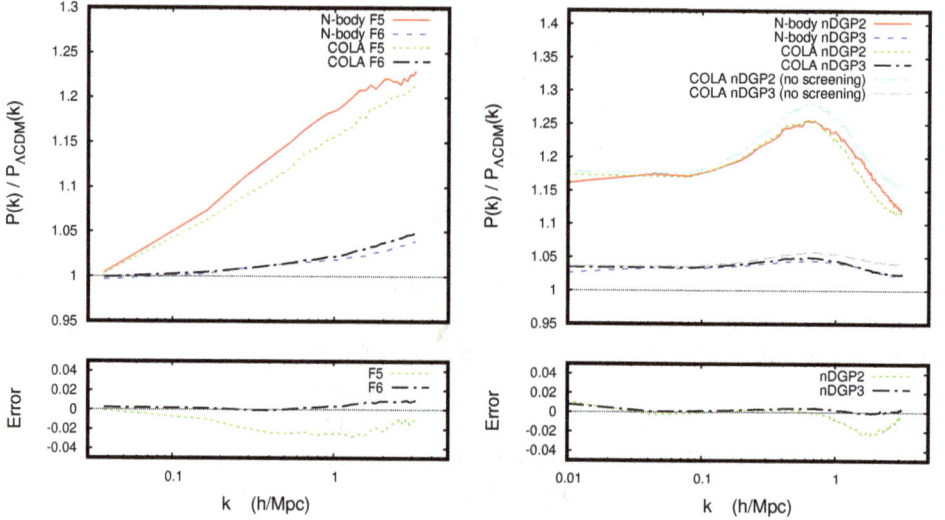

Fig. 2: *Left*: Comparisons between the predictions of the relative enhancement of the matter power spectra wrt ΛCDM, by MG-COLA and full N-body simulations for the Hu-Sawicki $f(R)$ model F5 and F6 (at $z = 0$). *Right*: The same as the left but for two nDGP models. The figure is adapted from Winther *et al.* (2017).[372]

2.3. *Rescaled simulations*

All the fast simulation approaches described so far involve tracking of particles throughout a simulation, and only differ in how the forces are calculated or how the particles are followed. The method discussed in this subsection is different in that it does not follow the movements of particles during a period for all models. Instead, it only needs a single simulation — a ΛCDM one which is designed to have specific cosmological parameters and output times. These outputs of particle snapshots can then be 'rescaled' following some physically-motivated recipe to get an *approximate particle distribution* for a target modified gravity model.

The basic idea is first described in Ref. 374, which shows that one can rescale an N-body particle snapshot to approximate simulation result of a ΛCDM model with different cosmological parameters, by changing the box size, particle mass and redshift of the original box to match the halo mass functions (or HMFs for short) in the two cosmologies, and then correcting the linear modes using the Zel'dovich approximation.

Consider a 'target' particle snapshot at redshift z' in a comoving box size of L', which one hopes to obtain by rescaling an 'original' simulation particle snapshot at redshift z with box size $L = L'/s$ (we follow the convention that all quantities in the 'target' cosmology are primed, and s is the rescaling factor of box size and particle coordinates). The first step is to find the correct z, s, since the original simulation is a full simulation that needs to be run with these specifications. These are determined by minimising

$$\delta_{\text{rms}}^2(s, z; z') = \frac{1}{\ln R_2 - \ln R_1} \int_{R_1'}^{R_2'} \frac{dR'}{R'} \left[1 - \frac{\sigma(s^{-1}R', z)}{\sigma'(R', z')}\right]^2, \tag{2}$$

where $\sigma(R, z)$ is the linear density field variance at smoothing scale R at z:

$$\sigma^2(R, z) = \frac{1}{4\pi} \int_0^\infty dk k^2 P_{\text{lin}}(k) \tilde{W}^2(kR), \tag{3}$$

in which P_{lin} is the linear matter power spectrum of the corresponding cosmology and $\tilde{W}(kR)$ is the Fourier transform of the window function (taken as a real-space spherical tophat window of radius R). The minimisation of Eq. (2) can ensure that the linear fluctuation amplitudes in the original and target cosmologies are as close as possible over a scale range $[R_1, R_2] = [s^{-1}R_1', s^{-1}R_2']$ at redshift $z = z(z')$. In excursion set theory (see below), the halo mass function (HMF) depends on $\sigma(R, z)$, and so this is equivalent to minimising the differences in the HMFs of the target and original cosmologies in the mass range $[M(R_1), M(R_2)]$.[374]

The rescaling consists of a scaling of box size according to

$$L \to L' = sL, \tag{4}$$

and accordingly particle coordinates, a redshift mapping

$$z \to z', \tag{5}$$

and a particle mass rescaling

$$m_p \to m_p' = \frac{\Omega_m' H'^2 L'^3}{\Omega_m H^2 L^3} m_p, \tag{6}$$

where m_p is the simulation particle mass, and $H = H(z)$ (and $H' = H'(z')$) is the Hubble expansion rate at z (and z') for the given cosmology.

While the minimisation of Eq. (2) is designed so that the target and original simulations have matched power spectrum at scales where the dimensionless linear power spectrum $\Delta_{\text{lin}}^2(k) \equiv k^3 P(k)/(2\pi^2) \sim 1$, the matching of matter power spectra is not automatically guaranteed at much larger scales. To overcome this, one can add additional displacements to the particles in the scaled simulation, which accounts for corrections for the difference on large scales (defined as wavenumbers $k < R_{\text{nl}}^{-1}$ with R_{nl} given by $\sigma'(R_{\text{nl}}, z') = 1$), resulting in (where a subscript k' means that the operation is in Fourier space):

$$\vec{x} \to \vec{x}' = s\vec{q} + D_+'(a')s\vec{d} + \delta\vec{x}', \quad \delta\vec{x}_{k'}' = \left[\left[\frac{\Delta_{\text{lin}}'^2(k', z')}{\Delta_{\text{lin}}^2(k = sk', z)}\right]^{1/2} - 1\right] D_+'(a')s\vec{d}_{k'}, \tag{7}$$

in which $D_+(a)\vec{d}(\vec{q})$ is the displacement field — the difference between the Eulerian ($\vec{x}(a)$ at time a) and the initial Lagrangian ($\vec{q}$) coordinates of a particle: $\vec{x}(a) = \vec{q} + D_+(a)\vec{d}(\vec{q})$ with $D_+(a)$ being the linear growth factor at a. In this approach, the phase of each Fourier mode of the density field $\delta(a) = -D_+(a)\nabla \cdot \vec{d}$ is preserved, while its amplitude is changed to match the target power spectrum. The peculiar velocity (again only on scales $k < R_{\rm nl}^{-1}$) is corrected using the linear growth rate

$$\vec{v}_{{\rm lin},\tilde{k}} \rightarrow \vec{v}'_{{\rm lin},\tilde{k}'} = s\frac{a'\dot{D}'_+(a')}{a\dot{D}_+(a)}\frac{f'(k',a')}{f(k=sk',z)}\left[\frac{\Delta'^2_{\rm lin}(k',z')}{\Delta^2_{\rm lin}(k=sk',z)}\right]^{1/2}\vec{v}_{{\rm lin},\tilde{k}'}, \quad (8)$$

where $\vec{v}_{\rm lin} = a\dot{D}_+(a)\vec{d} = aD_+(a)H(a)f(a)\vec{d}$, and $f(a) = d\ln D_+(a)/d\ln a$ is the linear growth rate. The nonlinear velocities are rescaled as

$$\vec{v}_{\rm nl} \rightarrow \vec{v}'_{\rm nl} = \left[\frac{aL}{a'L'}\frac{m'_p}{m_p}\right]^{1/2}\vec{v}_{\rm nl} = \left[\frac{\Omega'_m H'^2 L'^2(1+z')}{\Omega_m H^2 L^2(1+z)}\right]^{1/2}\vec{v}_{\rm nl}, \quad (9)$$

which comes from a spherical analogy $v^2(r) \propto M(r)/r$.

In Ref. 374 it is found that the rescaling has excellent performance, able to reproduce the matter power spectrum of the target cosmology to better than 0.5% on scales $k < 0.1h{\rm Mpc}^{-1}$ and 3% on $k < 1h{\rm Mpc}^{-1}$, halo masses and concentrations to below 10%, and halo positions and velocities to better than $0.09h^{-1}{\rm Mpc}$ and 5%.

The method can be improved to include a rescaling of halo internal structures[375] and generalised to rescaling halo rather than particle catalogues.[376, 377] Its extension to modified gravity scenarios is also straightforward,[269] with the linear power spectrum, growth rate and growth factor replaced by their modified gravity counterparts. Figure 3 shows that for ΛCDM and Hu-Sawicki $f(R)$ gravity (F4, F5, F6) it is capable of reproducing the nonlinear matter power spectrum and redshift monopole with 5% accuracy down to $k = 1h{\rm Mpc}^{-1}$.

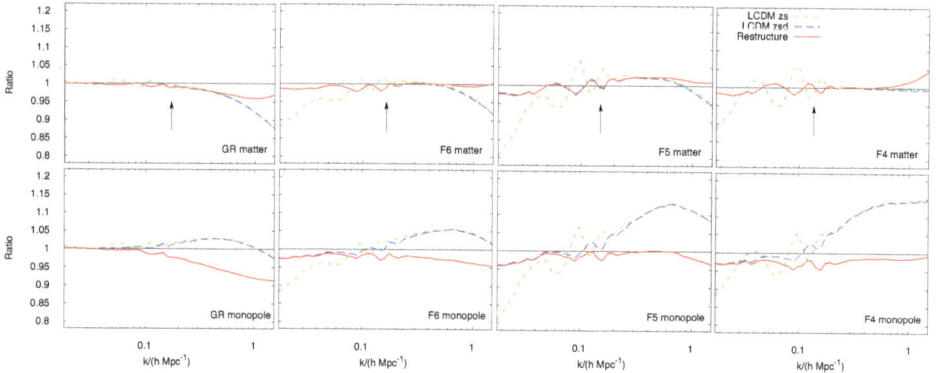

Fig. 3: *Upper panels*: The ratio between the predicted matter power spectra from rescaled and full simulations (at $z = 0$) for four models — from left to right ΛCDM, and Hu-Sawicki $f(R)$ model F6, F5, F4. The curves with different colours and line styles in each panel show the result of different stages of the rescaled simulation: rescaling redshift z and length s only (green shorted dashed), adding large-scale displacement field correction (blue long dashed) and full results by including also the rescaling of halo internal structures (red solid). *Lower panels*: The same as the upper panels for the the monopoles of the redshift space power spectra. This plot is adapted from Mead *et al.* (2015).[269]

3. Semi-Analytical Methods for Cosmological Observables

The fast and approximate simulation approaches described in the previous subsection have one thing in common — they produce actual particle snapshots. This is particularly useful if mock catalogues of galaxies and clusters are needed, for which one can obtain halo catalogues from the particle distributions and follow recipes such as halo occupation distribution[378,379] (HOD) to populate galaxies. Such catalogues are of great use when connecting theoretical models with observational data, the latter usually containing catalogues of galaxies.

In other situations, however, people are more interested in certain statistics that can be extracted from the large-scale distribution of galaxies, haloes or matter, the theoretical predictions for which do not necessarily require direct measurement from simulation. For example, when using cluster number counts to test a model, it is handy to have theoretical predictions of the counts (the simulation counterpart of which is the number of dark matter haloes as a function of halo mass, i.e., halo mass function, or HMF for short) which could be easily generated for arbitrary model parameter values; the theoretical HMF does not have to be measured from a simulation if it can be predicted in an alternative way.

Apparently, without tracking the full nonlinear physics of structure formation, these semi-analytical methods to calculate observables are approximate by nature. Their validity must be checked against full simulations and the physics unaccounted for in them are usually lumped together and described by free parameters which must then be calibrated by simulations. In this section we will discuss some of these methods. We will focus on three physical quantities that are fundamental to many observations — the HMF, matter power spectrum and two-point correlation function. For the latter two, there are approximate methods to predict them in both real and redshift spaces. We shall mention void abundance briefly in the end.

3.1. *Halo mass functions*

The Universe is known to host a 'cosmic web' of large-scale matter distribution, in which the majority of space has little matter and is occupied by the 'cosmic voids', whose boundaries are defined by rapid increases of matter and galaxy densities near filaments and sheets. Large nodes exist where filaments join, which are sites where large galaxy clusters are usually found. Such structures, seeded by the physics that took place at the primordial universe and shaped by the late-time evolution under the action of gravity, are complicated and often require full simulations to model.

Yet, if we are interested in certain simple statistics of the cosmic web, it is not always necessary to go a great length to run simulations. One primary example is the abundance of dark matter haloes in the universe, which can be analytically predicted by the simple Press-Schecter (PS) approach.[380] The PS method was extended later into the excursion set theory[381] and has motivated various fitting functions of the HMF that are calibrated by simulations[382–386] and used in theoretical studies.

A key premise, which is a reasonable physical assumption, in this approach is the connection between the dark matter haloes found in a late-time simulation snapshot and the small density peaks produced in the early universe as initial conditions. In the standard picture of cosmic structure formation, these initial density peaks grow by attracting more matter toward them through gravity, and thereby form the late-time matter clumps. To simplify the modelling, the density peaks are usually assumed to be spherical or ellipsoidal, so that their evolution toward singularities can be (semi)analytically followed. A specific point of this collapsing process, usually taken as the end of it, is associated to the time at which a dark matter halo is formed. Apparently, higher initial density peaks collapse earlier and so correspond to haloes which form at higher redshifts. Using this connection, the number counts of the haloes more massive than a given mass threshold can be obtained from the number densities of the density peaks higher than some threshold. Let's formulate this idea more quantitatively now, following the prescription of excursion set theory.[381]

3.1.1. *Spherical collapse and excursion set theory*

Let's consider a spherical tophat overdensity described by its density contrast $\delta_i \equiv \rho_i/\bar{\rho}(a_i) - 1$, where a subscript i denotes the initial time and $\bar{\rho}(a_i)$ is the mean matter density at a_i. The initial comoving size of this spherical region is R. Since $\delta_i \ll 1$, ρ_i is approximately the mean density at a_i, so that the enclosed mass is $M(R) \approx \bar{\rho}(a_i)(a_i R)^3 = \bar{\rho}_{m0} R^3$ with $\bar{\rho}_{m0}$ the background matter density today. We assume that during the evolution of this region there is no shell crossing and the density profile remains a tophat, a consequence of which is that the enclosed mass remains constant. Defining $y(t) = r(t)/a(t)R$ where $r(t)$ is the physical size of this patch at time t, the nonlinear evolution in ΛCDM is described by

$$y'' + \left[2 - \frac{3}{2}\Omega_m(N)\right]y' + \frac{1}{2}\Omega_m(N)\left(y^{-3} - 1\right)y = 0, \tag{10}$$

where a prime is the derivative wrt $N = \ln(a)$ and $\Omega_m(N)$ is the matter density parameter at N. The initial condition is taken as $y(a_i) = 1 - \delta_i/3$ and $y'(a_i) = -\delta_i/3$.

A dark matter halo of mass $M(R)$ is said to have formed at z_f if $r(z = z_f) = 0$ or $y(z_f) = 0$, and z_f is the halo formation redshift. For this to happen, the initial density contrast δ_i has to be tuned to the correct value $\delta_{i,c}$ (which often involves trial-and-error). In the literature, $\delta_{i,c}$ is usually extrapolated to today according to linear perturbation theory: $\delta_c \equiv \delta_{i,c}D_+(a = 1)/D_+(a_i)$ with $D_+(a)$ the linear growth factor at a. We shall follow the same convention by defining δ_c as the linearly extrapolated value of $\delta_{i,c}$. Throughout this section, we always use the extrapolated density contrast in equations and figures, unless otherwise stated. Apparently, δ_c is in general a function of R and z_f, $\delta_c(R, z_f)$, but this can be equivalently written as $\delta_c(M, z_f)$ thanks to the relation between R and M mentioned above. In ΛCDM, δ_c is independent of R or M — this can be most straightforwardly seen from Eq. (10), which does not depend on R; this does not hold true, for example, in the chameeon model, as we will see below.

To check if a spherical region with initial radius R or mass M has collapsed to form a dark matter halo by z_f, we only need to see if the initial density contrast inside it, $\delta(R)$, extrapolated to today (again using the linear growth rate), satisfies $\delta(R) \geq \delta_c(R, z_f)$. To calculate the abundance of these haloes, we need the probability distribution of $\delta(\vec{x}, R)$, which is the density contrast within a sphere of radius R centred at an arbitrary spatial location $\vec{x}$, given by

$$\delta(\vec{x}, R) = \int W(|\vec{x} - \vec{x}'|; R)\delta(\vec{x}')d^3\vec{x}' = \int \tilde{W}(k; R)\delta_{\vec{k}}e^{i\vec{k}\cdot\vec{x}}d^3\vec{k}, \qquad (11)$$

where $W(r; R)$ is a filter or window function of radius R and $\tilde{W}(k; R)$ its Fourier transform, and $\delta_{\vec{k}}$ is the Fourier transform of $\delta(\vec{x})$. Assuming $\delta_{\vec{k}}$ satisfies a Gaussian distribution, statistically it can be fully specified by its power spectrum $P(k)$ given by $(2\pi)^3\delta^{(3)}(\vec{k} + \vec{k}')P(|\vec{k}|) = \langle\delta_{\vec{k}}(\vec{k})\delta_{\vec{k}}^*(\vec{k}')\rangle$ where $\delta^{(3)}$ is the Dirac δ-function in 3D and $\langle\cdots\rangle$ denotes ensemble average. In this case we have

$$\langle\delta^2(\vec{x}, R)\rangle \equiv \sigma^2(R) \equiv S(R) = \int 4\pi k^2 P(k)\tilde{W}(k; R)dk, \qquad (12)$$

in which the variance $S(R) = \sigma^2(R)$ is defined, with $\sigma(R)$ the root-mean-squared (rms) density fluctuation within spherical window of size R. We still have the freedom to choose the smoothing window function. If W is a sharp k-space window,

$$\tilde{W}(k; R) = \Theta(1 - kR), \qquad (13)$$

where $\Theta(x)$ is the Heaviside function with $\Theta(x \geq 0) = 1$ and $\Theta(x < 0) = 0$, then in Eq. (11) any change of $\delta(\vec{x}, R)$ due to change of $R \to R - dR$ comes from extra k-modes that are newly brought into the integration. Because different k-modes in the linear spectrum are independent, the increment of $\delta(\vec{x}, R)$ due to an increase in R does not depend on the value of $\delta(\vec{x}, R)$ at the previous value of R (the Markov property). Note that, given initial power spectrum $P(k)$, $S(R)$ is a fixed monotonic decreasing function of R, and can be used interchangeably with R, M.

Because of its above properties, one can regard the change of $\delta(\vec{x}, R)$ with R or S as a Brownian motion, as schematically shown in Fig. 4. $S = 0$ corresponds to $R \to \infty$ so that $\delta(\vec{x}, R) \to 0$ is the overdensity of the whole universe, which means that the Brownian motion always starts at the origin (not shown in Fig. 4). As S increases, we are looking at smaller and smaller R, and $\delta(\vec{x}, R)$ fluctuate more and more strongly. S is effectively a 'time' variable for the Brownian motion.

The Gaussian smoothed density field $\delta(\vec{x}, R)$, or equally $\delta(\vec{x}, S)$ satisfies a Gaussian probability distribution

$$P(\delta, S)d\delta = \frac{1}{\sqrt{2\pi S}}\exp\left[-\frac{\delta^2}{2S}\right]d\delta. \qquad (14)$$

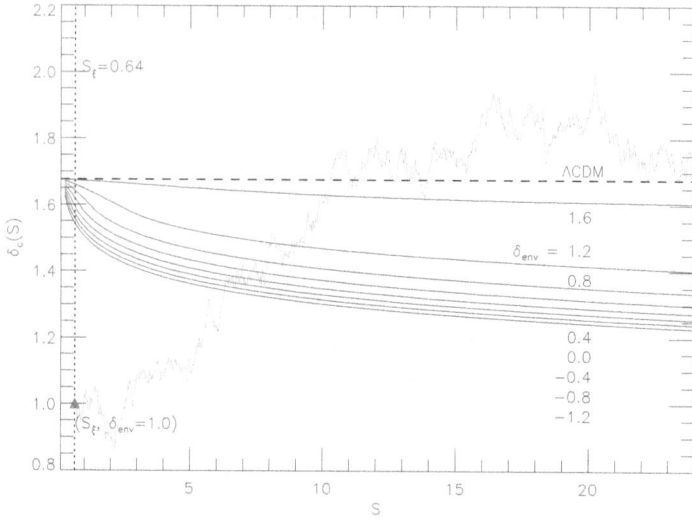

Fig. 4: An illustration of how the excursion set theory works. See the main text for a description. This figure is adapted from Li and Efstathiou (2012).[244]

At a given $\vec{x}$, a virialised dark matter halo within the mass range $[M - dM, M]$ forms at z_f *if and only if* the Brownian motion makes its *first* crossing of the barrier $\delta_c(M, z_f)$ (which is constant in M or S for ΛCDM; see the horizontal dashed line in Fig. 4) within $[S, S + dS]$. By its nature, the Brownian motion will cross this barrier infinitely many times at larger S which correspond to collapsed structures of smaller masses, but those are considered as the internal structures rather than main haloes. Similarly, if the Brownian motion crosses the barrier at smaller S, then the structure we look at is a substructure of some more massive halo. The probability density for this first crossing to take place inside $[S, S + dS]$ is given by

$$f(S, z_f)dS = \frac{\delta_c(z_f)}{\sqrt{2\pi}S^{3/2}} \exp\left[-\frac{\delta_c^2(z_f)}{2S}\right] dS, \tag{15}$$

and so the number density of haloes in the mass range $[M - dM, M]$ at z_f is

$$\frac{dn(M)}{dM}dM = \frac{\bar{\rho}_m(z_f)}{M}f(S, z_f)dS, \tag{16}$$

where $\bar{\rho}_m(z_f)$ is the mean matter density at z_f.

While Eq. (16) is generic, Eq. (15) has been derived under the assumption that the critical density for collapse, $\delta_c(M, z_f)$, only depends on z_f and not on M, R or S. In other words, the barrier that the considered Brownian motion has to cross is flat. Also, thanks to the spherical symmetry used in the derivation, $\delta_c(M, z_f) = \delta_c(z_f)$ does not depend on the large scale environment. Both assumptions are model specific and might not be valid in modified gravity models, as we shall see shortly.

3.1.2. *Halo bias*

Dark matter haloes are biased tracers of the underlying matter density field, which means that the halo number density contrast $\delta_h = n_h/\bar{n}_h - 1$, where n_h is the local halo number density and $\bar{n}_h$ its mean, is generally different from the matter density contrast δ. The two are related by the halo bias b, which depends on the halo mass (high mass haloes are more biased with $b > 1$), redshift and length scales considered.

Consider a halo of mass M which forms at z_f, and use the picture of a Brownian motion which crosses $\delta_c(z_f)$ between $[S, S + dS]$ with $S = S(M)$. This time let us assume that the halo forms in a large scale environment with initial density contrast δ_{env} and an initial comoving size ξ. In Fig. 4, this is equivalent to saying that the Brownian motion has passed through the point (S_ξ, δ_{env}), the red triangle, before crossing the barrier $\delta_c(z_f)$, in which S_ξ is the value of S corresponding to $R = \xi$ and δ_{env} is the initial density contrast enclosed in ξ extrapolated to today using the linear growth factor. Note that this necessarily means that $S_\xi < S$ and $\delta_{env} < \delta_c(z_f)$ as otherwise the environment itself would have collapsed to form a bigger halo.

The halo number density contrast satisfies[387]

$$(1 + \delta_h)f(S, z_f)dS = (1 + \delta_{env}^{z_f})f(S, z_f; S_\xi, \delta_{env})dS, \tag{17}$$

where $\delta_{env}^{z_f}$ is the linearly-extrapolated density contrast of the environment at z_f, and $f(S, z_f; S_\xi, \delta_{env})$ is the *conditional* probability density for the Brownian motion that has first passed through (S_ξ, δ_{env}) to cross $\delta_c(z_f)$ at S, which for a flat δ_c is:

$$f(S, z_f; S_\xi, \delta_{env}) = \frac{1}{\sqrt{2\pi}} \frac{\delta_c(z_f) - \delta_{env}}{(S - S_\xi)^{3/2}} \exp\left[-\frac{(\delta_c(z_f) - \delta_{env})^2}{2(S - S_\xi)}\right]. \tag{18}$$

Let us take the limit of large environmental regions, with $S_\xi \ll 1$ and $|\delta_{env}| \ll 1$, as a concrete example. With a Taylor expansion[388]

$$\delta_h = \sum_{n=0}^{\infty} \delta^n \frac{b_n}{n!} = b_0 + b_1 \delta_{env}^{z_f} + \cdots \tag{19}$$

where ellipsis represents higher-order terms neglected as a result of $\delta \ll 1$, we find b_1 as

$$b_1 = \left.\frac{d\delta_h}{d\delta_{env}^{z_f}}\right|_{\delta_{env}^{z_f}=0} = 1 + \frac{d\delta_{env}}{d\delta_{env}^{z_f}} \frac{f(S, z_f)df(S, z_f; 0, \delta_{env})}{d\delta_{env}} = 1 + g(z_f) \frac{\delta_c^2/S - 1}{\delta_c}, \tag{20}$$

in which we have used Eqs. (15), (17), (18), neglected the z_f-dependence of $\delta_c(z_f)$ and defined $g(z) \equiv d\delta_{env}/d\delta_{env}^{z_f} = D_+(z = 0)/D_+(z_f)$. As the coefficient of the linear term of the Taylor expansion, b_1 is called linear halo bias. Equation (20) makes explicit that b_1 depends on z_f and the halo mass M (or equivalently S). Note that the zeroth order term b_0 in the Taylor expansion vanishes as $\delta_h \to 0$ when $S_\xi \to 0$.

3.1.3. *Ellipsoidal collapse*

The prediction of excursion set theory, Eqs. (15), (16), provides a good starting point to model HMFs and generally shows qualitative agreement with simulations. However, it is not surprising that such a simple approach does not fully match simulation HMFs, especially at the low-mass end (where it overestimates halo abundance).

References 389, 390 suggest that the spherical collapse model used in excursion set theory is far from realistic, and should be replaced with a model in which the initial regions that form haloes at late times have ellipsoidal shapes. Unlike spherical collapse, the collapse of an ellipsoidal region depends on the surrounding shear field and the resulting critical density for collapsing at z_f now depends on the size of the region (or the mass enclosed in it), $\delta_{ec}(M, z_f)$ or $\delta_{ec}(S, z_f)$:

$$\delta_{ec}(S, z_f) = \sqrt{q}\delta_c(z_f) \left[1 + \beta \left(\frac{\delta_c(z_f)}{S}\right)^{-\alpha}\right], \tag{21}$$

in which $\alpha \approx 0.615$, $\beta \approx 0.485$ are derived from ellipsoidal collapse dynamics ($\alpha = \beta = 0$ in the spherical collapse limit), and $q \approx 0.7$ from normalising to simulations. It can be seen that as S increases, the barrier becomes higher, which means that it is harder for the Brownian motion to cross, leading to fewer low-mass haloes.

The first crossing probability density $f(S, z_f)$ in this case can be written as

$$f(S, z_f) = \frac{\sqrt{q}}{\sqrt{2\pi}} A \left[1 + q\frac{\delta_c^2}{S}\right]^{-p} \frac{\delta_c}{S^{3/2}} \exp\left[-\frac{q}{2}\frac{\delta_c^2}{S}\right], \tag{22}$$

where $p = 0.3$ and A is a normalisation factor which is determined by requiring that the integration of $f(S, z_f)$ over $S \in [0, \infty)$ is 1. In this equation we have again not explicitly written the z_f-dependence of $\delta_c(z_f)$ for simplicity.

Equation (16) can still be used to predict the halo abundance, with now the use of the $f(S, z_f)$ from Eq. (22). Meanwhile, the linear halo bias becomes

$$b_1 = 1 + g(z_f) \left[\frac{q\delta_c^2/S - 1}{\delta_c} + \frac{2p/\delta_c}{1 + (q\delta_c^2/S)^p}\right], \tag{23}$$

where $g(z_f)$ is the same as before and we again omit the z_f-dependence of δ_c.

3.1.4. *Excursion set theory in modified gravity*

The physical picture of excursion set theory and its generalisations to ellipsoidal collapse hold for modified gravity models. However, due to the fifth force, the non-linear evolution of a spherical or ellipsoidal tophat overdensity is in general different from the ΛCDM result discussed above. This means that the critical initial density required for such an overdensity to collapse at z_f would be different from the ΛCDM predictions. Indeed, although for ΛCDM spherical collapse δ_c depends only on z_f, in a modified gravity model it can depend on z_f, M and δ_{env}, as happens for chameleon-type theories. Because the analysis for Vainshtein-type models is simpler, in this subsection we shall focus on chameleons and only briefly mention some results for Vainshtein.

Before moving to the details, let us remark that even in the study of modified gravity models, we shall again extrapolate initial density contrasts to today using the linear growth factor of ΛCDM. This might sound counterintuitive, but one has to bear in mind that the linear extrapolation does not change anything in the initial density field apart from rescaling it by *some* constant. Strictly speaking, the extrapolation is not necessary as in the spherical or ellipsoidal collapse model what are needed are the distribution of *initial* density peaks and how they evolve *nonlinearly* under the action of gravity: linear theory has no role to play here apart from setting the initial condition. From a practical viewpoint, the extrapolation is convenient (for example, with it we will be dealing with values of $\delta_c(z_f)$ of order unity, and different people will agree on its value even though they would generally have different values of the linear density contrast at their *own*, *different*, initial redshifts, e.g., $z_i = 50, 100$) but not necessary. Unless otherwise stated, when talking about linear density contrast, we always mean the one linearly extrapolated to $z = 0$ using ΛCDM cosmology.

For chameleon models, the evolution equation (10) has to be modified to include the fifth force. Taking $f(R)$ gravity as an example, we have[244]

$$y_h'' + \left[2 - \frac{3}{2}\Omega_m(N)\right] y_h' + \frac{1}{2}\Omega_m(N)\left(y_h^{-3} - 1\right)y_h \left[1 + \frac{1}{3}\min\left\{1, 3\frac{\Delta r}{r}\right\}\right] = 0, \quad (24)$$

in which we have now used y_h instead of y, to denote that this is $r(t)/a(t)R$ for a halo under consideration. The expression in the squared brackets multiplied to the last term incorporates the effects of the fifth force. In $f(R)$ gravity, the fifth force can be between 0 and 1/3 the strength of the Newtonian gravity, but its maximum strength relative to standard gravity can be different in general chameleon models, for which the term in these square brackets should be changed accordingly. Here the fifth force effect has been modelled under the thin-shell approximation,[57, 58] with $\Delta r/r$ to be defined shortly. A more accurate expression for the fifth force may also be used:[191, 246] $\frac{1}{3}\min\left\{1, 3\frac{\Delta r}{r} - 3\left(\frac{\Delta r}{r}\right)^2 + \left(\frac{\Delta r}{r}\right)^3\right\}$.

To be more concrete, we assume that the initial region that forms a virialised dark matter halo at z_f is spherical with an initial comoving radius R and constant initial density δ_c (extrapolated to today using ΛCDM linear growth factor) inside. The spherical collapse takes place inside another, much larger, spherical region with initial comoving radius ξ and constant initial density δ_{env}. The latter is called the environment, and is used to model the environmental screening of the fifth force.

As the inner spherical overdensity evolves, the environment itself also evolves (either expands or contracts, depending on the sign of δ_{env}). We have to follow the co-evolution of these two systems. For the former we use Eq. (24) and for the latter we assume ΛCDM spherical evolution, similar to Eq. (10):

$$y_{\text{env}}'' + \left[2 - \frac{3}{2}\Omega_m(N)\right] y_{\text{env}}' + \frac{1}{2}\Omega_m(N)\left(y_{\text{env}}^{-3} - 1\right)y_{\text{env}} = 0, \quad (25)$$

where a subscript $_{\text{env}}$ is used to distinguish from y_h in Eq. (24). The use of ΛCDM equation for the evolution of the environment is justified if the initial comoving radius ξ is much larger than the range of the fifth force (or the Compton wavelength of the scalar field mediating the force). Meanwhile, ξ cannot be too big for it to be a faithful environmental definition (if ξ is too large then the environmental density will simply be the cosmic mean). In Ref. 244, $\xi = 8h^{-1}$Mpc is adopted, though later we will briefly mention other possibilities explored in the literature. Again, $\delta_{\rm env} < \delta_c$ so that the environment itself has not collapsed to form virialised halo by z_f.

At any late redshift $z \geq z_f$, the co-evolving system consists of a small spherical tophat with uniform density $\rho_h(z) = \rho_{m0} y_h^{-3}(1+z)^3$, embedded in a larger one with density $\rho_{\rm env}(z) = \rho_{m0} y_{\rm env}^{-3}(1+z)^3$. Let $f_{R,\rm env}$ and $f_{R,h}$ be the values of the scalar field f_R inside constant density fields with $\rho = \rho_{\rm env}$ and $\rho = \rho_h$ respectively, and Ψ_N be the Newtonian potential of the inner spherical patch at its edge r, then as described in Chapter 1, if $|f_{R,\rm env} - f_{R,h}| \ll |\Psi_N|$, the scalar field can go from its external environmental value $f_{R,\rm env}$ to its internal halo value $f_{R,h}$ within a distance Δr inside the edge of inner patch. This is known as the thin-shell[57,58] which for $f(R)$ gravity can be expressed as:

$$\frac{\Delta r}{r} = \frac{f_{R,\rm env} - f_{R,h}}{2\Psi_N}, \tag{26}$$

Using $\Psi_N = -\frac{4\pi G}{3}\rho_h(z)r^2 = -\frac{4\pi G}{3}\rho_{m0}R^2 y_h^{-1}(1+z)$ and

$$f_{R,h} = f_{R0}\left[\frac{\frac{\rho_h(z)}{\rho_{m0}} + 4\frac{\Omega_\Lambda}{\Omega_m}}{1 + 4\frac{\Omega_\Lambda}{\Omega_m}}\right]^{-2} \quad ; \quad f_{R,\rm env} = f_{R0}\left[\frac{\frac{\rho_{\rm env}(z)}{\rho_{m0}} + 4\frac{\Omega_\Lambda}{\Omega_m}}{1 + 4\frac{\Omega_\Lambda}{\Omega_m}}\right]^{-2} \tag{27}$$

for Hu-Sawicki $f(R)$ model with $n = 1$ in Eq. (26), we can express $\Delta r/r$ in terms of $z, R, y_h, y_{\rm env}$ and cosmological parameters such as $H_0, \Omega_m, \Omega_\Lambda$. Hence, Eqs. (24), (25) are coupled differential equations which need to be solved together. As in the standard case, through trial-and-error we can find the correct value of the initial density contrast for the inner patch to collapse at z_f, extrapolated to today using linear growth factor of ΛCDM. The result is written as $\delta_c(S, z_f, \delta_{\rm env})$, where the dependence on $S, \delta_{\rm env}$ is because the thin-shell condition above depends on $R, y_{\rm env}$. Recall that $\delta_{\rm env}$ is the initial density contrast in the environment that is extrapolated to today using the linear growth factor of ΛCDM — it is a single number rather than a function of time such as $y_{\rm env}$, and so more convenient to use (as the initial size of the environment ξ is fixed, given $\delta_{\rm env}$ we can calculate $y_{\rm env}$ at arbitrary times).

The solid lines in Fig. 4 are $\delta_c(S, z_f = 0, \delta_{\rm env})$ for a selection of $\delta_{\rm env}$ values as indicated in the legend. Two physical features are observed: (1) the barrier δ_c is lower at larger S (corresponding to smaller mass), meaning that the first crossing of Brownian motions is more likely to take place at smaller S, therefore larger haloes are more likely to form; (2) in higher-density environments (larger $\delta_{\rm env}$), the barrier is closer to the ΛCDM result, a feature of environmental screening.

The dependence of $\delta_c(S, z_f, \delta$env$)$ on S makes it impossible to find analytical solutions to the first-crossing probability $f(S, z_f, \delta_{\rm env})$ for a given $\delta_{\rm env}$. Furthermore, the dependence on $\delta_{\rm env}$ means that, unlike in ΛCDM, there is no unique barrier but infinitely many. These both add complications to the solution. In practice, one is interested in the overall, or environment-averaged, first-crossing probability density:

$$f(S, z_f) = \int_{-\infty}^{\delta_c^{\Lambda CDM}} q(\delta_{env}, \delta_c^{\Lambda CDM}, S_\xi) \times f(S, z_f; S_\xi, \delta_{env}) d\delta_{env}, \tag{28}$$

in which $f(S, z_f; S_\xi, \delta_{env})dS$ is the conditional probability that a Brownian motion that has passed through (S_ξ, δ_{env}) will cross the barrier $\delta_c(S, z, \delta_{env})$ between $[S, S + dS]$, which can be calculated numerically.[391] $q(\delta_{env}, \delta_c^{\Lambda CDM}, S_\xi)$ is the distribution of δ_{env}, or the probability density that a Brownian random motion passes through (S_ξ, δ_{env}) and never exceeds $\delta_c^{\Lambda CDM}$ at $S \leq S_\xi$ (as otherwise the environment itself would have collapsed). As the environment follows a ΛCDM evolution, we have[381]

$$q(\delta_{env}, \delta_c^{\Lambda CDM}, S_\xi) = \frac{1}{\sqrt{2\pi S_\xi}} \left\{ \exp\left[-\frac{\delta_{env}^2}{2S_\xi}\right] - \exp\left[-\frac{(\delta_{env} - 2\delta_c^{\Lambda CDM})^2}{2S_\xi}\right] \right\}, \tag{29}$$

for $\delta_{env} \leq \delta_c^{\Lambda CDM}$ and 0 otherwise. The integration in Eq. (28) can be performed numerically. Note that halo bias can also be calculated in this approach, though the calculation will be more involved and has not been done so far.

Once $f(S, z_f)$ is obtained, the calculation of the HMF follows the same expression as Eq. (16). Figure 5 shows an example of the derived HMF for a chameleon model (solid) and a ΛCDM model which only differs in the absence of a fifth force (dashed). The bottom panel shows the relative difference between the two models, which shows the expected result that they predict the same number of massive haloes, but the chameleon model predicts more intermediate-mass haloes due to the enhancement of gravity. Because more small haloes have merged to form

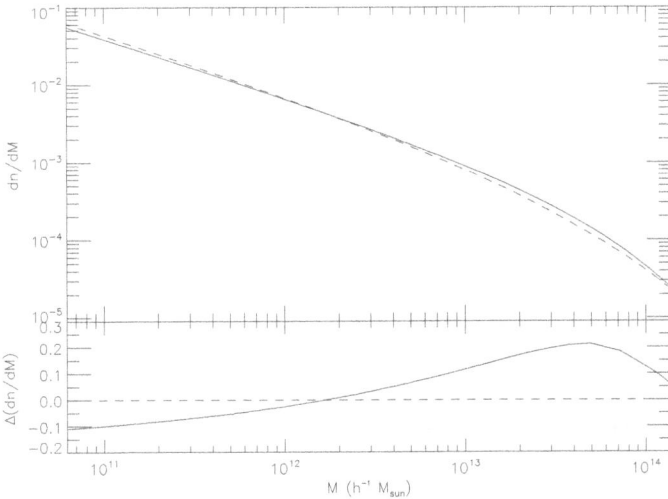

Fig. 5: An example showing the prediction of the HMF in chameleon model from the extended excursion set theory (see text for more details). This figure is adapted from Li and Efstahiou (2012).[244]

intermediate-mass haloes in the chameleon model, fewer are left at the low-mass end.

The method above has gone a great length in modelling the fifth force effect on the spherical collapse, especially its environmental dependence. Given that spherical collapse is at best an approximation, this part can be simplified. For example, one can eliminate the need for integrating over an environment distribution in Eq. (28) by using an average environment $\bar{\delta}_{\rm env}$ given by $\int \delta_{\rm env} q(\delta_{\rm env}, \delta_c^{\Lambda\rm CDM}, S_\xi) d\delta_{\rm env}$ to compute a mean barrier[191, 246] $\delta_c(S, z_f, \bar{\delta}_{\rm env})$, which will be a single barrier to calculate the first-crossing probability. Alternatively, one can calculate the mean of $\delta_c(S, z_f, \delta_{\rm env})$ as $\bar{\delta}_c(S, z_f) = \int \delta_c(S, z, \delta_{\rm env}) q(\delta_{\rm env}, \delta_c^{\Lambda\rm CDM}, S_\xi) d\delta_{\rm env}$ in which one again ends up with a single barrier. An even simpler treatment is to consider two limiting cases of the fifth force — zero and $1/3$ of the strength of standard gravity — which will lead to band of possible values of the HMF.[240]

On the other hand, the modelling of environment above can also be improved. The environments were taken as spherical regions whose initial comving (or Lagrangian) radius is $\xi = 8h^{-1}\rm Mpc$, and thus are called *Lagrangian environments*. However, depending on the initial density inside an environment, its size can become much larger or smaller than $8h^{-1}\rm Mpc$ at late times, while a good definition of environment should be neither (if the size of the environment is too small it is dominated by the halo under consideration; if the size is too large is simply approaches the cosmological background; neither is a good definition for our modelling). In the mean time, it is known[392] that the thin shell model works very well if the environment has a fixed Eulerian size of order $5 \sim 8h^{-1}\rm Mpc$, and therefore it is physically sensible to define environments as regions whose Eulerian size is similar to the Compton wavelength of the scalar field. References 245, 393 define the environments to have a physical (Eulerian) size of $\zeta = 5h^{-1}\rm Mpc$ at z_f, and the calculation then proceeds as in the case with Lagrangian environment. The main practical difference is in the way the physical density in the environment is calculated: instead of solving Eq. (25), the nonlinear density contrast in the environment, at arbitrary redshift z, $\Delta_{\rm NL}(z)$, is given approximately as,[394, 395]

$$\Delta_{\rm NL}(z) = \left[1 - \frac{\delta_{\rm lin}(z)}{\delta_c^{\Lambda\rm CDM}}\right]^{-\delta_c^{\Lambda\rm CDM}}, \tag{30}$$

in which $\delta_{\rm lin}(z)$ is the initial density contrast extrapolated to redshift z using the linear growth rate of ΛCDM. The probability distribution of $\delta_{\rm env}$ is also different because all environments now have the same final Eulerian radius $\zeta = 5h^{-1}\rm Mpc$, and some approximate analytical expressions can be found in Ref. 396. For details of the implementation and result of this method see Refs. 245, 393.

The fifth force effect can be more easily modelled in other modified gravity (e.g., Vainshtein) models, in which case the spherical collapse equation can be written as

$$y_h'' + \left[2 - \frac{3}{2}\Omega_m(N)\right] y_h' + \frac{1}{2}\Omega_m(N) \left(y_h^{-3} - 1\right) y_h \left[1 + \Gamma(N, y_h)\right] = 0, \tag{31}$$

in which $\Gamma(N, y_h) = \Gamma(z, \rho_h(z))$ incorporate the fifth force effect,[174, 251, 342] which depends only on the redshift z and the density inside the spherical tophat at z. There is no dependence on δ_{env} or R (or equivalently halo mass M), and the resulting $\delta_c(z_f)$ is a single flat barrier just like in ΛCDM. The first crossing probability can be obtained analytically and so the HMF (and halo bias[342]) is straightforward to find.

An approximation used in the above method to calculate $\delta_c\,(S, z_f, \delta_{\text{env}})$ is that haloes have a tophat density profile throughout their evolution history. It is known, however, that an initial tophat overdensity generally develops nontrivial inner profiles during spherical collapse in $f(R)$ gravity,[242] which increases with radius and has a spike at the surface. Neither of this and a tophat reflects the true shapes of dark matter haloes in simulations, which is why in the references mentioned above a tophat profile is assumed for its simplicity. An alternative method to calculate the collapse threshold was proposed in Ref. 243, where haloes are assumed to take mean initial density profiles by smoothing an initial Gaussian random field, and the density profiles are numerically evolved until collapse allowing the shape to evolve as well. The resulting collapse threshold, for a given model, depends on z and the mass enclosed in the halo, M; the effect of the environment is not explicitly included in the calculation, and it is suggested that the use of mean density profiles implicitly accounts for that. The method predicts the same behaviour that δ_c decreases as the halo mass M decreases, and once δ_c is available the calculation of the first-crossing probability and the HMF is the same.

3.1.5. *Simulation calibrations*

The modified excursion set theory with spherical collapse represents a greatly simplified picture of the complicated nonlinear physics of modified gravity, with assumptions such as sphericity and tophat density profile. Deviations from these assumptions affect the accuracy of predictions of the HMF already in ΛCDM, and they further affect the reliability of the thin shell calculation of the fifth force and chameleon screening.[397–399] One way to improve on this, instead of making the semi-analytical modelling more complicated and intractable, is to include parameterised corrections for which the parameters are calibrated by numerical simulations for a selected few models, and expect these calibrated fitting formulae to work for general models.

Reference 254 proposes to incorporate all the unaccounted-for physics in the spherical model to obtain a 'corrected' critical density (or barrier) for collapse, δ_c^{cor}:

$$\delta_c^{\text{cor}} = \epsilon(S, z_f; M_{\text{th}}^{(1)}, M_{\text{th}}^{(2)}, \eta, \vartheta, \chi)\delta_c(S, z_f, \delta_{\text{env}}^{\text{peak}}), \tag{32}$$

where $\delta_c(S, z_f, \delta_{\text{env}})$ is the environment-dependent barrier calculated using the most probable value of δ_{env}, $\delta_{\text{env}}^{\text{peak}}$, for simplicity. ϵ is a correction function given by

$$\epsilon(S, z_f; M_{\text{th}}^{(1)}, M_{\text{th}}^{(2)}, \eta, \vartheta, \chi) = \frac{1 + \left[\frac{M}{M_{\text{th}}^{(1)}}\right]^{\eta} \left[\frac{\delta_c^{\Lambda\text{CDM}}(S, z_f)}{\delta_c(S, z_f, \delta_{\text{env}}^{\text{peak}})}\right]^{\chi} + \left[\frac{M}{M_{\text{th}}^{(2)}}\right]^{\vartheta} \frac{\delta_c^{\Lambda\text{CDM}}(S, z_f)}{\delta_c(S, z_f, \delta_{\text{env}}^{\text{peak}})}}{1 + \left[\frac{M}{M_{\text{th}}^{(1)}}\right]^{\eta} + \left[\frac{M}{M_{\text{th}}^{(2)}}\right]^{\vartheta}},$$

$$\tag{33}$$

where $M = M(S)$ is the halo mass, and $\eta, \vartheta, \chi, M_{\rm th}^{(1)}, M_{\rm th}^{(2)}$ are free parameters. ϵ goes to 1 for strongly screened theories where $\delta_c(S, z_f, \delta_{\rm env}^{\rm peak}) \rightarrow \delta_c^{\Lambda \rm CDM}(S, z_f)$.

Using Hu-Sawicki $f(R)$ gravity as an example, Ref. 254 discovers an empirical expression for χ, $\chi = 0.5 - 0.2 \log_{10}(-10^5 f_{R0})$, and fits $\eta, \vartheta, M_{\rm th}^{(1)}, M_{\rm th}^{(2)}$ as functions of f_{R0} and z_f using a suite of Gpc-box $f(R)$ simulations. The resulting $\delta_c^{\rm cor}$ is then used in the Sheth-Tormen fitting formula of the HMF.

The left panels of Fig. 6 compare the spherical collapse threshold $\delta_c(S, z_f = 0, \delta_{\rm env}^{\rm peak})$ (blue) with the corrected threshold Eq. (32) (red). This demonstrates that the spherical collapse calculation captures the qualitative features of the corrected threshold, while differing quantitatively in fine details. The right panels of Fig. 6 show the HMF enhancements relative to ΛCDM for three values of f_{R0} in Hu-Sawicki $f(R)$ gravity, at three redshifts $z_f = 0$ (blue), 0.2 (red) and 0.5 (green), where the agreement between simulation results (shaded squares) and excursion set predictions using $\delta_c^{\rm cor}$ (lines) is found to be within 5% level, which is currently the

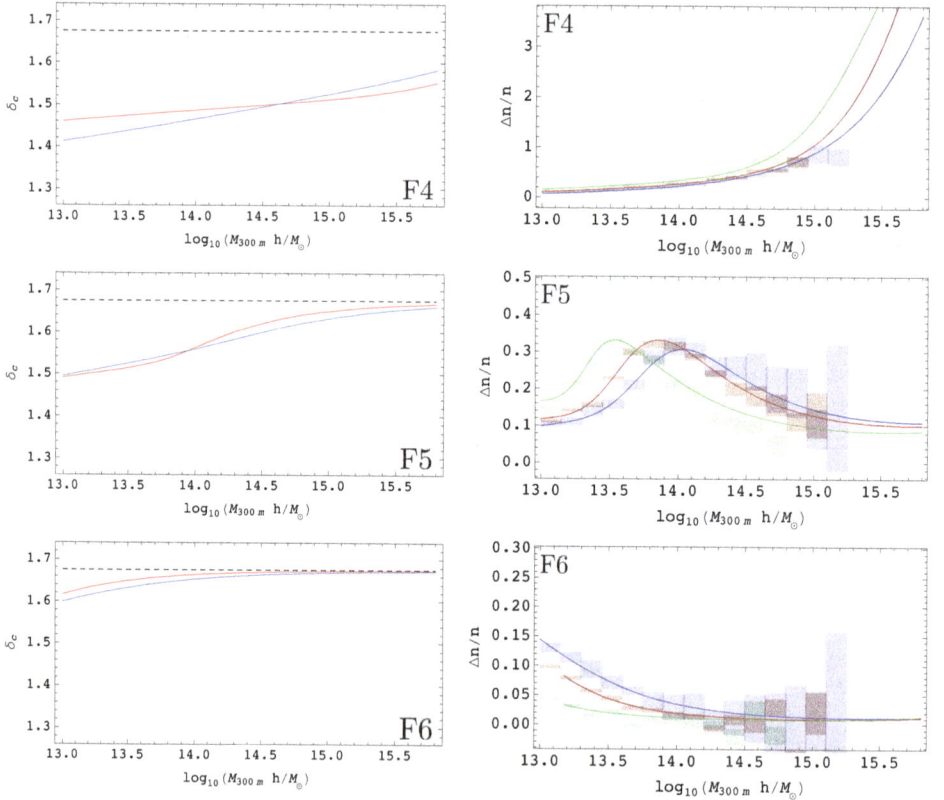

Fig. 6: An exampling showing how the simulation calibration improves the performance in predicting the HMFs for Hu-Sawicki $f(R)$ models. This figure is adapted from Cataneo *et al.* (2016).[254]

best theoretical model for halo mass function in chameleon-type models. Fitting formulae for Vainshtein models can be found in Ref. 253.

Calibrated HMFs are a crucial ingredient to test gravity using cluster observations such as abundance, as will be discussed in Chapter 6.

3.2. *Matter power spectra*

The matter power spectrum, $P(k)$, which is the Fourier-space counterpart of the two-point matter correlation function, $\xi(r)$, is an important statistic in cosmology. It measures the clustering power of matter at some given scale k (or particle pair separation r), and is sensitive to the presence of a modified gravity force, with different theories affecting different scales (chameleon models modify scales below the Compton wavelength of the scalar field, while in the DGP model (see Chapter 1) the modifications are above the Vainshtein radius). It is closely related to various other quantities, such as the weak lensing convergence or shear power spectrum and the redshift space matter or galaxy clustering, which are of importance in cosmological tests of models, making it one of the main focuses of studies of theoretical model predictions.

While linear perturbation theory can be used to efficiently and accurately compute $P(k)$ on large linear scales (e.g., $k < 0.1h\mathrm{Mpc}^{-1}$), on small scales it loses accuracy due to the nonlinear evolution of structures. At mildly nonlinear scales, bulk flow causes damping of linear signal and coupling of different Fourier modes (which are decoupled in linear theory), and these effects can be modelled by using higher-order perturbation theories[400] or removed using reconstruction techniques.[401–405] On even smaller scales where shell crossing takes place, the perturbation theory breaks down and has to be replaced by numerical simulations. The nonlinear effect is often worse in modified gravity models due to the inherently nonlinear nature of screening mechanisms, making it possible that linear theory fails whenever it predicts a modified gravity effect.[406]

Ideally, one would rely on full simulations to provide theoretical templates for $P(k)$ in modified gravity models, but as mentioned above, their cost is too high for continuous parameter space searches or covariance matrix estimation. (Semi)analytical models, with acceptable accuracy in certain regimes, can be more economic alternatives.

3.2.1. *Perturbation theory*

The growth of matter density contrast, $\delta = \delta(\vec{x}, t)$, is governed by the continuity and Euler equations,

$$\frac{\partial \delta}{\partial t} + \frac{1}{a}\vec{\nabla} \cdot [(1 + \delta)\vec{u}] = 0, \tag{34}$$

$$\frac{\partial \vec{u}}{\partial t} + H\vec{u} + \frac{1}{a}\vec{u} \cdot \nabla \vec{u} = -\frac{1}{a}\vec{\nabla}\Phi, \tag{35}$$

in which $\vec{u} = \vec{u}(\vec{x}, t)$ is the velocity field and Φ is the Newtonian potential given by the Poisson equation

$$\nabla^2 \Phi = 4\pi G \delta \rho \tag{36}$$

in which the density perturbation is defined as $\delta\rho(\vec{x}, t) = \rho(\vec{x}, t) - \bar{\rho}(t) = \bar{\rho}(t)\delta$ and $\bar{\rho}(t)$ is the mean matter density at time t. Note that for simplicity we have assumed that vorticity (the curl part of the peculiar velocity field) does not exist in the initial conditions and is not generated during the late-time evolution in perturbation theory (we do not consider the vorticity generated when structure formation enters the highly nonlinear regime, which requires higher order perturbation theory or full simulations to study). One can take the divergence of Eq. (35) to rewrite it as a scalar equation for $\theta \equiv \vec{\nabla} \cdot \vec{u}$.

In Fourier space, Eqs. (34), (35) become

$$\tilde{\delta}' + \tilde{\theta} + \frac{1}{(2\pi)^3} \int d^3\vec{k}_1 d^3\vec{k}_2 \delta_d^{(3)}(\vec{k} - \vec{k}_{12}) \alpha(\vec{k}_1, \vec{k}_2) \tilde{\theta}(\vec{k}_1, a) \tilde{\delta}(\vec{k}_2, a) = 0, \tag{37}$$

$$\tilde{\theta}' + \left[2 + \frac{H'}{H} \right] \tilde{\theta} + \frac{1}{2(2\pi)^3} \int d^3\vec{k}_1 d^3\vec{k}_2 \delta_d^{(3)}(\vec{k} - \vec{k}_{12}) \beta(\vec{k}_1, \vec{k}_2) \tilde{\theta}(\vec{k}_1, a) \tilde{\theta}(\vec{k}_2, a)$$
$$= \left[\frac{k}{aH} \right]^2 \tilde{\Phi}, \tag{38}$$

in which a prime is the derivative wrt $\ln(a)$, $k = |\vec{k}|$, $\vec{k}_{12} \equiv \vec{k}_1 + \vec{k}_2$, $\delta_d^{(3)}$ is the 3D Dirac-δ function, and a tilde denotes the Fourier transform of a quantity, e.g.,

$$\tilde{\delta}(\vec{k}) = \int d^3\vec{x} \exp(-i\vec{k} \cdot \vec{x}) \delta(\vec{x}), \quad \delta(\vec{x}) = \frac{1}{(2\pi)^3} \int d^3\vec{k} \exp(i\vec{k} \cdot \vec{x}) \tilde{\delta}(\vec{k}), \tag{39}$$

and we use shorthand $\tilde{\delta} = \tilde{\delta}(\vec{k}, a)$ for simplicity. α and β are functions encoding the mixing of different Fourier modes:

$$\alpha(\vec{k}_1, \vec{k}_2) = 1 + \frac{\vec{k}_1 \cdot \vec{k}_2}{k_1^2}, \quad \beta(\vec{k}_1, \vec{k}_2) = \frac{|\vec{k}_1 + \vec{k}_2|^2 (\vec{k}_1 \cdot \vec{k}_2)}{k_1^2 k_2^2}. \tag{40}$$

In the regime of smaller field, $\tilde{\delta} \ll 1$, a power series expansion can be done:

$$\tilde{\delta}(\vec{k}, a) = \sum_{n=1}^{\infty} \tilde{\delta}^{(n)}(\vec{k}, a), \quad \tilde{\theta}(\vec{k}, a) = \sum_{n=1}^{\infty} \tilde{\theta}^{(n)}(\vec{k}, a), \tag{41}$$

where $\tilde{\delta}^{(n)}$ and $\tilde{\theta}^{(n)}$ are the n-th order term. From Eq. (37) at first order we have:

$$\tilde{\delta}^{(1)}(\vec{k}, a) = D_+(a)\delta_0(\vec{k}), \quad \tilde{\theta}^{(1)}(\vec{k}, a) = -f(a)D_+\delta_0(\vec{k}) = -f(a)\tilde{\delta}^{(1)}(\vec{k}, a), \tag{42}$$

where D_+ is the (growing mode of the) linear growth factor normalised to $D_+ = 1$ today, $\delta_0(\vec{k})$ the initial density contrast extrapolated to the present day using linear theory, and $f = d\ln D_+/d\ln a$ is the linear growth rate.

In an Einstein-de Sitter (EdS) universe, one has $\Omega(a) = 1$ and $H'/H = -3/2$, and $D_+ = a$, $f = 1$. In this case, the time and spatial dependences of the solutions to Eqs. (37), (38) can be separated, as can be easily checked:

$$\tilde{\delta}(\vec{k}, a) = \sum_{n=1}^{\infty} a^n \tilde{\delta}^{(n)}(\vec{k}), \quad \tilde{\theta}(\vec{k}, a) = \sum_{n=1}^{\infty} a^n \tilde{\theta}^{(n)}(\vec{k}). \tag{43}$$

The higher-order perturbation terms can then be read recursively from Eqs. (37), (38), (42). For example, at second order we have

$$\tilde{\delta}^{(2)}(\vec{k}, a) = \frac{1}{(2\pi)^3} \int d^3\vec{k}_1 d^3\vec{k}_2 \delta_d^{(3)}(\vec{k} - \vec{k}_{12}) F_2(\vec{k}_1, \vec{k}_2; a) \tilde{\delta}_0(\vec{k}_1) \tilde{\delta}_0(\vec{k}_1), \tag{44}$$

$$\tilde{\theta}^{(2)}(\vec{k}, a) = \frac{1}{(2\pi)^3} \int d^3\vec{k}_1 d^3\vec{k}_2 \delta_d^{(3)}(\vec{k} - \vec{k}_{12}) G_2(\vec{k}_1, \vec{k}_2; a) \tilde{\delta}_0(\vec{k}_1) \tilde{\delta}_0(\vec{k}_1), \tag{45}$$

with

$$F_2(\vec{k}_1, \vec{k}_2; a) \equiv a^2 \left[\frac{5}{14} \left[\alpha(\vec{k}_1, \vec{k}_2) + \alpha(\vec{k}_2, \vec{k}_1) \right] + \frac{1}{7}\beta(\vec{k}_1, \vec{k}_2) \right], \tag{46}$$

$$G_2(\vec{k}_1, \vec{k}_2; a) \equiv -a^2 \left[\frac{3}{14} \left[\alpha(\vec{k}_1, \vec{k}_2) + \alpha(\vec{k}_2, \vec{k}_1) \right] + \frac{2}{7}\beta(\vec{k}_1, \vec{k}_2) \right]. \tag{47}$$

Even higher-order terms can also be derived straightforwardly, but we shall not go to the details here.

In non-EdS models, the separability of time and spatial dependencies is generally not exact, but for realistic ΛCDM models as a good approximation one can write

$$\tilde{\delta}(\vec{k}, a) = \sum_{n=1}^{\infty} F_1^n(a) \tilde{\delta}^{(n)}(\vec{k}), \quad \tilde{\theta}(\vec{k}, a) = -f(a) \sum_{n=1}^{\infty} F_1^n(a) \tilde{\theta}^{(n)}(\vec{k}), \tag{48}$$

in which $F_1(a) = D_+(a)$ is still scale independent. For modified gravity models, however, this is not necessarily true, and one has to refer to the more general expressions:

$$\tilde{\delta}^{(n)}(\vec{k}, a)$$
$$= \frac{1}{(8\pi^3)^{n-1}} \int d^3\vec{k}_1 \cdots d^3\vec{k}_n \delta_d^{(3)}(\vec{k} - \vec{k}_{1\ldots n}) F_n(\vec{k}_1, \cdots, \vec{k}_n; a) \tilde{\delta}_0(\vec{k}_1) \cdots \tilde{\delta}_0(\vec{k}_n),$$

$$\tilde{\theta}^{(n)}(\vec{k}, a) \tag{49}$$
$$= \frac{1}{(8\pi^3)^{n-1}} \int d^3\vec{k}_1 \cdots d^3\vec{k}_n \delta_d^{(3)}(\vec{k} - \vec{k}_{1\ldots n}) G_n(\vec{k}_1, \cdots, \vec{k}_n; a) \tilde{\delta}_0(\vec{k}_1) \cdots \tilde{\delta}_0(\vec{k}_n).$$

In addition, the modified gravity effects on matter clustering has to be taken into account, often through the modified Poisson equation, which becomes[165] now

$$-\frac{k^2}{(aH)^2} \tilde{\Phi}(\vec{k}, a) = \frac{3}{2}\Omega_m(a)\mu(k, a)\tilde{\delta}(\vec{k}, a) + S(\vec{k}, a), \tag{50}$$

where $\mu(k, a)$ parameterises the dependences of Newton's constant on space and time, while $S(\vec{k}, a)$ incorporates nonlinear source terms, which up to third order in the perturbations can be written as

$$
S(\vec{k}, a) = \frac{1}{(2\pi)^3} \int d^3\vec{k}_1 d^3\vec{k}_2 \delta_d^{(3)}(\vec{k} - \vec{k}_{12}) \gamma_2(\vec{k}, \vec{k}_1, \vec{k}_2; a) \tilde{\delta}_0(\vec{k}_1) \tilde{\delta}_0(\vec{k}_2)
$$
$$
+ \frac{1}{(2\pi)^6} \int d^3\vec{k}_1 d^3\vec{k}_2 d^3\vec{k}_3 \delta_d^{(3)}(\vec{k} - \vec{k}_{123}) \gamma_3(\vec{k}, \vec{k}_1, \vec{k}_2, \vec{k}_3; a) \tilde{\delta}_0(\vec{k}_1) \tilde{\delta}_0(\vec{k}_2) \tilde{\delta}_0(\vec{k}_3).
$$
$$(51)$$

In ΛCDM, one has $\mu(k, a) = 1$ and $\gamma_2 = \gamma_3 = 0$. Expressions for these quantities in chameleon and Vainshtein models can be found in, e.g., Refs. 165, 238, 407.

In perturbation theory, the matter and velocity (divergence θ) power spectra can be calculated from the above higher-order perturbation solutions to $\tilde{\delta}$ and $\tilde{\theta}$:

$$
P_{ab}(k) \delta_d^{(3)}(\vec{k} + \vec{k}) = (2\pi)^3 \langle \tilde{\delta}_a(\vec{k}) \tilde{\delta}_b(\vec{k}') \rangle,
$$
$$(52)$$

where $\{a, b\} = \{\delta, \theta\}$. Up to quadratic order in the linear matter power spectrum, the result is given by

$$
P_{ab}(k) = P_{ab}^{(1,1)}(k) + P_{ab}^{(1,3)}(k) + P_{ab}^{(2,2)}(k),
$$
$$(53)$$

where $P_{ab}^{(1,1)}(k)$ is the linear spectrum. The terms are given respectively by

$$
P_{\delta\delta}^{(1,1)}(k, a) = F_1^2(k, a) P_0(k),
$$
$$
P_{\delta\theta}^{(1,1)}(k, a) = F_1(k, a) F_1(k, a) P_0(k),
$$
$$(54)$$
$$
P_{\theta\theta}^{(1,1)}(k, a) = G_1^2(k, a) P_0(k),
$$

$$
P_{\delta\delta}^{(1,3)}(k, a) = \frac{6}{(2\pi)^3} \int d^3\vec{k}' P_0(k) P_0(k') F_1(k; a) F_3(\vec{k}, \vec{k}', -\vec{k}'; a),
$$

$$
P_{\delta\theta}^{(1,3)}(k, a) = \frac{3}{(2\pi)^3} \int d^3\vec{k}' P_0(k) P_0(k') G_1(k; a) F_3(\vec{k}, \vec{k}', -\vec{k}'; a)
$$
$$(55)$$
$$
+ \frac{3}{(2\pi)^3} \int d^3\vec{k}' P_0(k) P_0(k') F_1(k; a) G_3(\vec{k}, \vec{k}', -\vec{k}'; a),
$$

$$
P_{\theta\theta}^{(1,3)}(k, a) = \frac{6}{(2\pi)^3} \int d^3\vec{k}' P_0(k) P_0(k') G_1(k; a) G_3(\vec{k}, \vec{k}', -\vec{k}'; a),
$$

and

$$
P_{\delta\delta}^{(2,2)}(k, a) = \frac{1}{(2\pi)^3} \int d^3\vec{k}' P_0(|\vec{k} - \vec{k}'|) P_0(k') F_2^2(\vec{k} - \vec{k}', \vec{k}'; a),
$$

$$
P_{\delta\theta}^{(2,2)}(k, a) = \frac{1}{(2\pi)^3} \int d^3\vec{k}' P_0(|\vec{k} - \vec{k}'|) P_0(k') F_2(\vec{k} - \vec{k}', \vec{k}'; a) G_2(\vec{k} - \vec{k}', \vec{k}'; a), \quad (56)
$$

$$
P_{\theta\theta}^{(2,2)}(k, a) = \frac{1}{(2\pi)^3} \int d^3\vec{k}' P_0(|\vec{k} - \vec{k}'|) P_0(k') G_2^2(\vec{k} - \vec{k}', \vec{k}'; a).
$$

Following a lengthier calculation, the bispectrum can also be computed within the framework of perturbation theory using its definition,

$$B(\vec{k}_1, \vec{k}_2)\delta_d^{(3)}(\vec{k}_1 + \vec{k}_2 + \vec{k}_3) = (2\pi)^3 \langle \tilde{\delta}(\vec{k}_1)\tilde{\delta}(\vec{k}_2)\tilde{\delta}(\vec{k}_3) \rangle, \qquad (57)$$

though little work has been done on this (e.g., Refs. 166, 408).

In principle, higher-order corrections can be included in this framework straightforwardly, but this approach is known to have divergence problems on both small and large scales, the former being a consequence of perturbations becoming progressively nonlinear and the latter because oscillations on large scales at a given order can affect small-scale modes. Resummation techniques, such as the regularised perturbation theory,[409–411] overcomes the latter problem by including small-k contributions from all orders in perturbation theory, effectively introducing a damping on the oscillations of the power spectrum. This makes it possible to Fourier transform $P(k)$ to calculate the correlation function reliably, as well as alleviating small-scale divergence of the standard perturbation theory. This has been used in recent works to predict the nonlinear matter and velocity power spectrum in $f(R)$ and DGP models.[168, 407, 412–414] Another approach is the Lagrangian perturbation theory,[415–419] which has been extended to modified gravity models recently.[420]

Figure 7 shows a few examples of perturbation theory predictions for $P_{\delta\delta}$, $P_{\delta\theta}$ and $P_{\theta\theta}$ at $z = 1$, for three models (GR, nDGP N1 and Hu-Sawicki $f(R)$ gravity F4). The agreement with full simulation results is good down to scales $k \sim 0.1$-$0.2h$ Mpc^{-1}, below which perturbation theory fails to fully reproduce simulation results.

3.2.2. *Halo model and* HALOFIT

Moving to the fully nonlinear regime of structure formation, the frequent crossing of particle trajectories and virialised motions make it impossible to track the evolution of density perturbations perturbatively and analytically. One of the commonly used methods to predict the nonlinear matter power spectrum is the halo model.[258, 259]

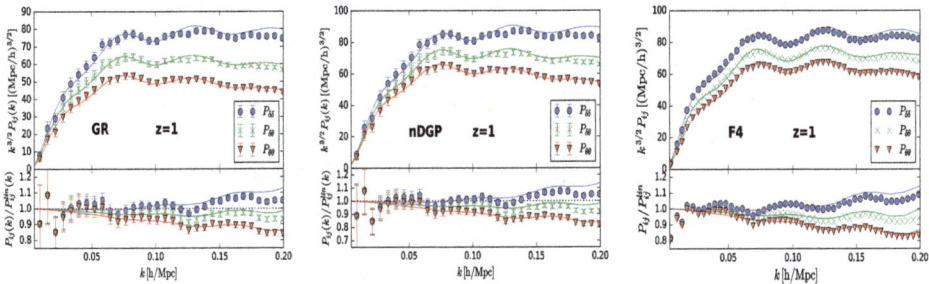

Fig. 7: Perturbation theory predictions of the matter and velocity power spectra (lines), plotted against full simulation results (symbols). The lower panels show the ratios of them. This figure is adapted from Bose *et al.* (2016)[168] (the left and right panels) and Bose *et al.* (2017)[413] (the middle panel).

The main premise of the halo model is that all matter in the universe is in bound structures (haloes). The matter 2-point correlation function becomes the sum of the contributions from the correlations between masses which belong to different haloes (the 2-halo term) and those in the same halo (the 1-halo term). The matter power spectrum can be written as (see Ref. 260 for a review)

$$P(k) = P_{1h}(k) + P_{2h}(k), \qquad (58)$$

with

$$P_{1h}(k) = \int dM \frac{M}{\bar{\rho}_{m0}^2} \frac{dn(M)}{d\ln M} |y(k, M)|^2 ,$$

$$P_{2h}(k) = \left[\int dM \frac{1}{\bar{\rho}_{m0}} \frac{dn(M)}{d\ln M} b_{\mathrm{lin}}(M) |y(k, M)| \right]^2 P_{\mathrm{lin}}(k), \qquad (59)$$

in which $dn/d\ln M$ is the HMF, $b_{\mathrm{lin}}(M)$ the linear halo bias, $P_{\mathrm{lin}}(k)$ the linear matter power spectrum and $y(k, M)$ the Fourier transform of the halo density profile truncated at halo radius and normalised such that $y(k \to 0, M) \to 1$.

Of the three main building blocks of the halo model, the halo mass function and linear halo bias have been discussed above in the frameworks of both ΛCDM and modified gravity. The halo density profiles in CDM cosmologies are known to follow the universal Navarro-Frenk-White (NFW) formula,[421]

$$\rho_{\mathrm{NFW}}(r) = \frac{\rho_s}{(r/R_s)(1 + r/R_s)^2}, \qquad (60)$$

where haloes are assumed to be spherical and r is the distance from the halo centre. ρ_s and R_s are the characteristic density and scale radius, which vary from halo to halo. The halo mass, is given by integrating Eq. (60) between $r = 0$ and the halo radius R_h:

$$M = 4\pi \rho_s \frac{R_h^3}{c^3} \left[\ln(1 + c) - \frac{c}{1 + c} \right] \qquad (61)$$

where $c = R_h/R_s$ is called the halo concentration. In practice, the halo radius R_h has no unique definition, and some common choices are the radius in which matter is virialised, or the radius enclosing an average density which is Δ times the critical or mean matter density at the halo redshift. In these latter cases, R_h is often written as R_Δ with mass $M_\Delta = \frac{4}{3}\pi\Delta\bar{\rho}R_\Delta^3$ and concentration $c_\Delta = R_\Delta/R_s$. c_Δ, M_Δ can be more conveniently used instead of ρ_s, R_s to characterise a halo's density profile, and a concentration-mass relation, $c_\Delta(M_\Delta)$, can be either fitted using simulations[422–425] or computed from physical modelling.[426] $y(k, M)$ for an NFW profile is given by

$$y(k, M) = 4\pi \rho_s R_s^3 \left[\frac{\sin(kR_s)}{M} \left(\mathrm{Si}[(1 + c)kR_s] - \mathrm{Si}(kR_s) \right) \right]$$

$$+ 4\pi \rho_s R_s^3 \left[\frac{\cos(kR_s)}{M} \left(\mathrm{Ci}[(1 + c)kR_s] - \mathrm{Ci}(kR_s) \right) - \frac{\sin(ckR_s)}{M(1 + c)kR_s} \right], \qquad (62)$$

where c is the concentration, $\mathrm{Si}(x) \equiv \int_0^x dx' \sin(x')/x'$, $\mathrm{Ci}(x) \equiv - \int_x^\infty dx' \cos(x')/x'$.

To a good approximation, the c_Δ-M_Δ relation in ΛCDM cosmologies follows a power law, which depends on cosmological parameters and redshift. In modified gravity models, the prediction further depends on the behaviour of gravity. In models with Vainshtein mechanisms, such as Galileons and the nDGP model (with realistic parameters), the screening strongly suppresses the fifth force inside haloes, so that the halo density profile is not appreciably affected.[253, 300] In other models, such as chameleons, the screening efficiency depends on various factors including model parameters, redshift, halo mass and environment, leaving nontrivial imprints to the density profiles of haloes. While earlier works indicate that the NFW profile of Eq. (60) is generally still a good description for these models,[173, 248] recent higher-resolution simulations show that the c_Δ-M_Δ relation is no longer a simple power law.[367] Note that the deviation from a simple power-law in the c_Δ-M_Δ relation has been considered in Ref. 248 as a way to incorporate the chameleon screening effect, by promoting the coefficient of the power to a function of halo mass. More recently, based on numerical simulations, Ref. 369 has more systematically studied the effects of chameleon $f(R)$ gravity on the halo concentration, and found a universal description of these effects which has weak or no dependence on the Hu-Sawicki $f(R)$ model parameter, redshift, halo mass and cosmological parameters such as Ω_m.

The halo model approach can be straightforwardly applied to modified gravity scenarios with the use of the relevant physical quantities, such as $c_\Delta(M_\Delta)$, for the latter. Therefore, at the basic level its application in modified gravity does not require any extension (unlike, for example, halo mass function or the perturbation theory prediction of the power spectra). Some of its applications in different modified gravity models can be found in Refs. 175, 191, 253. But as a simple model, the quantitative agreement of its prediction with simulations is relatively poor, in particular at intermediate scales where the transition between the 2-halo and 1-halo terms takes place, which is true even when looking at the relative difference between modified gravity and ΛCDM,[253] though the inaccuracy of the halo model affects both modified gravity and ΛCDM so that its effects cancel to certain extent when looking at the relative difference between the models.

There are different ways to improve on this, including simple phenomenological fixes such as using an interpolated P_{lin} between the modified gravity and ΛCDM $P_{\text{lin}}(k)$ in Eq. (59) to partially compensate the underestimation of the nonlinear screening effect on relatively large scales (Fig. 8). This is similar in philosophy to the methods used in Ref. 427, 428, which use the predictions of perturbation theory to replace $P_{\text{lin}}(k)$ in the 2-halo term. A more elaborated approach is to include new parameters to the default halo model to account for missing physical effects such as halo exclusion and nonlinear damping of small-scale power spectrum, as well as promoting some physical quantities to adjustable parameters to increase the freedom and flexibility of the model prediction,[429] where the transition between the 1- and 2-halo terms is also modified so that the total $P(k)$ is no longer a simple sum of the two. The approach has been shown to work well for a range of models

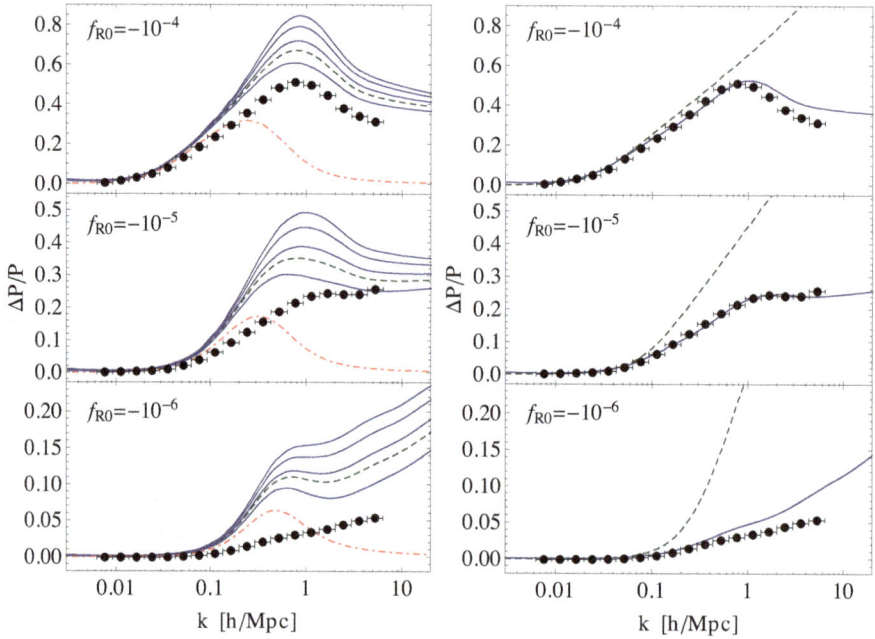

Fig. 8: Halo model predictions of the matter power spectrum enhancements wrt ΛCDM, for three variants of the Hu-Sawicki $f(R)$ model as shown in the legends. *Left panels*: the blue curves come from the use of HMFs predicted by the extended excursion set theory predictions for $f(R)$ gravity, with different assumed values of $\delta_{\rm env}$; for the green dashed $\delta_{\rm env}$ is taken as the mean $\langle\delta_{\rm env}\rangle$. *Right panels*: The halo model is improved by using an interpolated $P_{\rm lin}(k)$ between the modified gravity and ΛCDM linear power spectra in the two-halo term; the green dashed line is linear theory prediction here. In all panels black symbols are simulation results at $z = 0$. This figures is adapted from Lombriser *et al.* (2014)[191]

including chameleons and nDGP.[263] The most popular extension to the halo model is perhaps HALOFIT,[430, 431] which takes a similar approach by using free parameters mimicking the missing physical effects to gain more flexibility, and tuning these free parameters using simulations. HALOFIT and its variations for non-standard cosmologies (e.g., massive neutrinos[432]) is implemented in standard Boltzmann codes such as CAMB,[433] but it is not tested for modified gravity models and a naive application to the latter can lead to results very different from simulations;[406] later, in Ref. 434 the author extends it to the case of chameleon $f(R)$ gravity, and the resulting fitting function is accurate to 6% on scales up to $k = 1h{\rm Mpc}^{-1}$ and 12% in $1 < k < 10h{\rm Mpc}^{-1}$.

The nonlinear matter power spectrum predicted by the halo model or HALOFIT can be a good starting point in the prediction of other observables. An example is the weak lensing convergence or shear angular power spectrum, which is a weighted integration of $P(k)$ along the line of sight:

$$C_{\kappa\kappa}(\ell) = \int_0^{\chi_s} d\chi \frac{W(\chi)^2}{\chi^2} P_{\delta\delta}\left(k = \frac{\ell}{\chi}, z(\chi)\right), \tag{63}$$

where χ is the comoving distance from the observer, χ_s for the lensing source, $z(\chi)$ the redshift corresponding to comoving distance χ, and $W(\chi) = \frac{3}{2}\Omega_m H_0^2 \frac{\chi}{\chi_s}(\chi_s - \chi)[1 + z(\chi)]$ is a weak lensing kernel for a spatially flat universe; the integration accumulates the lensing effects of structures between the source and observer. This has been used in, e.g., Refs. 435, 436, using the nonlinear $P_{\delta\delta}(k)$ from either simulations or (generalised) HALOFIT. Other weak lensing statistics have been studied in, e.g., Refs. 247, and lensing tests of gravity will be covered in more detail in Chapter 5.

3.3. *Redshift space distortions (RSD)*

In real observations, apart from the nonlinear evolution of structures and biasing of tracers with respect to the underlying dark matter field, there is also another complication in the confrontation with theories — redshift space distortions, the fact that we directly measure the redshifts of galaxies, which are not equivalent to their radial distance due to peculiar velocities. However, if modelled accurately, RSD can be a useful probe which is sensitive to the velocity field and the law of gravity.

Let $\vec{u}$ be the peculiar velocity of a particle, then its position in redshift space, $\vec{s}$, is relate to its true position in configuration space $\vec{r}$, as

$$\vec{s} = \vec{r} + \frac{1}{aH}(\hat{z}\cdot\vec{u})\hat{z} = \vec{r} + \frac{u_z}{aH}\hat{z}, \tag{64}$$

where $\hat{z}$ is the unit vector in the line-of-sight (LOS) direction, chosen as the z axis for simplicity. The redshift-space density contrast $\delta^z(\vec{s})$ is related to its configuration-space counterpart, $\delta^r(\vec{r})$, by mass conservation,

$$[1 + \delta^z(\vec{s})]\,d^3\vec{s} = [1 + \delta^r(\vec{r})]\,d^3\vec{r} \;\Rightarrow\; \delta^z(\vec{s}) = \left|\frac{\partial\vec{s}}{\partial\vec{r}}\right|^{-1}[1 + \delta^r(\vec{r})] - 1, \tag{65}$$

where $d^3\vec{s}$ and $d^3\vec{r}$ are volume elements. The Fourier transform of $\delta^z(\vec{s})$ is

$$\tilde{\delta}^z(\vec{k}) = \int d^3\vec{s}\, e^{-i\vec{k}\cdot\vec{s}}\delta^z(\vec{s}) = \int d^3\vec{r}\left[\delta^r(\vec{r}) - \frac{\nabla_z u_z}{aH}\right]e^{-i\left(\vec{k}\cdot\vec{r} + \frac{k_z u_z}{aH}\right)}, \tag{66}$$

where $|\partial\vec{s}/\partial\vec{r}| = 1 + \nabla_z u_z/aH$ is used. The redshift-space power spectrum reads

$$P^z(\vec{k}) = \int d^3\vec{x}\, e^{-i\vec{k}\cdot\vec{x}}\left\langle e^{-ik_z[u_z(\vec{r}) - u_z(\vec{r}')]}[\delta^r(\vec{r}) - \nabla_z u_z(\vec{r})][\delta^r(\vec{r}') - \nabla_z u_z(\vec{r}')]\right\rangle, \tag{67}$$

where $\vec{x} = \vec{r} - \vec{r}'$.

By rewriting Eq. (67) in terms of cumulants, treating the resulting expression perturbatively and neglecting terms with third or higher powers of the linear matter power spectrum, one gets the 1-loop result for the redshift-space power spectrum by Taruya, Nishimichi and Saito (TNS; Ref. 437)

$$P_{\mathrm{TNS}}^{z}(\vec{k}) = D_{\mathrm{FoG}}(k_z \sigma_v) \left[P_{\delta\delta}(k) - 2\mu^2 P_{\delta\theta} + \mu^4 P_{\theta\theta} + A(\vec{k}) + B(\vec{k}) + C(\vec{k}) \right], \quad (68)$$

where D_{FoG} is a phenomenological factor describing the exponential damping of P^z due to virial motions of particles at small scales, $\mu = k_z/h$ and A, B, C are correction terms whose detailed expressions are not presented here. These terms represent higher-order interactions between the density and velocity fields, neglecting which Eq. (68) reduces to the previous results of Ref. 438. If one further neglects the random-motion-induced Finger-of-God effect (the factor D_{FoG}) and adopts linear theory predictions for P_{ab}'s with $a, b = \delta, \theta$, then Kaiser's formula[439] is recovered:

$$P_{\mathrm{Kaiser}}^{z}(\vec{k}, a) = \left[F_1^2(k, a) - 2\mu^2 F_1(k, a) G_1(k, a) + \mu^4 G_1^2(k, a) \right] P_0(k), \quad (69)$$

where F_1, G_1 are the same as used in the previous section.

The redshift-space two-point correlation function can be obtained as Fourier transform of $P^z(\vec{k})$, or directly in the configuration space, following the streaming model.[440, 441] Reference 414 applies the streaming model to modified gravity models, and find their generalised streaming model (GSM) prediction agrees with the Fourier transform of the TNS power spectrum reasonably well, see Fig. 9.

3.4. *Void abundances*

Cosmic voids[442] are a relatively new cosmological probe, believed to be particularly useful for testing modified gravity models with screening mechanisms, due to such mechanisms not working in low-density regions. Unlike large galaxy clusters, cosmic

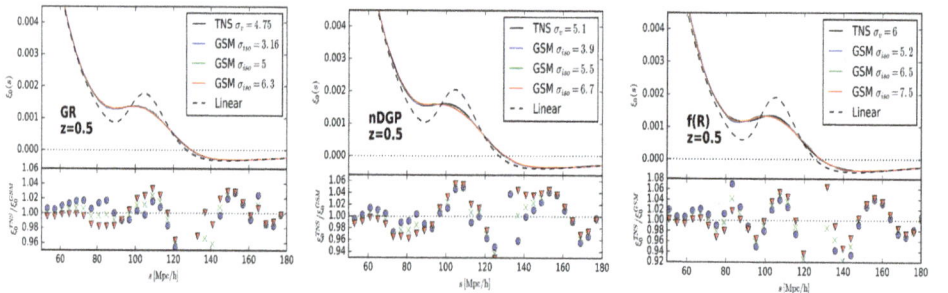

Fig. 9: The comparison of the redshift space correlation function monopole for three models (from left to right: GR, nDGP and $f(R)$ gravity F5, all at $z = 0.5$) predicted by the GSM model and by Fourier transforming the power spectrum from the TNS model. Agreements between both approaches are good, as are the agreements with simulation results. This figure is adapted from Bose and Koyama (2017).[414]

voids primarily develop at sites where the primordial density is low, where matter is evacuated and attracted to their surroundings. Voids and haloes, however, do share an important similarity, namely both form from extreme regimes of the nearly Gaussian distribution of the initial density field ($\delta < 0$ and $\delta > 0$ respectively), so that some of their statistical properties are analytically tractable. Following the same principle of using excursion set theory to predict halo abundance, one can connect the observed abundance of voids to the counts of initial density field with $\delta < \delta_v < 0$, where δ_v is some threshold (similar to δ_c for haloes) to define void formation.

In Ref. 443, in the framework of standard gravity, the authors model voids as spherical tophat underdensities (similar to the case of haloes, which are modelled as spherical overdensities in the simple treatments) which expand over time. The matter evacuated from this region accumulates at its edge, forming an overdensity ridge which can be a defining feature of voids. The assumption of spherical tophat may sound like an oversimplified assumption, but Ref. 443 shows that different underdensity profiles evolve towards it over time. The threshold δ_v which (like in the halo case) is often taken as the value of $\delta_{v,\text{ini}}$ extrapolated to the present day using linear theory, can be chosen arbitrarily as a free parameter for void definition, but a natural choice is $\delta_v = -2.81$ which in an EdS universe is a scale-independent value at which shell crossing (the event that different layers of the void profile cross each other) happens today. The corresponding value of the nonlinear overdensity inside the spherical tophat at today is $\Delta \approx -0.8$, and the latter is often (but not always) used as the defining criterion for numerical algorithms for void finding. The evolution of the underdensity from $\delta_{v,\text{ini}}$ at initial time to $\Delta = -0.8$ at late times is governed by the growth equation, Eq. (10), for ΛCDM, again.

An important feature of the model used in Ref. 443 is the proper accounting for voids-in-voids (VIV) and voids-in-clouds (VIC) effects. The former says that subregions in a void can satisfy a chosen void definition but should not be counted as independent voids (they are subvoids), and the latter says that a region of space satisfying the void definition must not be considered as viable voids at late times should they happen to live in larger overdense regions which would have collapsed to form haloes (and so squashed the underdense region). In the language of excursion set theory, these are respectively equivalent to saying a void is formed if a random walk crosses δ_v at S_*, with initial Lagrangian radius of $R(S_*)$ and final radius $R_v \approx 1.7R(S_*)$ ($1.7 \approx (1+\Delta)^{-\frac{1}{3}}$ with $\Delta = -0.8$), but two situations should be excluded: (i) VIV — if the random walk cross δ_v at different values of S then only the first crossing (at the smallest S) counts, and (ii) VIC — if the random walk also crosses δ_c (the halo formation threshold) at smaller S then it does not correspond to a viable void. The void abundance predicted is given by

$$n_v(M) = \frac{\bar{\rho}_m}{M} \mathcal{F}(S, \delta_v, \delta_c) \frac{dS}{dM}, \tag{70}$$

where M is the mass enclosed in the void radius R (recall that R, S, M can be used interchangeably), and $\mathcal{F}(S, \delta_v, \delta_c)$ is the probability density for a random walk to first cross δ_v at S and not cross δ_c at smaller S (where $\mathcal{D} \equiv \delta_v/(\delta_c - \delta_v)$, for a derivation see Ref. 443):

$$F(S, \delta_v, \delta_c) = \sum_{j=1}^{\infty} \frac{j^2 \pi^2 \mathcal{D}^2}{\delta_v^2} \frac{\sin(j\pi\mathcal{D})}{j\pi} \exp\left[-\frac{j^2 \pi^2 \mathcal{D}^2}{2\delta_v^2/S}\right]. \tag{71}$$

The treatment of VIC in the model described in Eqs. (70), (71) is approximate: it eliminates the possibility of forming voids of any size inside regions which are destined to form haloes, but does not properly account for the effects that voids residing in slightly overdense regions survive being crushed but nevertheless their sizes get squeezed (or grow less) because of the contraction of their surroundings. An indication of the unphysical-ness of this approximation is that in this model the predicted void volume fraction (void volume as a fraction of the total volume of the universe) is ~ 2. Reference 444 shows that this problem can be alleviated by replacing the single second barrier for the random walk, δ_c, with ones corresponding to the Eulerian volumes that the surroundings of the voids-to-be would attain: as the voids cannot grow larger than their surroundings, their sizes should be set to be equal to these Eulerian volumes rather than by the S value where the random walk crosses δ_v. The voids predicted in this way are generally smaller, so that a smaller volume fraction (~ 1.2 though still unphysical) can be obtained. It is also noticed that improvements can be achieved by using correlated random walk steps.

The extension of the semi-analytical models for void formation to modified gravity models can be done similarly as in the halo case. In Ref. 256 the model in Eqs. (70), (71) is extended to chameleon-type models in a conceptually straightforward way: instead of having two constant (scale-independent) barriers δ_v, δ_c, the authors (i) introduce scale- and environment-dependences in δ_v following the method for δ_c (see above), (ii) numerically compute the modified $\mathcal{F}(S, \delta_v, \delta_c)$ — now becoming $\mathcal{F}[S, \delta_v(S, \delta_{env}), \delta_c(S, \delta_{env}), \delta_{env}]$, (iii) average over the distribution of δ_{env} to find $\mathcal{F}_{ave}(S)$ and (iv) apply Eq. (70) to calculate void abundance. In Ref. 257 this work is further generalised to use the Eulerian model of Ref. 444 to predict void abundance.

Unlike for halo abundance, semi-analytical models for void abundance, despite their successes in qualitatively predicting the behaviour, are still quite far from being used in precision cosmological tests. Theoretically, the assumptions of spherical evolution an excursion set are perhaps an oversimplified description of the evolution and mergers of voids embedded in the complicated cosmic web. Observationally, voids are found to have different and mostly irregular shapes rather than being spherical, adding another layer of complexity as how to define voids and their boundaries, and different void-finding algorithms can return very different void abundances even when applied to the same density field.[445, 446] Furthermore, the above theoretical models only apply to voids found in the dark matter density field, while in reality voids are usually found from a distribution of tracers (such as galaxies), and galaxy bias offers a freedom which can affect the void abundance; this can be translated to a degeneracy — if the two-point galaxy correlation functions of two different models (e.g., ΛCDM and modified gravity) are tuned to agree with each other, this more or less fixes the abundance and distribution of voids identified from the galaxy fields of the two models to be very similar to each other, even though their underlying dark matter density fields can be different.[446] For these reasons, the models for void abundance are less widely used than the other techniques mentioned in this review.

Chapter 7 will contain a more depth review of gravity tests using void observables.

4. Summary and Conclusions

In this chapter we have reviewed a range of approximate simulation techniques and (semi)analytical methods to predict various cosmological observables. Most of the methods described here have previously been used in the studies of ΛCDM, but their application to modified gravity models can involve nontrivial extensions depending on specific properties of the model being studied. This is one of the main points underlying the developments of these methods: while N-body simulations can appear to be a black box to users, analytical methods can help us track the highly nonlinear physics underlying (modified) gravity; their successes not only offer physical insights into the interpretation of simulation predictions, but are also potentially useful in assisting the development of (model specific) simulation algorithms.

From a practical point of view, the main advantage of approximate and analytical methods is their efficiency. At moderate and controlled loss of accuracy, these methods are usually orders of magnitude faster than full simulations, therefore allowing quick delineation of the model parameter space and serving as guidances for full simulation explorations. This is particularly important considering that we will need to study a large number of theoretical models and compare their predictions with a wide variety of cosmological observables.

Of course, these methods still cannot completely replace full N-body simulations, because they are approximate in nature. Simulations will be needed to assess their performance and range of validity and applicability, as well as calibrate their free parameters. An optimal combination of simulations and fast methods will be an effective tool to reliably test cosmological models and gravity in cosmology.

Acknowledgments

The author is supported by the European Research Council (ERC-StG-716532-PUNCA) and UK STFC Consolidated Grants ST/P000541/1, ST/L00075X/1.

Chapter 5

Large-Scale Structure Probes of Modified Gravity

Catherine Heymans[1] and Gong-Bo Zhao[2]

[1] *Institute for Astronomy, University of Edinburgh, Royal Observatory,*
Blackford Hill, Edinburgh, EH9 3HJ, UK
heymans@roe.ac.uk

[2] *National Astronomy Observatories, Chinese Academy of Science,*
Beijing, 100012, P.R. China &
Institute of Cosmology & Gravitation, University of Portsmouth,
Dennis Sciama Building, Portsmouth, PO1 3FX, UK
gbzhao@nao.cas.cn

Observations of the evolution of large-scale structures in the Universe provides unique tools to confront Einstein's theory of General Relativity on cosmological scales. We review weak gravitational lensing and galaxy clustering studies, discussing how these can be used in combination in order to constrain a range of different modified gravity theories. We argue that in order to maximise the future information gain from these probes, theoretical effort will be required in order to model the impact of beyond-Einstein gravity in the non-linear regime of structure formation.

Keywords: Observational cosmology; gravitational lensing; galaxy clustering; surveys.

1. Introduction

There is a multitude of potentially viable beyond-GR theories that introduce a variety of new scalar-field degrees of freedom. These new fields can couple to the matter fields in the Universe, modifying their gravitational interactions through a 'fifth force'. How these beyond-GR theories differ from Einstein's famed cosmological constant could be considered somewhat ambiguous, as in principle a 'cosmological constant' can be placed on either side of Einstein's field equations to modify them. With what we'll term 'modified gravity' theories, the modification is made to the 'curvature' side of the equations, the source of the gravitational field. With what we'll term 'dark energy' theories, uncoupled scalar fields, are seen as a new energy component modifying the 'stress-energy' tensor. Modified gravity theories, dark energy theories and the cosmological constant, are therefore all modifications of GR. Where the difference between these theories is unambiguous however is how they impact the expansion history of the Universe and the growth of structures over time.[27] Only the theories that couple to the matter field will contribute scale-dependent alterations to the linear growth of density fluctuations, as illustrated in Fig. 1. As such the observed growth of large-scale structures over time will be inconsistent with the growth predicted from the Hubble expansion alone. In order to confront a range of modified gravity theories and to separate them from dark energy theories, one therefore needs a combination of observations to probe both the expansion history and the growth of structures. In addition, in order to distinguish between modified gravity scenarios where theories can be broadly classed by their predictions for how matter bends space and time, and how the gravitational constant G evolves, we also need observations that can test the 'curvature of space' (the curvature potential Φ) independently from the 'curvature of time' (the Newtonian potential Ψ).

Over the coming decade a number of large-scale cosmological surveys are coming online. The imaging surveys, the Large Synoptic Survey Telescope (LSST) and Euclid, will survey over 10,000 square degrees, mapping the evolution in the growth of large-scale structures of dark matter using a technique called weak gravitational lensing. The spectroscopic surveys, the Dark Energy Spectroscopic Instrument (DESI) and Euclid, will survey similar areas, mapping the 3-D distribution of galaxies across cosmic-time. These facilities used in combination with high-precision measurements of the early-Universe from the exquisite Planck observations of the cosmic microwave background (CMB) will provide a wealth of observational evi-

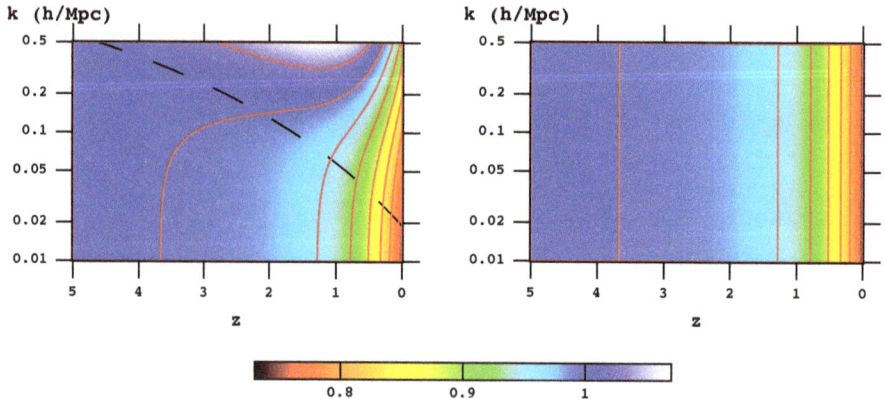

Fig. 1: The evolution of the growth factor for the linear matter power spectrum $\Delta_m(k, z)$, showing $[(1 + z)\Delta_m(k, z)]/[6\Delta_m(k, z = 5)]$ as a function of redshift z and scale k. The left panel corresponds to the Hu-Sawicki $f(R)$ model[152] where the $R + \Lambda$ term in the action of GR is replaced by a function of the Ricci scalar, $f(R)$. The promotion of R to $f(R)$ introduces an additional degree of freedom making the growth of structures scale-dependent in comparison to the scale-independence of the ΛCDM case shown in the right panel. For this analysis the $f(R)$ function is set so that at the background level, $f(R) \simeq R + \Lambda$, with an effective dark energy equation of state $w_{\text{eff}} = -1$ and an amplitude $|f_{R0}| = 10^{-4}$. Figure reproduced from Ref. 90.

dence to confront a range of different theories. In this chapter we describe three large-scale structure probes; baryon acoustic oscillations, redshift-space distortions and weak gravitational lensing, highlighting how, when used in combination, these probes can break degeneracies in constraining a range of beyond-GR theories.

2. Large-Scale Structure Probes

2.1. *Gravitational lensing*

Large-scale structures of dark matter gravitationally deflect light rays as they travel through the Universe, coherently distorting the distant galaxy images that we observe. This is often referred to as 'cosmic shear' (for a review see Ref. 447). Photons of light from different galaxies can act as relativistic particles for a large-scale cosmological gravity experiment. In its cosmological trajectory, the light travels through as much time as space, and hence, lensing is sensitive to the curvature of both space and time. As lensing probes both the universal expansion of the Universe and structure formation it is a powerful tool for the study of both standard and beyond-GR cosmological models.

Figure 2 shows the expected signal-to-noise of a measurement of the cosmic shear power spectrum P_κ for a future deep, wide-area 15,000 square degree survey like the Large-Survey Synoptic Telescope (LSST[448]). P_κ is the power spectrum of the projected surface mass density along the line of sight and is related to the matter power spectrum P_δ through

$$P_\kappa(\ell) = \int_0^{\chi_H} Q(\chi) P_\delta(\ell/\chi, \chi), \tag{1}$$

where χ is the comoving radial distance and $Q(\chi)$ is a redshift-weight that combines information about the depth of the weak lensing survey with the lensing efficiency of foreground structures.[447] As $Q(\chi)$ peaks at very roughly half the redshift of the lensed background galaxy sample, for this $z > 1$ LSST forecast from Ref. 449, we can see that the lensing measurement has its strongest constraining power in the deeply non-linear regime around physical scales of $k \sim 1h\,\mathrm{Mpc}^{-1}$.

In the lower panel of Fig. 2 we compare the cosmic shear power spectrum from Ref. 450 for two $f(R)$ gravity models[152] and an example Dilaton model.[59] Here the non-linear behaviour of the non-GR models has been calculated using a phenomenological halo model approach.[238] We can see that the most significant deviations from GR for these models are in the non-linear regime, where the weak lensing observations are very sensitive. This therefore strongly motivates the theoretical study of modifications to gravity in the non-linear regime, but with a strong caveat. Both observational systematics and the uncertain impact of baryon feedback on the dark matter non-linear matter power spectrum are likely to be significant on these scales[451, 452]

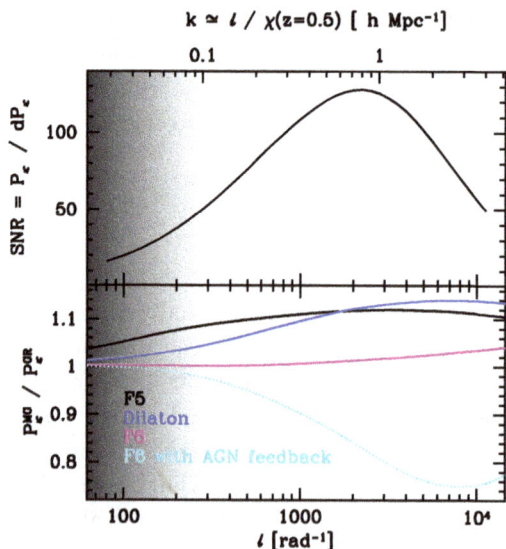

Fig. 2: Weak lensing is most sensitive to structures in the non-linear regime. In the upper panel, the forecasted signal-to-noise ratio for the shear power spectrum $P_\kappa(\ell)$ measured from a $z > 1$ sample of galaxies in an LSST-like survey. The upper axis illustrates the physical k-scales which the projected weak lensing signal is most sensitive to. The shaded box indicates the transition from a linear (black) to a non-linear (white) regime. The lower panel shows the difference between the modified gravity and GR weak lensing power spectrum for $f(R)$ gravity with $|f_{R0}| = 10^{-5}$ (F5 - black), and with $|f_{R0}| = 10^{-6}$ (F6 - pink), and an example Dilaton model (blue). Baryon feedback also impacts the non-linear power spectrum on these scales. The dotted line compares a dark-matter only GR power spectrum with an F6 cosmology that also includes strong AGN feedback. The data used to create this figure was taken from Refs. 449, 450.

2.2. *Baryon acoustic oscillations and redshift space distortions*

Observationally, scale-dependent structure growth is a smoking gun of modified gravity, and it can in principle be probed by large spectroscopic galaxy surveys that map the Universe from the distant past to the present epoch. By measuring redshifts from spectra of a large number of galaxies, redshift surveys produce three-dimensional maps of the Universe in redshift space. The statistical quantities of the clustering of the galaxies can then be used to confront a range of cosmological models.

Galaxy clustering measurements are often separated into two regimes. Large-scale real-space clustering measurements are used to detect an enhanced galaxy clustering signal at $\sim 150h^{-1}$ Mpc that was imprinted in the baryonic matter distribution as a result of photon-baryon interaction in the early Universe. This fixed 'baryon acoustic oscillation' length scale in the distribution of galaxies acts as a standard ruler with which to probe the geometry and background expansion of the Universe.[7] Redshift-space clustering measurements look at the relative distortion in the clustering signal along and perpendicular to the line-of-sight that arises from the peculiar motions of galaxies.[453] These 'redshift space distortions' (RSD) are driven on all scales by local gravitational potentials which means that these observations are directly related to gravity on cosmological scales and thus are a powerful probe of gravity models. There is however a caveat that the galaxies are a biased tracer of the total matter distribution and this bias needs to be accounted for when comparing observations to theoretical models.

3. Observational Constraints on Modified Gravity from Combined Probe Analyses

Individually both RSD and weak lensing observations can provide constraints on beyond-GR models of gravity (see for example a recent RSD analysis in Ref. 454), but their true power comes in their combination.[455] One approach taken by many so far is to look for consistency with GR in these observations by determining joint constraints on phenomenological modified gravity models to observations of RSD and weak lensing, in combination with geometry probes such as BAO and the cosmic microwave background (CMB).

One example of a phenomenological modified gravity model is to alter the Newtonian potential Ψ and curvature potential Φ in the Friedmann-Robertson-Walker metric with a spatially scale-independent modification that scales with time in proportion to the effective dark energy density[151, 232, 455, 457, 458];

$$\Psi_{\mathrm{MG}}(k, a) = [1 + \mu_0 \, \eta(a)] \, \Psi_{\mathrm{GR}}(k, a), \tag{2}$$

$$\Psi_{\mathrm{MG}}(k, a) + \Phi_{\mathrm{MG}}(k, a) = [1 + \Sigma_0 \, \eta(a)] \, [\Psi_{\mathrm{MG}}(k, a) + \Phi_{\mathrm{GR}}(k, a)], \tag{3}$$

where $\eta(a) = \Omega_\Lambda(a)/\Omega_\Lambda(a = 1)$. For the case of GR, $\Sigma_0 = \mu_0 = 0$. Figure 3 shows current constraints for this modified gravity parameterisation from Ref. 232 (panel C) and Ref. 9 (panel D). Both show constraints using weak lensing measurements from the Canada-France-Hawaii Telescope Lensing survey (CFHTLenS,[459] panel A showing the tomographic shear correlation function $\xi_\pm(\theta)$) and a compilation of redshift space distortion measurements (panel B showing $f\sigma_8$, where f and σ_8 are the logarithmic growth rate and the root mean square of the density fluctuation on a scale of 8 Mpc/h, respectively, see Ref. 456 for details). Focussing on panel C we see that RSD (shown in green) is only sensitive to changes in the curvature of time μ_0, as galaxies are non-relativistic tracers of the gravitational potential. Weak lensing (shown in red), however is sensitive to changes in both of the Bardeen potentials, as lensing is a relativistic tracer. The combination of these two probes breaks the degeneracy between these two phenomenological modified gravity parameters and when combined with BAO[460] and WMAP CMB observations[461] there is good consistency found with GR. Panel D revisits this analysis using updated CMB constraints from Planck[9] finding some tension with the GR-prediction. This tension however stems from differences between the probes even in a standard flat ΛCDM analysis.[462] When combining Planck with weak lensing

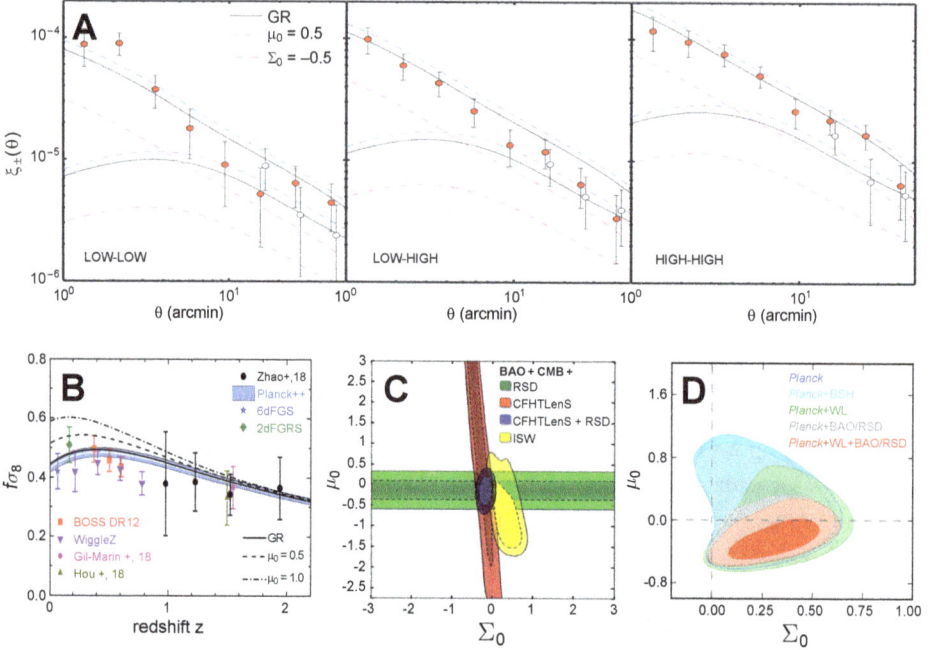

Fig. 3: A compilation of observational constraints. The impact of varying the parameterised modified gravity parameters μ_0 and Σ_0 can be seen for weak gravitational lensing observables in panel A. Here the tomographic shear correlation function $\xi_+(\theta)$, closed circles, and $\xi_-(\theta)$, open circles, has been measured from the CFHT Lensing Survey (CFHTLenS) within a low (left) and high (right) redshift bin. The shear correlation between the two redshift bins is also shown (centre panel). A compilation of redshift space distortion measurements is shown in panel B, compared to two different μ_0 values. RSD is insensitive to changes in Σ_0. For the details of the different surveys shown in panel B see Ref. 456. Joint constraints on Σ_0 and μ_0 are shown adopting CMB measurements from WMAP (panel C) and Planck (panel D, where BSH in the inset refers to the analysis that includes constraints from BAO, Supernovae and a prior on the Hubble constant). Figures reproduced from Refs. 9, 232, 456.

data from the Dark Energy Survey[3, 463] the results are found to be GR-consistent with $\mu_0 = -0.07^{+0.19}_{-0.32}$ and $\Sigma_0 = 0.018^{+0.059}_{-0.048}$. As RSD and weak lensing observations improve in both precision and accuracy over the coming decade it will be interesting to see whether consistency or tension arises between these different and independent probes.

The 'gravitational slip' statistic, E_G,[150, 464] provides an alternative statistic to discriminate between different modified gravity theories by taking ratios of weak lensing, clustering and RSD measurements. Whilst a promising probe for modifications to gravity, Ref. 465 argue that as E_G primarily constrains Ω_m, all systematic sources of tension between the CMB and large-scale structure probes need to be ruled out before E_G can be used as a reliable consistency test of GR. In addition Refs. 466 and 467 show that whilst the approach of taking ratios of observables reduces the impact of observational uncertainties relating to galaxy bias, it is still a significant source of error, reducing the arguably already limited constraining power of this statistic. The future focus is therefore on joint cosmological parameter inference over all large-scale structure and CMB observables.

A range of different phenomenological modified gravity parameterisations exist[200, 468–470] in addition to well tested software made publicly available in order to to carry out these analyses.[206, 471] It is however challenging to map any phenomenological constraints to the underlying theories that we wish to confront. Development in this area has been led by Refs. 472, 473 that link the μ, Σ parameterisation from Eqs. (2) and (3), to general Horndeski theories, finding a series of consistency conditions that can discriminate against broad classes of theories. There has also been recent focus to use observations to directly constrain the parameters for Horndeski gravity models.[211, 474, 475] All these developments are however, limited to the linear regime, where large-scale structure observations are least constraining (see Fig. 2).

Observational constraints have so far primarily focused on analysing the two-point statistics of the density field, but there is strong motivation to look to higher-order statistics[476, 477] including density-weighted two-points statistics.[267, 478, 479] We have argued the importance of analysing departures from GR in the non-linear regime. It therefore naturally follows that as this is also the regime where the density field becomes non-Gaussian, information will be lost by only analysing two-point statistics. Higher-order statistics can be challenging to model and measure observationally and can also be subject to a different set of systematics,[480] but encouraging early results[481] promotes the future importance of higher-order statistics to enhance modified gravity constraints.

4. Conclusions

Over the next decade we will see the results of the upcoming suite of 'Stage-IV' experiments that have been designed to constrain the dark energy equation of state parameter to the percent level. DESI has been designed so its RSD measurements will be able to distinguish between a range of modified gravity theories at a high significance, and in combination with improved CMB observations from the Simon's Observatory, and weak lensing measurements from Euclid and LSST, a range of sophisticated GR tests can and will be performed on cosmological scales. Harnessing the full power of large-scale structure probes of gravity will, however, require significant theoretical developments to in order to accurately model modifications to gravity into the non-linear regime.

Acknowledgments

We thank Dragan Huterer and Joachim Harnois-Deraps for kindly providing the data used to create Fig. 2. CH and GBZ are supported by the European Research Council (ERC) under the European Union's Horizon 2020 research and innovation programme, grant agreement numbers 647112 "GLOBE" (CH) and 646702

"CosTesGrav" (GBZ). GBZ also acknowledges support from NSFC Grants 11720101004, 11673025 and 11711530207, the National Basic Research Program of China (973 Program) (2015CB857004), and a Royal Society Newton Advanced Fellowship.

Chapter 6

Tests of Gravity with Galaxy Clusters

Matteo Cataneo

Institute for Astronomy, University of Edinburgh, Royal Observatory,
Blackford Hill, Edinburgh, EH9 3HJ, UK
matteo@roe.ac.uk

David Rapetti

Center for Astrophysics and Space Astronomy, Department of Astrophysical and
Planetary Science, University of Colorado, Boulder, CO 80309, USA
NASA Ames Research Center, Moffett Field, CA 94035, USA
Universities Space Research Association, Mountain View, CA 94043, USA
David.Rapetti@colorado.edu

Changes in the law of gravity have far-reaching implications for the formation
and evolution of galaxy clusters, and appear as peculiar signatures in their mass-
observable relations, structural properties, internal dynamics, and abundance.
We review the outstanding progress made in recent years towards constraining
deviations from General Relativity with galaxy clusters, and give an overview of
the yet untapped information becoming accessible with forthcoming surveys that
will map large portions of the sky in great detail and unprecedented depth.

Keywords: Modified gravity; structure formation; cosmology.

1. Introduction

Gravity has a central role in the formation of galaxy clusters, the most massive
bound structures in the universe.[482] These astrophysical objects emerge from the
coherent infall of matter towards the highest peaks of the primordial density fluctuations and, subsequently, evolve through a combination of accretion and hierarchical
merging. Modifications of the law of gravity can have dramatic consequences for
the growth of structure across different scales, ranging from astrophysical systems
to the large-scale structure of the universe. Galaxy clusters are at the crossroads of
these two regimes, which makes them ideal laboratories to test theories of gravity
affecting the distribution of matter on cosmic scales while recovering the standard
predictions on small scales. Hence, modifications to General Relativity (GR) have
profound implications for the formation and evolution of galaxy clusters, as well
as for their properties. Their abundance, gravitational potentials, shape, and other
bulk properties are all sensitive to the presence of a fifth force. Thanks to their different components — gas, stars, and dark matter — galaxy clusters can be observed

with a variety of techniques and in a broad range of wavelengths, thus providing us with a wealth of data that are key to discriminate among the numerous alternatives to GR.

In the following we will review various tests of gravity that use galaxy clusters as a probe for signatures beyond GR. In Secs. 2 and 3 we summarize constraints on modified gravity derived from cluster counts and cluster-mass estimates, two observables that have been widely employed over the past decade and helped rule out substantial deviations from standard gravity. In Sec. 4, we present preliminary studies using the gravitational redshift measured in galaxy clusters as a test of gravity. Finally, in Sec. 5, we discuss recently proposed tests, some of which will require data from the next generation of large volume surveys.

2. Cluster Abundance

The abundance of galaxy clusters as a function of mass and redshift is a highly sensitive probe of both cosmic expansion history and growth of structure formation, making this an excellent test for departures from GR. In this section we review the leading studies that have employed cluster abundance (CA) data to either examine the consistency of GR at large scales with observations or constrain specific models of modified gravity.

2.1. Surveys at different wavelengths

Future and ongoing galaxy cluster surveys in multiple wavelengths should continue to provide key insights into cosmological gravity. Here we briefly present only the surveys that have been or are about to be utilized for this task, and which will thus be featured in the following subsections. There are various physical mechanisms that allow us to detect galaxy clusters in different parts of the electromagnetic spectrum. In optical, the observable employed is the number of galaxies identified as members of a cluster (richness) through the so-called Red Sequence method,[483] which is based on the observed fact that cluster galaxies are generally older than those in the field. Optical cluster surveys have been built from the Sloan Digital Sky Survey (SDSS)[484, 485] and the Dark Energy Survey (DES).[486]

In X-ray, the strong gravitational pull exerted by the large mass in clusters heats the gas to high virial temperatures of 10^7–10^8 K, at which the diffuse intracluster medium (ICM) emits X-ray photons through primarily collisional processes.[487] Using mainly X-ray flux, spectral hardness, and spatial extent as observables, X-ray cluster surveys are built with a relatively straightforward selection function. Examples are the ROSAT Brightest Cluster Sample (BCS),[488] which covered the northern hemisphere up to $z < 0.3$ above a flux limit (F_X) of 4.4×10^{12} erg s^{-1} cm^{-2} (0.1–2.4 keV), the ROSAT-ESO Flux-Limited X-ray Galaxy Cluster Survey (REFLEX),[489] covering the southern hemisphere with $z < 0.3$ and F_X(0.1–2.4 keV)

$> 3 \times 10^{-12}$ erg s^{-1} cm^{-2}, and the Massive Cluster Survey (MACS),[490] which extended this work to higher redshifts ($0.3 < z < 05$) and slightly fainter fluxes [for Bright MACS, $F_X(0.1$–$2.4\,\mathrm{keV}) > 2 \times 10^{-12}$ erg s^{-1} cm^{-2}].

Other X-ray cluster catalogs, covering significantly smaller areas than those above from the ROSAT All-Sky Survey (RASS), have also been constructed based on serendipitous discoveries from pointed observations of the ROSAT mission, such as the 400 Square Degree ROSAT Position-Sensitive Proportional Counter (PSPC) Galaxy Cluster Survey (400sd).[491]

Using the Sunyaev–Zel'dovich (SZ) effect, through which clusters are seen as shadows in the Cosmic Microwave Background (CMB) when its photons scatter off electrons in the ICM, the South Pole Telescope (SPT), the Planck satellite mission, and the Atacama Cosmology Telescope (ACT) have also built various SZ cluster surveys.[492–494]

2.2. *Observational constraints on the consistency with GR*

This subsection describes observational tests of the consistency of GR at cosmic scales with cluster number count data and other complementary measurements, assuming that only the mean background density and its linear perturbations could evolve differently than GR, while nonlinear structure formation occurred as in GR. An additional assumption taken in these analyses was the preservation of the GR property of having the same spatial and temporal gravitational potentials, leading to a unique, underlying cluster mass regardless of the measurement technique used.

Following the cluster abundance analysis of Mantz *et al.*,[495] which presented the first constraints on dark energy from a cluster counts experiment,[a] Rapetti *et al.*[497] employed a popular model of deviations from the growth of structure of GR to report also the first constraints from this experiment on the cosmic linear growth index, γ.[155, 156, 498, b]

This parameter allows deviations from GR of the linear growth rate of density perturbations on large scales, $g(a)$, as a function of the scale factor, a, in the form of a power law as follows:

$$g(a) \equiv \frac{d \ln \delta}{d \ln a} = \Omega_m(a)^\gamma, \tag{1}$$

in which the definition of $g(a)$ is based on $\delta \equiv \delta\rho_m/\rho_m$, the ratio of the comoving matter density fluctuations, $\delta\rho_m$, with respect to the cosmic mean, ρ_m. $\Omega_m(a) = \Omega_m a^{-3} E(a)^{-2}$ is the evolving mean matter density in units of the critical density of the universe, with Ω_m being its present-day value and $E(a) \equiv H(a)/H_0$ the evolution parameter, where $H(a)$ is the Hubble parameter and H_0 its present-day value. $E(a)$ parameterizes the cosmic expansion history such as

$$E(a) = [\Omega_m a^{-3} + (1 - \Omega_m)a^{-3(1+w)}]^{1/2}, \tag{2}$$

---

[a]These results were independently confirmed soon after by Vikhlinin *et al.*[496]
[b]See Sec. 2.3 for details on the first $f(R)$ gravity constraints using this probe.[499]

and w is a kinematical parameter that usually represents the dark energy equation-of-state. For analyses with no assumption on the origin of the late-time cosmic acceleration, w can be used to conveniently and generally fit expansion history data instead of associating it with a fluid component such as dark energy, matching the expansion of ΛCDM when $w = -1$. In a similar fashion, GR is recovered when $\gamma \simeq 0.55$.[c] It is worth noting, however, that even though for this w modeling there are no dark energy perturbations, the γ parametrization of linear growth is then required to account for any additional density fluctuations beyond those predicted by GR at subhorizon scales, including those relevant for the integrated Sachs–Wolfe (ISW) effect of the CMB.[497, 501]

Equations (1) and (2) thereby model the growth and expansion histories, respectively, with γ and w parameterizing simultaneously the phenomenological departures from GR and ΛCDM. Given this linear modeling, the number of dark matter halos as a function of mass and redshift can be obtained as

$$n_\Delta(M, z) = \frac{\bar{\rho}_m}{M} \frac{d \ln \sigma^{-1}}{d \ln M} f(\sigma, z), \tag{3}$$

where $f(\sigma, z)$ is the multiplicity function fitted to N-body simulations. The variance of the linear matter density field convolved with a top hat window function of radius R, enclosing a mass $M = 4\pi R^3 \bar{\rho}_m / 3$, in which $\bar{\rho}_m$ is the mean background density, can be calculated as

$$\sigma^2(R, z) = \int \frac{d^3 k}{(2\pi^3)} P_L(k, z) |W(kR)|^2. \tag{4}$$

$P_L(k, z) \propto k^{n_s} T^2(k, z_t) D(z)^2$ is the linear matter power spectrum as a function of wavenumber, k, and redshift, z, n_s the scalar spectral index of the primordial fluctuations, $T(k, z_t)$ the matter transfer function at a redshift z_t, $D(z) \equiv \delta(z)/\delta(z_t)$ the growth factor of linear perturbations normalized at z_t, and $W(kR)$ the Fourier transform of the window function. Thus, the halo mass function (HMF), $n_\Delta(M, z)$, combines both the linear and nonlinear descriptions of cosmic structure formation.

Figure 1 shows the first measurements on the linear w, γ modeling obtained from cluster abundance data.[497] The latter provided low-z constraints on the evolution of the amplitude of the linear matter power spectrum, conventionally parameterized with $\sigma_8 = \sigma(R = 8\,h^{-1}\,\mathrm{Mpc}, z = 0)$, while data from the high multipoles of the anisotropies power spectrum of the CMB strongly constrained this amplitude at high-z (when the universe was decelerating), and their low multipoles added a relatively weak constraint at low-z from the ISW effect at large scales. Additional measurements on the expansion history at low-z came from SNIa and cluster gas mass fraction (f_{gas}) datasets. The CMB, SNIa, and f_{gas} data also helped breaking degeneracies and constraining additional parameters of the overall cosmological model that otherwise would have been poorly constrained.

---

[c]This value is, however, only acceptable as a GR reference for the current level of constraints. At higher accuracy, the growth index of GR has small redshift and background parameter dependencies.[500]

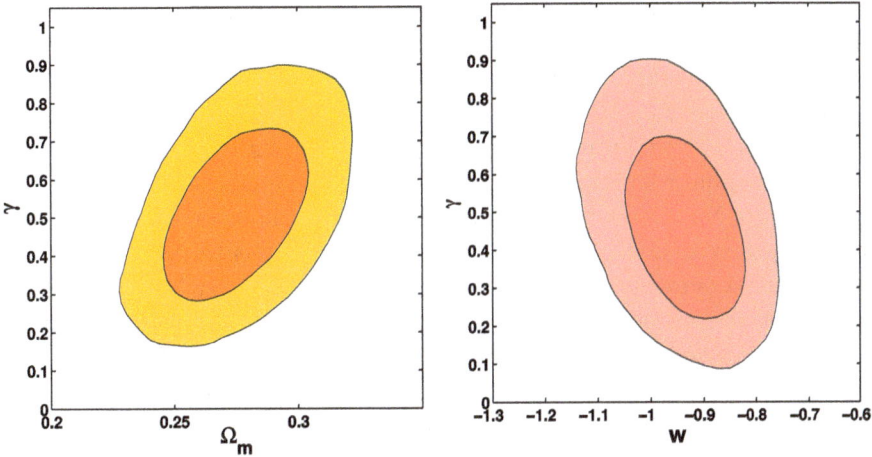

Fig. 1: Figures taken from Ref. 497 showing the first constraints (68.3% and 95.4% confidence regions) from a combination of cluster abundance, CMB, Supernovae Type Ia (SNIa), and $f_{\rm gas}$ data on a phenomenological model using γ and w (right panel) to allow deviations from GR and ΛCDM, demonstrating consistency with both at the same time. The left panel shows also the correlation of the linear growth index γ with another parameter of this model, the mean matter density, Ω_m.

These initial results were in good agreement with both GR and ΛCDM, as shown in Fig. 1. Reassuringly for both experiments, Reyes *et al.*[464] also found consistency of the Standard Model with independent, noncluster datasets using a different contemporaneous test based on a parameter, E_G, that combines measures of large-scale gravitational lensing, galaxy clustering, and structure growth rate.[d] Recent results of this test[465, 467] continue to be largely consistent with GR+ΛCDM despite not statistically significant hints of tensions,[e] with various studies suggesting the need for further modeling of observational systematics and theoretical uncertainties.[462, 502]

The X-ray cluster survey data used for the first constraints on γ came from the aforementioned BCS, REFLEX, MACS, and 400 sd samples. This work also employed a mass–luminosity relation calibrated with hydrostatic masses from pointed ROSAT PSPC, and RASS X-ray observations[503] at low-z, assuming a self-similar evolution and a generic, linearly evolving scatter, as well as applying a correction for the bias due to the assumption of hydrostatic equilibrium. As a consistency check, these results were compared to others from weak-lensing data free of that assumption. In the γ analysis, an HMF based on GR simulations, from Jenkins *et al.*,[383] was employed to describe the nonlinear structure formation. Hence, this analysis tested only linear density deviations from GR through the calculation of Eq. (4) while nonlinearities were assumed to be standard. Note also that when

---

[d]Using the E_G measurement of Reyes *et al.*[464] together with other data sets, Lombriser[201] also constrained w and γ finding again consistency with the concordance model.
[e]Amon *et al.*[465] pointed out that the current tension in Ω_m also impacts those in E_G.

allowing the mean curvature energy density to be free, this work found negligible covariance between Ω_k and γ.

In the next generation of these cluster studies, a series of papers[504–507] constrained departures from the standard cosmological model with up to a factor of 2–3 improvements[504] with respect to the previous results of Mantz *et al.*[495] using the same survey data. This analysis incorporated X-ray follow-up data from ROSAT or the Chandra X-ray Observatory (with a certain overlap between them, useful for testing purposes) spanning over the same redshift range as the survey data, up to $z \lesssim 0.5$. The measurements of cluster properties such as X-ray luminosity, average temperature, and gas mass obtained from the follow-up data were used to constrain luminosity–mass and temperature–mass scaling relations.

In this analysis, the gas mass data was used in the role of a total mass proxy. This was motivated by the fact that it can be measured with very little bias independently of the dynamic state of the clusters, unlike the total mass via hydrostatic equilibrium. The latter was the method employed previously to calibrate masses, forcing the use of relatively large uncertainties to accommodate the hydrostatic bias. The new analysis ultimately also relied on hydrostatic equilibrium to relate the gas mass to the total mass, but it did so through f_{gas} clusters,[508] which include only hot, massive, dynamically relaxed objects with minimal bias due to nonthermal pressure. For this purpose, however, using only the six lowest redshift clusters ($z < 0.15$) from Allen *et al.*[508] was sufficient to constrain this relation while avoiding direct constraints on cosmic expansion by not employing the high redshift objects of the sample. The modeling of systematic uncertainties utilized in the f_{gas} experiment[508] was also included in the new abundance analysis with the improved mass calibration.

In fact, a major innovation of this work was to model all the datasets described above into a global likelihood analysis able to provide robust constraints on both cosmological and astrophysical parameters at the same time, accounting for selection effects, covariances, and systematic uncertainties. This was a pioneer development for the utilization of cluster abundance measurements to test cosmology, including gravity at large scales. A simultaneous and self-consistent analysis of both cosmology and mass-observable scaling relations[504, 505] allows to properly take into account Malmquist and Eddington biases present in all surveys. To visualize this key concept, it is helpful to utilize the following cartoon from Mantz *et al.*[505] (see also the review of Allen *et al.*[487]). As depicted in Fig. 2, near the threshold this fictitious survey will preferentially include higher luminosity objects within the scatter (Malmquist bias; see the top-left panel of the figure), and this effect will be larger for a distribution skewed towards lower luminosity, less-massive objects (Eddington bias; see the bottom-left panel), as it is the case for the mass function of galaxy clusters. It is therefore crucial for cluster abundance surveys to model the sample selection and cluster-mass function together with the mass-observable scaling relations into a single likelihood function. This is currently the benchmark methodology employed in the field for robust constraints on the cosmic growth of structure.

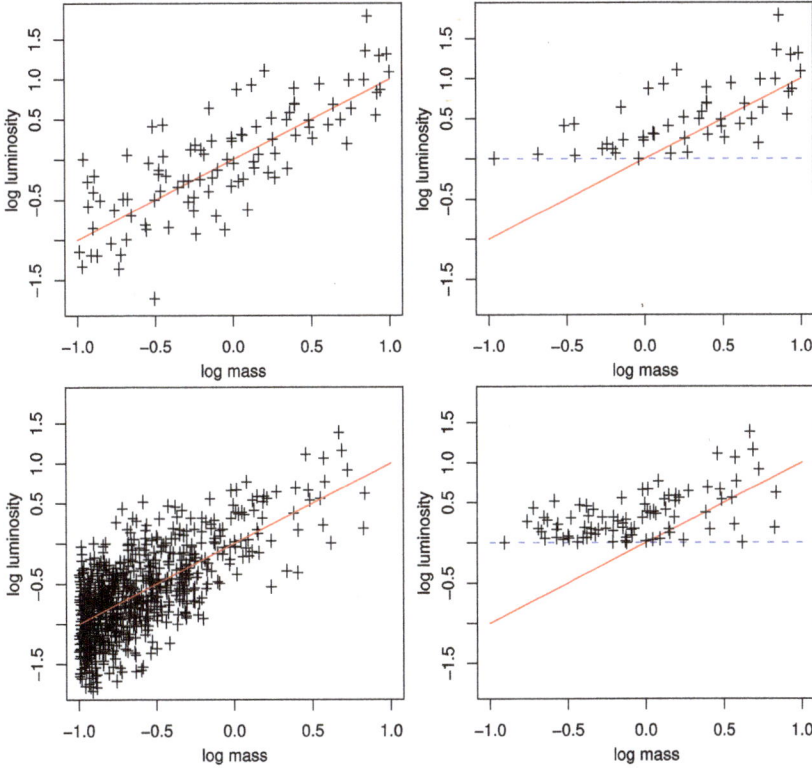

Fig. 2: Figures from Ref. 505 illustrating the importance of self-consistently and simultaneously fitting cosmological and mass-observable scaling relation parameters to avoid Malmquist and Eddington biases. In this cartoon, the red line is a fictitious underlying scaling relation from which simulated clusters (black crosses) are generated either uniformly (top panels) or exponentially (bottom panels) in log-mass. The dashed, blue lines represent a luminosity threshold. When the latter is applied in the right panels, fitting the remaining data without accounting for the full distribution of objects (shown in the left panels) given by both sample selection and halo mass function will bias the answer with respect to the true scaling relation. This is particularly clear in the bottom panels.

Rapetti *et al.*[506] utilized the innovative cluster analysis to simultaneously constrain the cosmic expansion and growth histories as parameterized by the kinematical parameter w and the growth index γ, respectively, as described by Eqs. (1) and (2). The left panel of Fig. 3 shows the results obtained using survey data from BCS, REFLEX, and MACS, which are tighter than those from the previous analysis, particularly considering that the 400 sd sample was not used in the new analysis. These results also include CMB, SNIa, f_{gas}, and BAO measurements. Another update in the new cluster analysis was the use of the then-more-recent halo mass function from Tinker *et al.*,[386] which was still based on GR, but accounted also for redshift evolution of the fitting parameters. In addition to a multivariate normal prior for all the mass function parameters, a systematic uncertainty reflecting physical effects not included in the simulations, such as the presence of baryons or possible exotic dark energy properties, was also added by scaling the covariance matrix that had been obtained from fitting the simulations. However, results were shown to be insensitive to changes in the mass function relative to the dominant errors due to uncertainties in the mass calibration. It was also verified that the additional systematic parameter added to account for residual evolution of the mass function was essentially uncorrelated with γ.

The incorporation of follow-up observations covering the full redshift range of the survey data was made possible by the new internally consistent method. Importantly, this redshift coverage allowed to directly test for evolution in the scaling relations, which is especially relevant for the analysis of the growth index. It is key for such work to examine potential correlations between γ and any astrophysical evolution parameters. A model with flat ΛCDM for the background expansion, a constant γ parametrization for the structure growth rate, and two additional free parameters to allow departures from self-similarity and redshift evolution in the scatter of the luminosity–mass relation revealed weak correlations between γ and those astrophysical evolution parameters. The constraints on γ corresponding to the blue contours in the right panel of Fig. 3, for which the additional evolution parameters are free to vary, are only $\sim 20\%$ weaker than those from the gold contours of the self-similar, constant scatter model. As found in Ref. 505 for a GR plus flat

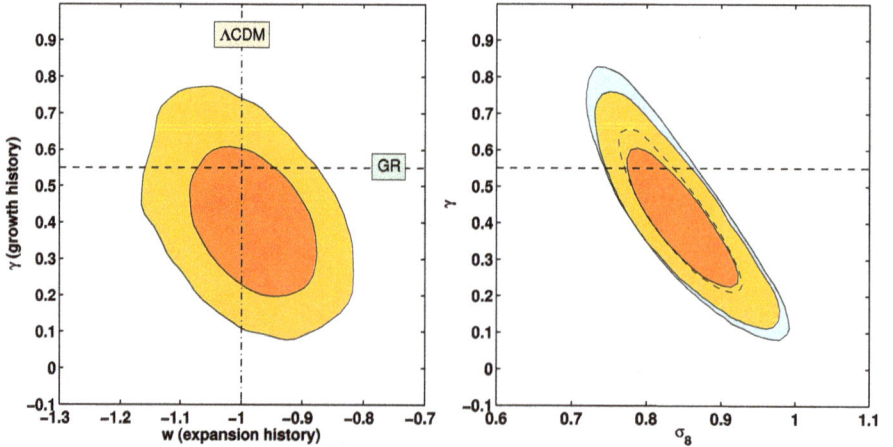

Fig. 3: Figures from Ref. 506 displaying robust, joint measurements from a combination of cluster abundance, CMB, SNIa, $f_{\rm gas}$, and Baryon Acoustic Oscillations (BAO) data on (w, γ) for a flat $\gamma+w$CDM model (left panel) and on (σ_8, γ) for a flat $\gamma+\Lambda$CDM model (right panel). The gold contours (at the 68.3% and 95.4% confidence levels) assume self-similar evolution and constant scatter, while the blue contours (right panel) show a small increase on the constraints when a parameter for departures from self-similarity and another for redshift evolution in the scatter of the luminosity–mass relation vary freely. The tight correlation between σ_8 and γ promises significant improvements by adding independent, precise measurements on σ_8.

ΛCDM model, the Deviance Information Criterion (DIC)[509] indicated that the minimal self-similar and constant scatter model remained a valid description of the data even when γ was included as a parameter in the analysis.[506]

Together with the aforementioned robustness of this analysis, another key finding was a tight correlation between σ_8 and γ such as that $\gamma(\sigma_8/0.8)^{6.8} = 0.55^{+0.13}_{-0.10}$, with a correlation coefficient of $\rho = -0.87$ for the case of $w = -1$ (ΛCDM; see the right panel of Fig. 3). This tight correlation appears when combining cluster abundance with particularly CMB data, due to its strong constraints on σ_8 at high redshift. This suggested that the incorporation of data with independent, precise constraints on σ_8 should be able to break this degeneracy and obtain significantly stronger results on γ.

By adding galaxy clustering data on redshift space distortions (RSD) and the Alcock–Paczynski (AP) effect to the cluster plus CMB data analysis, Rapetti *et al.*[510] obtained indeed much tighter constraints on γ, as shown by the gold contours in Fig. 4. For the $\gamma+\Lambda$CDM model, the left panel of the figure includes also the results on the (σ_8, γ) plane for each individual experiment, demonstrating the required agreement between datasets in order to combine them. The right panel shows constraints on the (w, γ) plane for the $\gamma+w$CDM model and the different combinations of dataset pairs, showing an excellent consistency with GR and ΛCDM. Studies using galaxy clustering and other cosmological probes but not including cluster data have also been finding good agreement with the Standard Model.[232, 511, 512]

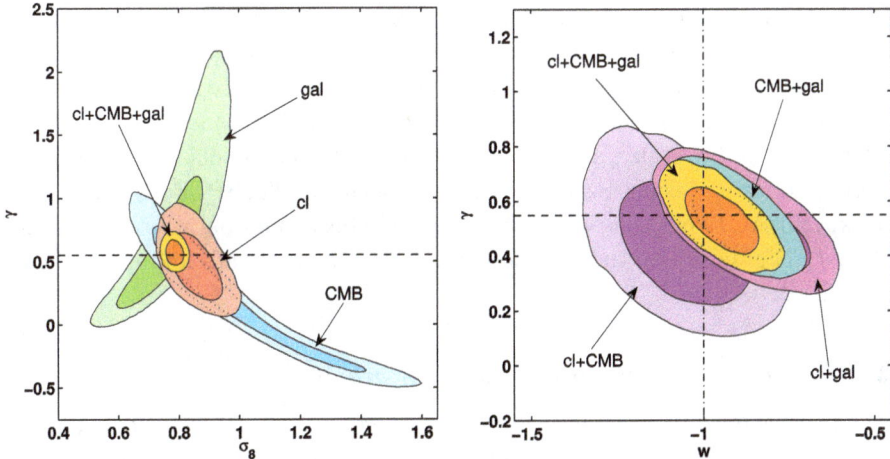

Fig. 4: Figures from Ref. 510 where the complementary degeneracies in the (σ_8, γ) plane between galaxy clustering data (RSD+AP) and clusters (abundance + f_{gas}) or CMB data provide tight constraints on these parameters when all these data is combined (gold contours; 68.3% and 95.4% confidence levels), while remaining consistent with GR ($\gamma \sim 0.55$) and ΛCDM ($w = -1$). Importantly, the consistency between the individual, independent datasets allows their combination.

To overcome the dominant systematic uncertainty when using an f_{gas} mass calibration, the bias in estimating total masses due to assuming hydrostatic equilibrium,
the Weighing the Giants (WtG) project employed instead high-quality weak-lensing
data to calibrate cluster masses.[501, 513–516] To incorporate these new data, an additional self-consistent part of the likelihood function was implemented, which led to
improved cosmological constraints, including for the $\gamma+w$CDM model.[501,f] Since
then, weak lensing has become the standard technique to calibrate masses[517–521] in
cluster abundance studies.[501, 522–524] However, subsequent SZ cluster count analyses
from the Planck Collaboration[525] still used hydrostatic equilibrium mass measurements from XMM Newton X-ray observations, which might have introduced some
of the observed tension between these and the corresponding Planck CMB results,
as well as with other cosmological datasets, as indicated by a WtG weak-lensing
mass calibration analysis of Planck clusters.[526] Other similar studies, however, were
performed with varying results.[527, 528] Hence, the follow-up Planck full mission data
set study on the reported tension between CMB and cluster constraints presented
various results according to the different cluster-mass calibrations adopted from
external weak-lensing analyses.[493] The latest work on this topic from the Planck
Collaboration provides further insights into this tension,[529] showing again no dis-
crepancy when adopting the WtG mass scaling, and agreement now with a recent
reanalysis of CMB-cluster lensing data by Zubeldia and Challinor,[897] even though
remaining discrepancies still exist with other weak-lensing studies.

Using SZ cluster data from the SPT 720 square-degree survey[531] together with a
velocity-dispersion-based mass calibration and X-ray follow-up observations of sample objects, Bocquet et $al.$[530] performed an independent, simultaneous analysis of
cosmology and scaling relations, and found consistency with GR and ΛCDM when
allowing the growth index γ and the dark energy equation-of-state parameter w to
vary. As shown in Fig. 5, for a $\gamma+\Lambda$CDM model this work reproduced the previously found strong degeneracy between σ_8 and γ (top-left panel), and reported a
weak correlation between γ and the species-summed neutrino mass, Σm_ν (top-right
panel). Also, using various SZ cluster surveys, such as SPT, SPTPol (polarization),
Planck, and ACTPol, Mak et $al.$[532] forecasted constraints on a modified gravity
theory, $f(R)$ gravity (see the next subsection for further details and measurements
on this model).

2.3. Constraining alternative models of gravity

Beyond consistency tests as those described above, using galaxy cluster counts as a
function of mass and redshift to observationally constrain modified gravity models
requires not only to calculate the linear behavior of the model but also to compute

---

[f]Note that when using the γ parametrization to test GR, the only deviations explored are those
in which the masses obtained from weak-lensing and dynamical mass measurements equate, as in
GR. As discussed in Sec. 3, in general these masses differ for modified gravity models, although
there are also cases for which these can be exactly or approximately the same (see Sec. 2.3).

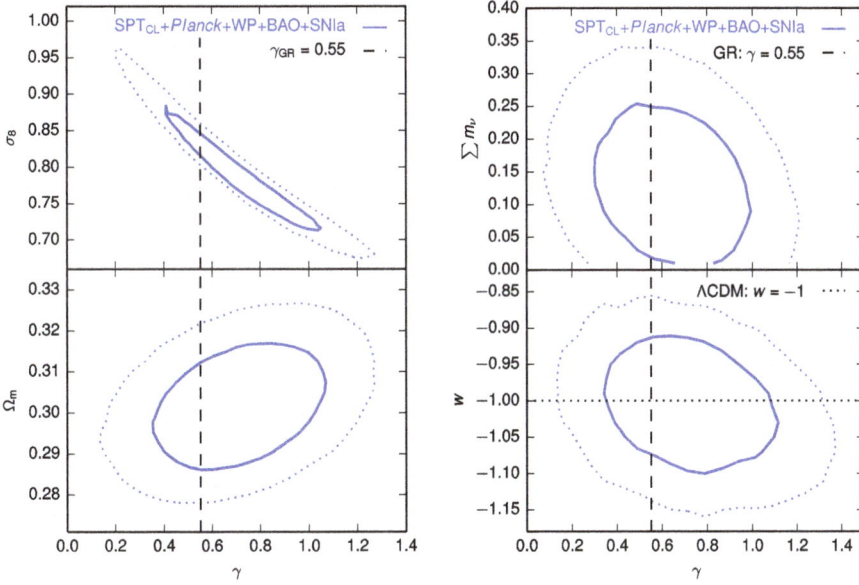

Fig. 5: Figures from Ref. 530 with SPT constraints (68% and 95% confidence contours) on the $\gamma+\Lambda$CDM model, where the strong degeneracy between σ_8 and γ was independently found (top-left panel), as well as on the consistency with both GR and ΛCDM when simultaneously allowing γ and w to be free for the $\gamma+w$CDM model (bottom-right panel). The top-right panel shows constraints on yet another extended model with γ and the species-summed neutrino mass Σm_ν free to vary.

its relevant nonlinear effects on structure formation. The goal is to build an accurate HMF that adequately incorporates the dependencies of the cosmological parameters of interest to perform likelihood analyses. Full N-body simulations are the present, ultimate benchmark with which to validate HMF modeling (see Chapter 3 for details). Since these are computationally expensive, however, alternative, faster approaches has been pursued in the literature depending on the aimed precision, such as fitting procedures and approximate methods (see Chapter 4 for details).

A well-studied alternative to GR at large scales is a simple modification of the Einstein–Hilbert action obtained by substituting the Ricci scalar R with a nonlinear function of itself, $f(R)$. This is in fact a special case of the more general scalar–tensor theory of Brans–Dicke when $\omega_{\mathrm{BD}} = 0$ — for further information on this and other cosmological modified gravity models, see Chapter 1. The fifth force carried by the added scalar degree of freedom, the scalaron field $f_R = df/dR$, has a range of interactions determined by the Compton wavelength of the field, $\lambda_c = (3df_R/dR)^{1/2}$. As long as this scale is smaller than that of the horizon (H^{-1}), the additional force enhances the growth of structure by a factor of 4/3 at scales below λ_c, while above this scale GR is recovered.

Currently viable cosmic gravity models possess nonlinear screening mechanisms to suppress the modification of GR in high density environments, such as the Solar

System, wherein gravity has been measured to agree with GR at high precision. For $f(R)$ gravity, the so-called chameleon mechanism provides such property. Popular forms of $f(R)$ able to evade local constraints are the Hu-Sawicki (HS)[72] and designer[90, 93] models. Pending on a closer examination of systematic uncertainties, constraints on $f(R)$ gravity at galactic scales exist that are somewhat tighter than those achievable by cosmological probes. Cluster counts, however, have been shown to be able to explore $f(R)$ as an effective theory of gravity at cosmic scales all the way down to ~ 1–$20\, h^{-1}\,\mathrm{Mpc}$, allowing to investigate the critical transition from linear to nonlinear scales when the field is of the order of the cluster gravitational potential.

The original functional form of the HS class of models is

$$f(R) = -2\Lambda \frac{R^n}{R^n + \mu^{2n}}, \tag{5}$$

with free parameters Λ, μ^2, and n. This model does not strictly contain a cosmological constant, but in the high-curvature regime, $R \gg \mu^2$, it can be approximated as

$$f(R) = -2\Lambda - \frac{f_{R0}}{n} \frac{\bar{R}_0^{n+1}}{R^n}, \tag{6}$$

where $\bar{R}_0 \equiv \bar{R}(z = 0)$ is the present background R, and the value of the field today, $f_{R0} = -2n\Lambda\mu^{2n}/\bar{R}_0^{n+1}$, is then used as the free parameter of the model that controls the strength of the modification of GR as well as the scale, which for a flat ΛCDM expansion corresponds to a present-day value of

$$\lambda_{c0} \approx 29.9 \sqrt{\frac{|f_{R0}|}{10^{-4}} \frac{n+1}{4 - 3\Omega_m}}\, h^{-1}\,\mathrm{Mpc}. \tag{7}$$

Note also that it follows from this expression that larger values of n will correspond to weaker constraints on f_{R0} from the data.[533, g] On the other hand, the designer class of models is commonly parameterized as a function of the dimensionless Compton wavelength squared in Hubble units,

$$B_0 \equiv \frac{f_{RR}}{1 + f_R} R' \frac{H}{H'} \bigg|_{z=0} \approx 2.1\Omega_m^{-0.76} |f_{R0}|, \tag{8}$$

where $f_{RR} = df_R/dR$ and $' \equiv d/d\ln a$.

While the background expansions of both families of $f(R)$ models above mimic closely or exactly, respectively, that of the cosmological constant, they produce detectable scale-dependent, linear growths of structure, allowing strong tests of GR at large scales. To fully describe the nonlinear part of the HMF in terms of $f(R)$ parameters, such as f_{R0} or B_0, as well as the other relevant cosmological parameters, N-body simulations are presently the tool of choice. After a breakthrough

---

[g]Ferraro *et al.*[534] also previously obtained this behavior based on the HS ($n = 1$) results of Schmidt *et al.*,[499] by demonstrating the ability to rescaling n to the value of interest from the originally calibrated HS ($n = 1$) HMF of Schmidt *et al.*[240]

in performing such calculations[292] others continued this work to include model extensions and/or provide larger and higher-resolution simulations.[173, 284, 293] Even though these computations became then common practice, exhaustive explorations of such parameter spaces are still prohibitively time-consuming. Schmidt *et al.*,[240] however, combined the spherical collapse approximation and the Sheth–Tormen (ST) prescription[382] into a less expensive semianalytic approach that conservatively matched simulation results. For the ST HMF, one can write the comoving number density of halos per logarithmic interval of the virial mass M_v as

$$n_{\Delta_v} \equiv \frac{dn}{d\ln M_v} = \frac{\bar{\rho}_m}{M_v} \frac{d\ln\nu}{d\ln M_v} \nu f(\nu), \tag{9}$$

where $\nu = \delta_c/\sigma(M_v)$ and δ_c are the peak height and density thresholds, respectively, and the multiplicity function $f(\nu)$ is given by the expression

$$\nu f(\nu) = A\sqrt{\frac{2}{\pi}a\nu^2}[1 + (a\nu^2)^{-p}] \exp\left[-\frac{a\nu^2}{2}\right]. \tag{10}$$

Using cluster abundance measurements from the 400 sd sample[491, 496] together with CMB and other cosmological datasets, Schmidt *et al.*[499] obtained the first results on $f(R)$ gravity using a cluster counts experiment, leading to the tightest cosmological constraints on the HS model at the time, $|f_{R0}| \lesssim 1.3 \times 10^{-4}$ at the 95.4% confidence level (used throughout hereafter), as shown in the top panels of Fig. 6. This work rescaled σ_8 at a fixed pivot mass mapping the modifications of gravity into GR by matching the ST HMF for $f(R)$ to a GR HMF,[386] when analyzing both cluster abundance and CMB data.

Instead of renormalizing σ_8, a follow-up, improved analysis by Cataneo *et al.*[533] implemented the same ST HMF modeling, calibrated with a GR HMF,[386] into the full WtG likelihood function of Mantz *et al.*[501] This combined survey, scaling relations (X-ray), and mass calibration (weak-lensing) data, to properly account for all convariances, including those between astrophysical and cosmological parameters, systematic uncertainties, and observational biases. For this $f(R)$ analysis, no changes in the weak-lensing mass calibration were required because weak-lensing and dynamical masses only deviate by a factor on the order of f_{R0}, which is negligible. Cataneo *et al.*[533] calibrated the new $f(R)$ HMF, n_Δ, by multiplying $n_\Delta|_{\text{Tinker}}$, i.e. the HMF of Eq. (3) for the GR $f(\sigma, z)$ fitted by Tinker *et al.*,[386] by a pre-factor that contains the deviations from GR via the ratio of the ST HMF in $f(R)$ to the ST HMF in GR,

$$n_\Delta = \left(\frac{n_\Delta^{f(R)}}{n_\Delta^{\text{GR}}}\bigg|_{\text{ST}}\right) n_\Delta|_{\text{Tinker}}. \tag{11}$$

Figure 7 shows the constraints obtained by the new $f(R)$ analysis,[533] which represented about an order of magnitude improvement with respect to the previous, $\log_{10}|f_{R0}| < -4.79$ (for the combination with the WMAP[535, 536] data; and -4.73

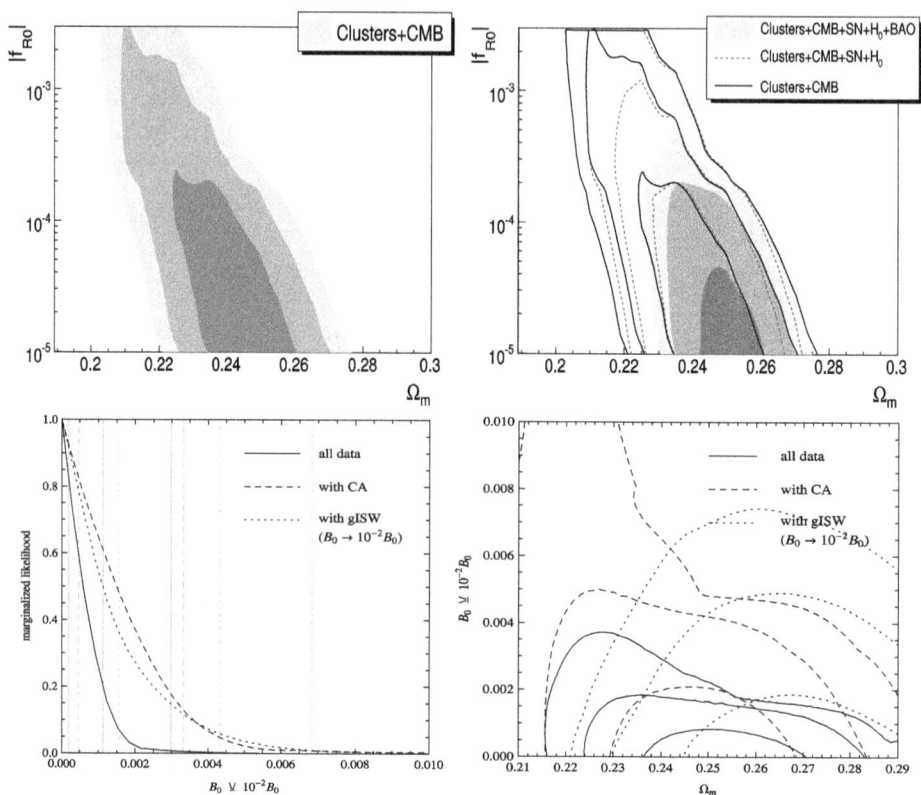

Fig. 6: Figures from Ref. 499 (top panels) and Ref. 252 (bottom panels), with the first constraints on HS (with $n = 1$; top) and designer (bottom) $f(R)$ gravity models using cluster count data together with CMB and additional cosmological datasets. The top panels show 68.3%, 95.4%, and 99.7% confidence contours, and the bottom panels, 1D (left) and 2D (right) marginalized 68%, 95%, and 99% confidence levels. The latter are from either a combination of CMB, SNIa, and BAO datasets or this plus additional measurements from galaxy-ISW cross-correlations (gISW; note that for illustration purposes this constraint was increased by a factor of 100) or CA.

with Planck[537] data), and still are the gold standard in the field.[h] These results started entering the intermediate-field regime ($|f_{R0}| \sim 10^{-5}$) where the scalaron amplitude becomes comparable to the Newtonian potentials of massive halos. However, in order to significantly benefit from upcoming cluster data including well-calibrated lower-mass objects a more accurate modeling of the chameleon screening mechanism was required. For this purpose, using different methods and simulations, Cataneo et al.[254] and Hagstotz et al.[255] derived new HMFs which, importantly,

---

[h]Using SZ clusters from Planck with the WtG mass calibration, Peirone et al.[538] found results that strongly depended on the HMF employed, highlighting the need for a robust HMF. At the conservative end of their range, these constraints are similar enough to those from the X-ray analysis of Cataneo et al.,[533] albeit being based on a different HMF.

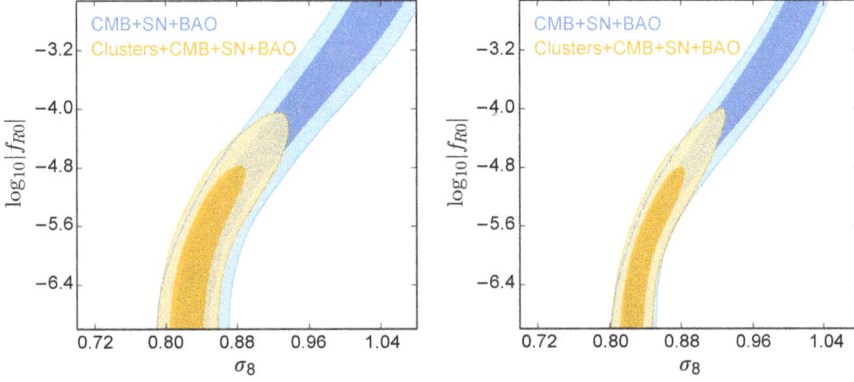

Fig. 7: Figure from Ref. 533 with constraints (68.3% and 95.4% confidence contours) that represented about an order of magnitude improvement with respect to those on the HS ($n = 1$) model in Fig. 6, and entered the intermediate-field regime where the scalaron is comparable to the Newtonian potential of large clusters. In the legend, clusters refer to cluster abundance plus f_{gas} data.[539] The difference between the two panels is the combination with CMB data, which in both cases includes SPT[540–542] and ACT[543] plus either WMAP (Wilkinson Microwave Anisotropy Probe; left) or Planck CMB+Planck gravitational lensing+WMAP polarization (ePlanck) (right) data. In the same work, similar constraints were obtained for the designer model.

are consistent. Both works also used their respective results to forecast small-field regime ($|f_{R0}| \sim 10^{-6}$) constraints for a cluster survey such as that ongoing for DES.

For the designer model, Lombriser *et al.*[252] employed optical data from the MaxBCG catalog[484] of SDSS, together with other complementary cosmological datasets, to obtain the first cluster abundance constraints on B_0, as shown in the bottom panels of Fig. 6. Consistently, these results, which can be translated to $|f_{R0}| \lesssim 2 \times 10^{-4}$, were only slightly weaker than the first on the HS ($n = 1$) model.[499] For this initial study of the designer model, modifications of gravity were included only in calculating the linear component σ of $n_\Delta|_{Tinker}$, since the nonlinear description of GR was considered accurate enough for observational constraints in the large-field regime ($|f_{R0}| \gtrsim 10^{-4}$), in which the effects of the nonlinear chameleon mechanism to the HMF can be neglected. Afterwards, Cataneo *et al.*[533] accounted for both linear and nonlinear effects by using an $f(R)$ HMF calculated via Eq. (11), as required in the intermediate-field regime entered due to having improved the initial results for this model also by about an order of magnitude.

Using cluster abundance data from the 400 sd sample[491, 496] together with CMB and additional cosmological datasets, Chudaykin *et al.*[544] constrained, in the presence of a sterile neutrino of the order of eV, yet another $f(R)$ gravity model,[88]

$$f(R) = R + \lambda R_s \left[\left(1 + \frac{R^2}{R_s^2} \right)^{-n} - 1 \right] \tag{12}$$

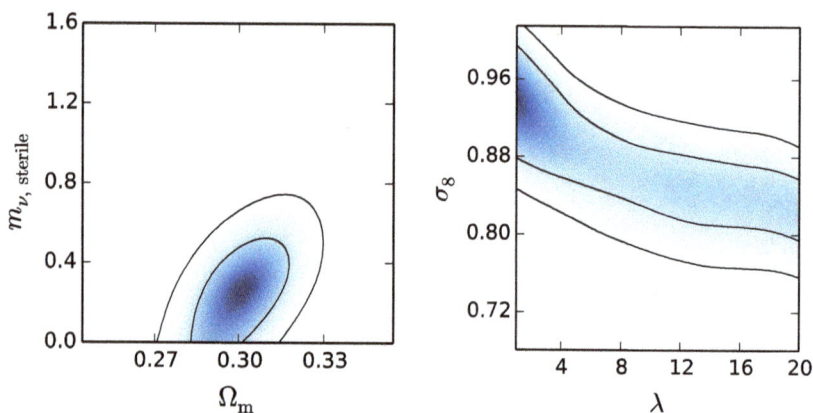

Fig. 8: Figures from Ref. 544 showing constraints (65% and 95% confidence contours) from cluster abundance, CMB (ePlanck+SPT+ACT), BAO, H_0, and lensing potential data for a Starobinsky $f(R)$ gravity model, assuming one massive sterile neutrino and three massless active neutrinos.

with the appropriate correction and conditions for the model to be viable at Solar System densities,[544] and where n, λ, and R_s are model parameters, from which $n = 2$ was chosen to be fixed. Results from this analysis are shown in Fig. 8.

3. Mass Estimates and Cluster Profiles

In theories of modified gravity cluster masses inferred from dynamics can differ from their (in general inaccessible) true masses, a combination of dark matter, gas, and stars.[250, 545, 546] Dynamical masses can be obtained by observing the velocity dispersions of cluster galaxies, the X-ray properties of the hot ionized intracluster gas or the SZ effect. For a subclass of theories of gravity the mass deduced from gravitational lensing is unaffected by the modifications and will match GR predictions. Only within such particular models the lensing mass does indeed correspond to the true mass of the cluster. More typically, deviations from standard gravity also produce changes in the lensing mass. In addition, screening mechanisms restore GR on nonlinear scales and make departures of dynamical and lensing masses from the true mass depend on the true mass itself, on the scale considered and on the environment surrounding the cluster.

The sheer complexity of the system inevitably requires some simplifications when comparing theoretical predictions to the data, which will impact our conclusions to an extent that must be assessed *a posteriori*. Simplifying assumptions often include: the quasistatic approximation (QSA), in which time derivatives are neglected and clusters are treated as virialized systems; intracluster gas in hydrostatic equilibrium; spherical symmetry; Navarro–Frenk–White (NFW) density profiles[547] for the host dark matter halos.[i]

---

[i]Several works showed that modified gravity does not qualitatively change the shape of the dark

The starting point for any mass estimate are the Poisson equations for the dynamical potential Ψ and the lensing (or Weyl) potential Φ_{lens},

$$k^2 \Psi = -4\pi G_{\text{matter}} a^2 \delta\rho, \tag{13}$$

$$k^2 \Phi_{\text{lens}} = -4\pi G_{\text{light}} a^2 \delta\rho, \tag{14}$$

where G_{matter} and G_{light} are the effective gravitational constants for nonrelativistic matter and light, respectively, and $\delta\rho$ is the matter density excess with respect to the background. In most alternative theories of gravity the two modified constants are in fact not constant at all, and can be arbitrary functions of both time and space. Nonetheless, in Horndeski gravity (see Chapter 1) these assume a relatively simple form in the linear subhorizon regime, i.e.[548, 549]

$$G_{\text{matter}}(a, k) = h_1 \left(\frac{1 + k^2 h_5}{1 + k^2 h_3} \right) G, \tag{15}$$

$$G_{\text{light}}(a, k) = h_6 \left(\frac{1 + k^2 h_7}{1 + k^2 h_3} \right) G, \tag{16}$$

where h_{1-5} are functions of time only, $h_6 = h_1(1 + h_2)/2$, and $h_7 = (h_5 + h_2 h_4)/(1 + h_2)$. The functions $h_i(a)$ are completely determined by the Lagrangian functions $K(\phi, X)$ and $G_{3-5}(\phi, X)$ (see Chapter 1), and explicit expressions can be found in Ref. 548. Standard gravity is recovered for $h_{1,2} = 1$, $h_{3-5} = 0$. The evolution of linear perturbations in many popular modified gravity theories can be readily described by Eqs. (15) and (16) (see Chapter 2 for more details). A notable example are viable $f(R)$ gravity models where

$$\frac{G_{\text{matter}}}{G} = \frac{1 + 4 f_{RR} \left(\dfrac{k}{a} \right)^2}{1 + 3 f_{RR} \left(\dfrac{k}{a} \right)^2}, \tag{17}$$

$$\frac{G_{\text{light}}}{G} = 1, \tag{18}$$

implying $\Phi_{\text{lens}} = \Psi_N$, with Ψ_N being the standard Newtonian potential. On the other hand, for the Cubic Galileon model of Ref. 344 one has $G_{\text{matter}} = G_{\text{matter}}(a)$, and $\Phi_{\text{lens}} = \Psi \neq \Psi_N$ in the absence of anisotropic stress.[550]

Moving to smaller scales requires either solving highly nonlinear differential equations for an isolated halo or running expensive cosmological simulations (see e.g. Refs. 248 and 551). The reason being that clusters are nonlinear objects, and screening mechanisms must be in action to guarantee that on small scales modifications of gravity are suppressed. As a matter of fact, if screening conditions are satisfied there exists a screening scale R_{scr} such that for radii $R \ll R_{\text{scr}}$ gravity is back to GR, i.e. $G_{\text{matter}} = G_{\text{light}} = G$, whereas for $R \gg R_{\text{scr}}$ the linear prediction

---

matter halos, and on scales below the virial radius the NFW profiles match reasonably well the averaged halo profiles measured in simulations.[173, 240, 248, 261, 333]

of Eq. (15) and Eq. (16) apply. The transition at intermediate scales can depend on details such as the cluster mass, density profile, shape, and environment.

For a collisional system, such as the intracluster gas, observable effects induced by modifications of gravity can be quantified from the equation for hydrostatic equilibrium, that under the assumption of spherical symmetry reads (see e.g. Ref. 552)

$$\frac{dP}{dr} = \rho_{\text{gas}} \frac{d\Psi}{dr}, \tag{19}$$

where ρ_{gas} is the gas mass density, and Ψ receives contributions beyond the standard Newtonian potential. The total pressure P can be decomposed into thermal and nonthermal components. For an ideal gas with temperature T_{gas} the thermal pressure $P_{\text{therm}} = n_{\text{gas}} k_B T_{\text{gas}}$, with $n_{\text{gas}} = \rho_{\text{gas}}/\mu m_p$ denoting the number density of gas particles with mean molecular weight μ. From Eq. (19) it is straightforward to obtain the thermal mass as

$$M_{\text{therm}}(<r) = -\frac{kr T_{\text{gas}}}{\mu m_p G} \left(\frac{d \ln n_{\text{gas}}}{d \ln r} + \frac{d \ln T_{\text{gas}}}{d \ln r} \right). \tag{20}$$

The gas temperature and profile are linked to the observed projected X-ray temperature and surface brightness, respectively.[553] Alternatively, measurements of the temperature difference of the CMB photons caused by the SZ effect provide information on the gas thermal pressure, that in combination with X-ray surface brightness data can be used to estimate the thermal mass.[554] Nonthermal pressure generated by magnetic fields, cosmic rays, bulk motion, etc. can be a source of systematic uncertainty if not properly accounted for.[482] A common approach consists in defining the nonthermal contribution as a scale-dependent fraction of the total pressure, i.e.

$$P_{\text{nonthermal}} \equiv g(r) P(r), \tag{21}$$

where the function $g(r)$ is calibrated against hydrodynamical simulations.[555]

The trajectories of photons traveling from distant galaxies are perturbed by the presence of foreground massive galaxy clusters according to Eq. (14). Measurements of the convergence profile of galaxy clusters (or quantities closely related to it, such as the tangential shear) probe how gravity interacts with light. In fact, the lensing convergence is derived from the projected lensing potential ψ as[447, 556]

$$\kappa(\theta) = \frac{1}{2} \nabla_\perp^2 \psi, \tag{22}$$

where $\nabla_\perp^2 \equiv \partial_{\theta_x}^2 + \partial_{\theta_y}^2$ is the two-dimensional Laplacian on the plane of the sky, and

$$\psi \left(\theta = \frac{r}{D_L} \right) = \frac{D_{LS}}{D_L D_S} \frac{2}{c^2} \int_{-D_L}^{D_{LS}} dl \, \Phi_{\text{lens}}(r, l), \tag{23}$$

with D_L being the angular diameter distance between the observer and the lens (i.e. the cluster), and D_S, D_{LS} denote, respectively, the angular diameter distances to the source, and between the lens and the source.

Integration of Eq. (14) gives

$$\kappa(\theta) = \frac{\Sigma(\theta)}{\Sigma_c},\tag{24}$$

where we have used the surface mass density definition

$$\Sigma\left(\theta = \frac{r}{D_L}\right) \equiv \frac{1}{4\pi G}\int dl\nabla^2\Phi_{\text{lens}},\tag{25}$$

and have introduced the critical surface mass density

$$\Sigma_c \equiv \frac{c^2}{4\pi G}\frac{D_S}{D_{LS}D_L}.\tag{26}$$

In theories of gravity that do not modify the lensing potential one recovers the standard GR expression

$$\Sigma(\theta) \equiv \int dl\delta\rho(\theta D_L,l).\tag{27}$$

Observations provide the convergence profiles of galaxy clusters [Eq. (24)], which can then be compared to the predictions obtained from Eq. (22). For the purpose of testing generic deviations from GR it is key that lensing mass reconstructions avoid any assumption about the distribution of matter in the cluster, i.e. the analysis should be nonparametric.[557]

Alternatively, one could use the shear profiles by measuring the average deformation of background galaxy shapes around foreground galaxy clusters. The tangential shear profile can then be defined as the excess surface mass density of the cluster halo, $\Delta\Sigma(\theta)$, normalized to the critical surface mass density, i.e.

$$\langle\gamma_t\rangle(\theta) = \frac{\Delta\Sigma(\theta)}{\Sigma_c},\tag{28}$$

where

$$\Delta\Sigma(\theta) = \bar{\Sigma}(<\theta) - \langle\Sigma\rangle(\theta),\tag{29}$$

with $\bar{\Sigma}$ denoting the mean surface mass density in a circular aperture of angle θ, and $\langle\Sigma\rangle$ is the average surface mass density computed in a narrow annulus at the edge of the aperture.

In recent years, a growing number of studies have employed cluster profiles to detect potential signatures of modified gravity. Lombriser *et al.*[247] were the first to employ lensing data in search of departures from GR in the context of $f(R)$ gravity. They measured the stacked cluster-galaxy lensing signal $\Delta\Sigma$ generated by selected MaxBCG galaxy clusters and background galaxy sources in the SDSS imaging data.[558] Their cluster sample covered the redshift range of $0.1 < z < 0.3$, and the analysis focused on scales $0.5 \leq r\,h\,\text{Mpc}^{-1} \leq 25$, corresponding to the outskirt of a typical cluster and beyond. Because of the relatively shallow potential, this region experiences the largest modifications of gravity allowed, producing an excess signal associated with more infalling matter compared to standard gravity. In combination with BAO and SNIa distance measurements, as well as information from the CMB anisotropies, the shear lensing data in Ref. 247 provided the upper bound $|f_{R0}| < 3.5 \times 10^{-3}$ at the 95% confidence level.

Narikawa and Yamamoto[559] tested a generalized Galileon model with surface mass density data obtained from stacking strong- and weak-lensing measurements for four high-mass clusters (A1689, A1703, A370, and Cl0024+17).[560, 561] The authors allowed for modifications on large linear scales through a parameter μ, such that $G_{\text{light}} = G(1 + \mu)$, and parameterized the transition scale to standard gravity set by the Vainshtein screening (see Chapter 1) as

$$r_V = 13.4\epsilon^{2/3} \left(\frac{M}{10^{15} \, M_\odot/h} \right)^{1/3} h^{-1} \, \text{Mpc}, \tag{30}$$

where the stacked mass M and the dimensionless parameter ϵ are both constrained by the data. Newtonian gravity is recovered on all scales in the limits $\epsilon \to \infty$ or $\mu \to 0$. In their analysis, Narikawa and Yamamoto modeled the cluster-mass distribution with various density profiles, deriving constraints on μ and ϵ largely consistent across the different cases. Figure 9 shows the allowed region of parameter space in the $\{\ln \epsilon, \mu\}$ plane for predictions assuming an NFW profile. The lensing data in Refs. 560 and 561 cover scales of $r \lesssim 5 \, h^{-1} \, \text{Mpc}$, thus for large screening radii, i.e. $\epsilon \gg 1$, the linear deviation μ remains unconstrained. On the other hand, for $\epsilon \ll 1$ linear departures on cluster scales are limited to a small range around $\mu = 0$.

Relying on lensing information only, Barreira *et al.*[344] also looked into possible deviations from standard gravity in the context of the Cubic Galileon model, and extended their analysis to nonlocal gravity cosmologies as well. Differently from Narikawa and Yamamoto,[559] they used individual radially-binned lensing conver-

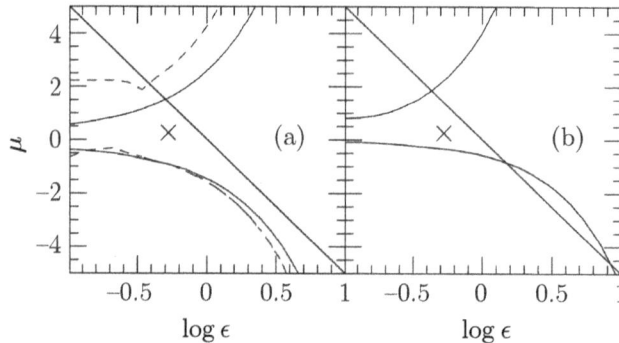

Fig. 9: The 68% confidence regions for the modified gravity parameters pair $\{\mu, \epsilon\}$ describing a generalized Galileon model. The cross marks values for the Cubic Galileon model.[112] *Left*: Constraints from stacked surface mass density data (solid curve), together with those using data for the logarithmic slope $d \ln \Sigma / d \ln r$ (dashed curve). *Right*: Same as left panel using the differential surface mass density $\Delta\Sigma_+ \equiv \Delta\Sigma/(1 - \kappa)$. In all panels theoretical predictions assume the NFW profile. Figure taken from Ref. 559.

gence profiles for 19 X-ray selected galaxy clusters from the Cluster Lensing and Supernova Survey with the Hubble Space Telescope (CLASH).[562–564] These data have the advantage that the lensing signal reconstruction makes no assumption on the mass distribution of the clusters, a desirable feature for applications aimed to test departures from GR. Barreira *et al.* found that within the virial radius ($\sim 1\,h^{-1}\,\mathrm{Mpc}$) both modifications of gravity under consideration had no measurable impact on the mass and concentration parameters describing the halo profiles, thus showing that these observations cannot distinguish these particular models from ΛCDM.

Instead of lensing measurements, Terukina and Yamamoto[565] opted for a complementary approach based exclusively on temperature profiles data of the Hydra A cluster in search of chameleon force effects. The presence of an attractive fifth force changes the gas distribution inside the cluster, with potentially observable effects on the X-ray surface brightness profiles. Assuming hydrostatic equilibrium and a polytropic equation-of-state for the gas component, Terukina and Yamamoto compared their temperature profile predictions against data reduced from Suzaku X-ray observations of the Hydra A cluster out to the virial radius.[566] For an NFW dark matter halo profile and the scalar field coupling $\beta = 1$ they obtained the upper bound $\phi_{\mathrm{BG}} < 10^{-4}\,M_{\mathrm{pl}}$,[j] where mass, concentration, and temperature at the cluster center and background amplitude of the scalar field were all allowed to vary simultaneously in their analysis. However, in $f(R)$ gravity the coupling strength is too weak ($\beta = 1/\sqrt{6}$), and the relatively large uncertainties in the data preclude any meaningful constraint.

Deviations from standard gravity for theories predicting $\Phi_{\mathrm{lens}} \neq \Psi$ can be strongly constrained with the combination of lensing and dynamical measurements.[545, 546] Terukina *et al.*[265] pioneered such analysis employing observations of the X-ray surface brightness and temperature profiles,[567–569] the SZ effect,[570] as well as weak-lensing mass and concentration priors[571, 572] of the Coma cluster. As in Ref. 565, the authors searched for signatures of a chameleon fifth force by modeling the gas and dark matter distribution under spherical symmetry and hydrostatic equilibrium. Moreover, they considered the effects of the nonthermal pressure component to be largely negligible. Although following studies (e.g. Ref. 573) also adopted this last assumption, and explored the potential implications of relaxing it, there is no consensus on the extent of the systematics associated with nonthermal pressure support, which if not properly accounted for can lead to spurious constraints on modified gravity parameters (cf. Refs. 574 and 546). Cluster asphericity and substructures are sources of systematic uncertainty too,[482] capable of biasing significantly our conclusions on departures from GR. An interesting aspect of the analysis performed in Terukina *et al.* is that the combination of multiwavelength observations helped break degeneracies between the parameters describing the cluster profiles for mass, gas, temperature, and those pertaining to the chameleon force.

---

[j]See Chapter 1 for details on chameleon models.

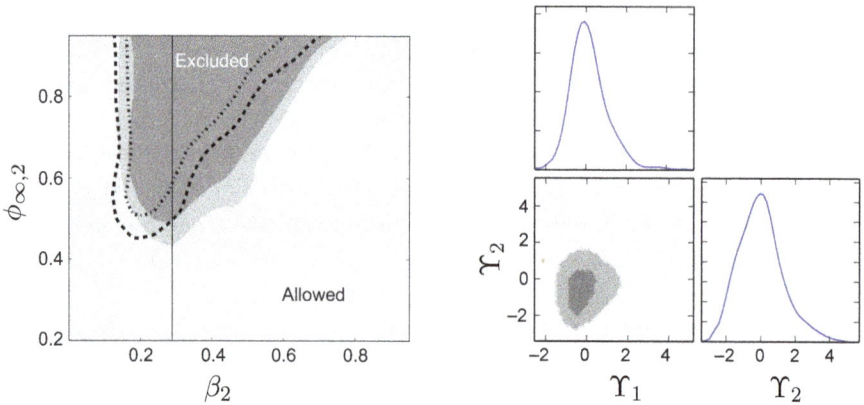

Fig. 10: *Left:* Wilcox *et al.*[573] 95% (light gray region) and 99% confidence level (mid-gray region) boundaries for the rescaled chameleon parameters $\beta_2 \equiv \beta/(1+\beta)$ and $\phi_{\infty,2} \equiv 1 - \exp(-\phi_{\mathrm{BG}}/10^{-4} M_{\mathrm{pl}})$. The overlapped dashed and dot-dashed lines represent the corresponding boundaries found by Terukina *et al.*[265] The vertical line marking $\beta = 1/\sqrt{6}$ shows the constraints for $f(R)$ gravity. Figure taken from Ref. 573. *Right:* Posterior distributions for the beyond Horndeski parameters Υ_1 and Υ_2, together with their combined 68% and 95% contours. GR corresponds to the point $\{\Upsilon_1, \Upsilon_2\} = \{0,0\}$. Figure taken from Ref. 266.

The dashed lines in the left panel of Fig. 10 delimit the excluded region in the rescaled $\{\beta, \phi_{\mathrm{BG}}\}$ plane obtained by Terukina *et al.* with the additional assumption that the Coma cluster is an isolated system. The vertical line corresponds to $f(R)$ gravity implying the upper bound $|f_{R0}| \lesssim 6 \times 10^{-5}$ at the 95% confidence level. Performing a similar analysis Ref. 575 investigated modifications of gravity in generalized Cubic Galileon models. However, despite the more recent X-ray data[576] and updated lensing information[577] employed, these models were only loosely constrained.

With access to a larger cluster sample, Wilcox *et al.*[573] and Sakstein *et al.*[266] implemented a different strategy later validated in Ref. 578. The Coma cluster is notoriously nonspherical[579–581] and is located at low redshift ($z \approx 0.02$), two facts that can weaken the robustness and efficacy of the derived modified gravity constraints. The method developed by Wilcox *et al.*, and already suggested in Terukina *et al.*, relies on stacked X-ray surface brightness and shear profiles of 58 X-ray selected clusters in the redshift range of $0.1 < z < 1.2$ and temperature range of $0.2\,\mathrm{keV} < T_{\mathrm{gas}} < 8\,\mathrm{keV}$. The dynamical information on these objects was obtained from the XMM Cluster Survey (XCS),[582–584] and the Canada–France–Hawaii Telescope Lensing Survey (CFHTLenS)[459, 585] provided the complementary lensing information for the same systems. In addition, Wilcox *et al.* explored the mass dependence of the screening mechanism in chameleon models [including $f(R)$ gravity] by splitting their cluster sample into two bins with a temperature threshold of $T_{\mathrm{gas}} = 2.5\,\mathrm{keV}$. The joint constraints on the chameleon parameters from the combined cluster subsamples are shown in the left panel of Fig. 10, where hydrostatic equilibrium, spherical symmetry, isothermality, and negligible nonthermal pressure were assumed. Also in this case the amplitude of the $f(R)$ scalar field is constrained to $|f_{R0}| \lesssim 6 \times 10^{-5}$ at the 95% confidence level. Sakstein *et al.* applied the same data and method to a subclass of beyond Horndeski theories that breaks the Vainshtein screening inside extended objects, and parameterized it with the dimensionless quantities Υ_1 and Υ_2, where the former measures changes in the motion of nonrelativistic particles and the latter affects exclusively light propagation.[k] GR is recovered for $\Upsilon_1 = 0$ and $\Upsilon_2 = 0$, and $\Upsilon_1 < 0$ (> 0) is equivalent to enhanced (suppressed) gravity. The joint posterior distribution as well as the two marginalized posteriors for Υ_1 and Υ_2 are shown in the right panel of Fig. 10.

More recently, Salzano *et al.*[586] considered a particular subset of beyond Horndeski theories with $\Upsilon_1 = \Upsilon_2 = \Upsilon$ (Ref. 131) characterized by a mismatch between the dynamical and lensing potentials, and inactive Vainshtein screening inside large astrophysical systems. Taking advantage of these features, Salzano *et al.* selected the 20 most relaxed and symmetric galaxy clusters observed by both the X-ray Chandra telescope and the Hubble Space Telescope within CLASH,[587] and constrained the modified gravity parameter Υ quantifying the deviation from standard gravity. Interestingly, for this model one always has $\Upsilon > 0$ leading to weaker gravity for physically motivated dark matter profiles, with $\Upsilon = 0$ being GR. Under the same assumptions of previous studies they found the upper limit $\Upsilon < 0.16$ at the 95% confidence level, clearly consistent with no deviations from GR.

Finally, dynamical mass profiles can also be inferred from the motion of cluster galaxies, an approach followed by Pizzuti *et al.*,[588, 589] who compared the kinematic and lensing measurements of dynamically relaxed clusters obtained during the CLASH[562] and CLASH-VLT[590] observing campaigns. The authors searched for deviations from the standard relation $\eta \equiv \Phi/\Psi = 1$ (Chapters 1 and 2), finding $\eta(r_{200c}) = 1.01^{+0.31}_{-0.28}$, a result fully consistent with GR predictions. They also extended their analysis to Yukawa-like interactions with a free range parameter λ and a coupling constant fixed to $\beta = 1/\sqrt{6}$. This choice effectively mimics the fifth force generated by linearized fluctuations of the scalar field f_R in $f(R)$ gravity characterized by a mass $\bar{m}_{f_R} \sim 1/\lambda$, where the overbar denotes the background value.[l] In their most recent analysis, using data for the MACS J1206.2-0847 cluster Pizzuti *et al.* derived the upper limit $\lambda < 1.61$ Mpc at the 90% confidence level, 20% tighter than their previous result. Including a simplified implementation of the chameleon screening relaxes this bound to $\lambda < 20$ Mpc, or $|f_{R0}| \lesssim 5 \times 10^{-5}$, in agreement with studies based on properties of the intracluster gas.

---

[k]See Chapter 1 for details on these modified gravity theories and their parametrization.
[l]Equivalently, one can see this as an unscreened $f(R)$ gravity model.

4. Gravitational Redshift

Any metric theory of gravity predicts that a photon with wavelength λ_{em} emitted from within a gravitational potential well Ψ experiences an energy loss when leaving such potential. Then, in a static universe and in the weak-field limit, an observer at rest with respect to the source of the gravitational field measures the *gravitational redshift*

$$z^{gr} = \frac{\Delta\lambda}{\lambda_{em}} \approx \frac{\Delta\Psi}{c^2}, \tag{31}$$

where $\Delta\lambda$ is the wavelength difference between the observed and emitted photons, and $\Delta\Psi = \Psi(\mathbf{x}_{obs}) - \Psi(\mathbf{x}_{em})$. For a cluster of mass $\sim 10^{14} M_\odot$ the gravitational redshift $cz^{gr} \approx 10\,\text{km/s}$,[591–593] a tiny value compared to other redshift contributions. In fact, neglecting the evolution of the metric potentials, in our universe the total redshift z^{tot} of a photon emitted at the location $\mathbf{x}$ and observed at the origin of the reference frame can be written as

$$1 + z^{tot} = (1 + z^{cosmo}) \left\{ 1 + \frac{1}{c^2}\left[\Psi(0) - \Psi(\mathbf{x})\right] + \frac{\hat{\mathbf{n}} \cdot \mathbf{v}}{c} + \frac{v^2}{2c^2} \right\}. \tag{32}$$

Here, z^{cosmo} is the cosmological redshift associated with the background expansion, the second nontrivial term in curly brackets represents the Doppler shift along the line of sight $\hat{\mathbf{n}}$ due to the peculiar velocity $\mathbf{v}$ of the object emitting the photon, and the remaining kinetic term is known as transverse Doppler shift.[594] For typical galaxy clusters the gravitational and transverse Doppler shifts are of the same order of magnitude, and both are two orders of magnitude smaller than the longitudinal Doppler shift.

Measurements of these second-order effects are usually expressed in terms of total redshift difference between a *satellite* galaxy (S) and the *central* galaxy (C) in a cluster, i.e.[595, 596]

$$\Delta_S \equiv c\left(\frac{z_S^{tot} - z_C^{tot}}{1 + z_C^{cosmo}}\right) \approx \frac{\Psi(\mathbf{x}_C) - \Psi(\mathbf{x}_S)}{c} + \hat{\mathbf{n}} \cdot \mathbf{v}_S - \hat{\mathbf{n}} \cdot \mathbf{v}_C + \frac{v_S^2}{2c} - \frac{v_C^2}{2c}, \tag{33}$$

where now the gravitational redshift is negative, which can be interpreted as the blueshift of photons emitted by the satellite galaxy and observed at the location of the central galaxy. Averaging over the velocity of satellite galaxies with their phase-space distribution $f(\mathbf{x}_S, \mathbf{v}_S)$, after projecting along the line of sight one obtains the cluster velocity shift profile

$$\langle \Delta_S \rangle(R) = \frac{\int d\chi_S \int d^3 v_S \Delta_S f(\mathbf{x}_S, \mathbf{v}_S)}{\int d\chi_S \int d^3 v_S f(\mathbf{x}_S, \mathbf{v}_S)}$$

$$= \left\langle \frac{\Psi(\mathbf{x}_C) - \Psi(\mathbf{x}_S)}{c} \right\rangle + \left\langle \frac{v_S^2}{2c} \right\rangle, \tag{34}$$

where R is the distance from the central galaxy projected on the plane of the sky. In Eq. (34) we have assumed that the central galaxy has negligible cluster-centric velocity compared to the satellite galaxies, and also ignored the additional shift produced by the transformation from rest-frame coordinates to light-cone coordinates known as the past light-cone effect.[595, 596, m]

For a sample of galaxy clusters with a mass distribution dn/dM (i.e. the cluster-mass function) the ensemble average profile reads

$$\Delta(R) \equiv \frac{\int dM \Sigma(R) \frac{dn}{dM} \langle \Delta_S \rangle}{\int dM \Sigma(R) \frac{dn}{dM}}, \tag{35}$$

where

$$\Sigma(R) \equiv \int d\chi_S \int d^3 v_S f(\mathbf{x}_S, \mathbf{v}_S) \tag{36}$$

is the surface density profile of satellite galaxies.

Wojtak *et al.*[597] were the first to report a nearly 3σ detection of the velocity shift $\Delta(R)$ generated by galaxy clusters selected from the SDSS3 Data Release 7 (Ref. 598) and the associated Gaussian Mixture Brightest Cluster Galaxy catalog.[599] Key to this detection is the large number of galaxy clusters with member galaxies and interlopers having spectroscopically measured velocities. This is necessary to reduce the contamination from peculiar velocities and remove the effect of irregularities in individual clusters.[n] In their analysis Wojtak *et al.* identified the cluster centers and redshifts with the positions and redshifts of the brightest cluster galaxies. Their large cluster sample also helped control the impact of this approximation.[o]

Theoretical predictions for the velocity shift require knowledge of the mean gravitational potential profile and of the cluster-mass function [see Eqs. (33)–(35)]. Wojtak *et al.*[597] modeled the former using an NFW density profile and the latter as a power law, and then used the velocity dispersion profile of the composite cluster to constrain their free parameters. The resulting signal was interpreted as entirely due to gravitational redshift, ignoring the transverse Doppler contribution in Eq. (34). Their measurements and GR predictions are shown in the left panel of Fig. 11. Later, Zhao *et al.*[594] included the missing term which resulted in an overall upward shift of the GR prediction corresponding to the difference between the solid black line and the dotted blue line in the right panel of Fig. 11. However, it was only with Kaiser[595] and Cai *et al.*[600] that all terms in the velocity shift profile — gravitational redshift, special relativistic contributions, and past light-cone effects — were correctly implemented.

---

[m] Furthermore, the peculiar velocity of galaxies modulates their surface brightness through relativistic beaming. In flux-limited surveys this causes a bias of the redshift distribution of the selected galaxies comparable to z^{gr}. See Ref. 595 for details.
[n] See Ref. 600 for a detailed analysis of the systematics introduced by this assumption.
[o] For an updated analysis see Ref. 601.

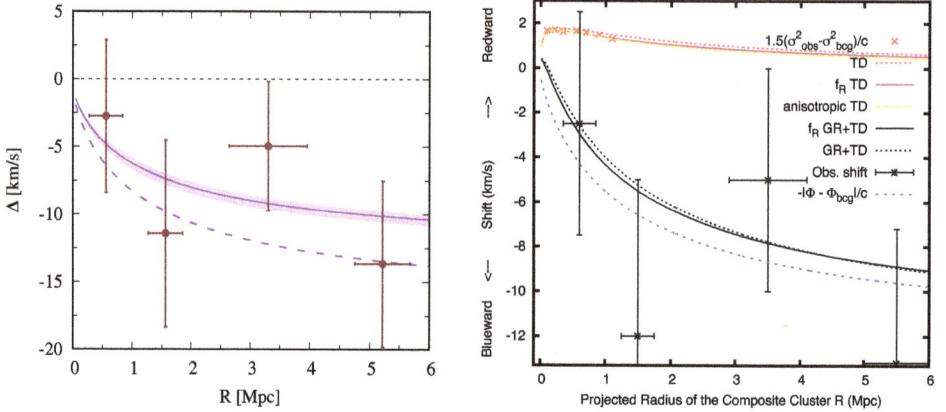

Fig. 11: *Left*: Stacked velocity shift profiles in galaxy clusters showed in the rest frame of their brightest cluster galaxies. Measurements (points with errorbars) at different projected radii R correspond to the mean velocity of the observed satellite galaxies velocity distributions. The solid line denotes the gravitational redshift prediction for GR, and the dashed line shows the same for unscreened $f(R)$ gravity assuming the observed range of cluster masses remains unchanged in Eq. (35). Figure credit: Radowslaw Wojtak. *Right*: Gravitational redshift (dotted blue line) and transverse Doppler (dotted magenta line) profiles in GR. Their sum is represented by the dotted black line. Data points and error bars match those in the left panel. NFW profiles are used to describe the mass distribution within clusters, and the halo mass function is approximated as $dn/dM \sim M^{-7/3}$. The integrated mass range is $M = (0.11–2) \times 10^{15} M_\odot$ in standard gravity, and is $M = (0.09–1) \times 10^{15} M_\odot$ in $f(R)$ gravity with $G_{\text{matter}} = 4G/3$. Figure taken from Ref. 594.

Wojtak *et al.*[597] also employed their data to test departures from standard gravity, as shown by the dashed line in the left panel of Fig. 11 for a fully unscreened $f(R)$ model. This prediction, however, is based on the false premise that the velocity dispersion of the galaxies in their sample tracks the Newtonian potential Ψ_N of their parent cluster. As a matter of fact, the potential inferred from the kinematics of galaxies matches the same potential responsible for the gravitational redshift since both physical processes are governed by the time component of the metric Ψ. Therefore, if only dynamical information for a narrow range of cluster masses is available, these observations simply probe the validity of the weak equivalence principle[34, 594, 595] and cannot discriminate among alternative theories of gravity. This point was explicitly verified in Ref. 594 where the gravitational potential in unscreened $f(R)$ gravity was computed as $4\Psi_N/3$, and the integration boundaries in Eq. (35) were adjusted to smaller halo masses compared to those used in the GR calculation. This reflects the fact that what is measured is the dynamical potential $\Psi \sim G_{\text{matter}} M$. The degeneracy between G_{matter} and the dynamical masses is visible in the right panel of Fig. 11, where the solid black line corresponding to $f(R)$ gravity is very similar to the GR prediction (dotted black line).[p] Nonethe-

---

[p]For a thorough study of the gravitational redshift in $f(R)$ and symmetron models see Ref. 602.

less, supplementary information can break this degeneracy, effectively promoting the velocity shift profile to a test of gravity. For example, changes in $\Delta(R)$ occur due to modifications to the cluster-mass function and halo profiles.[594] Moreover, in some theories of gravity $\Phi_{\text{lens}} \neq \Psi$ and lensing data could be used to define the integration boundaries in Eq. (35).

As a concluding remark note that gravitational redshifts can also in principle be extracted from X-ray spectra of the intracluster medium, with the advantage that owing to the large number of particles only a relatively small number of systems is required to keep the noise within acceptable levels.[592, 596] Although such measurements are beyond the reach of current X-ray observatories,[603, 604] the excellent spectral resolution of future X-ray telescopes will make this possible.[605]

5. Future Cluster Probes

We conclude this chapter by shortly reviewing tests of gravity probing physical processes in and around galaxy clusters, as well as their dynamical and structural properties, that have not yet been consistently applied to observations. The motion of galaxies and gas in proximity or beyond the virial radius of a galaxy cluster can provide powerful diagnostics of modified gravity theories, in that these regions are only marginally affected by screening mechanisms. In addition, the presence of a fifth force can induce changes in the rotation and shape of galaxy clusters. Some of these probes can already take advantage of available data (e.g. gas mass fractions), while others require high-quality measurements from the next generation of X-ray, SZ, imaging, and spectroscopic surveys to produce competitive constraints on infrared modifications of GR.

5.1. *Cluster gas mass fraction and mass-observable scaling relations*

The observed fraction of the X-ray emitting gas mass to the total mass (f_{gas}) in hot, luminous, massive, dynamically relaxed clusters has traditionally been employed to constrain the expansion history of the universe out to $z \approx 1$ as well as the mean matter content at $z = 0$, i.e. Ω_m.[508, 539] However, a modified gravity force could affect the temperature of the intracluster medium used to measure the total mass, leading to an inferred mass larger than the true mass. The X-ray luminosity of a cluster depends instead on its gas density, and thus its gas mass, which is proportional to the true cluster mass times the ratio of underlying background densities of matter species, baryons and dark matter, Ω_b/Ω_m. Modifications of gravity that obey the equivalence principle leave this quantity unchanged, allowing a more direct connection between the X-ray luminosity and the true total mass of a cluster. Therefore, if such a modified gravity scenario occurs but one assumes ΛCDM, constraints on Ω_b/Ω_m obtained from f_{gas} data at low-z would be in tension with those, for example, from CMB observations at high-z, where GR is restored (see further details in Li *et al.* (2016)).[606]

The mass–temperature and mass–luminosity scaling relations can also be altered by modified gravitational forces. The extent of the deviations from standard gravity scaling relations depends on the modification of the background, the coupling between matter and the scalar field, and possibly the mass of the cluster and its environment. At present, results from hydrodynamical simulations have been investigated for a subset of scalar–tensor theories.[359, 574, 607]

5.2. *Cluster galaxies kinematics*

The peculiar velocities of galaxies carry valuable information on the nature of the gravitational forces shaping the large-scale structure of the universe[406, 608–610] (see also Chapter 5 and references therein). Measurements of the coherent infall motion of galaxies towards massive galaxy clusters give access to tests of gravity on scales 2–$20\,h^{-1}\,\mathrm{Mpc}$, hence exploring the transition between the linear and nonlinear regimes. In particular, the joint probability distribution of the galaxies' projected positions and velocities probes the dynamical potential Ψ of the cluster, which can then be compared to the lensing potential Φ_{lens} obtained from weak-lensing observations. Predictions for this phase-space distribution have been either developed from a semianalytical approach based on the halo model[611, 612] or directly calibrated against cosmological simulations, therefore including in the latter case the full information on cluster-galaxy cross-correlations in redshift space.[613, 614] With their large cluster and galaxy samples, future Stage IV overlapping imaging[235, 615, 616] and spectroscopic surveys[617, 618] will greatly improve the signal-to-noise ratio of these kinematic observables, which in turn will allow stringent constraints on the properties of nonminimally coupled scalar fields. However, one should keep in mind that both modeling strategies for the phase-space distribution were validated against dark matter-only cosmological simulations, using halos or particles as proxy for the galaxies. Thus, a solid understanding of the impact of baryonic physics on the infall motion of galaxies, as well as of its variation across different galaxy populations, will be crucial to avoid undesired systematic uncertainties,[619] which ultimately can only be assessed with the aid of hydrodynamical simulations.

The random motion of cluster galaxies on the scales of 0.3–1 virial radii also has information on gravity. Starting from the phase-space configuration of these galaxies methods such as the escape velocity edges can be applied to reconstruct the dynamical potential profile of the parent galaxy cluster.[620] One can then compare the inferred averaged potentials for two separate cluster samples, one for high-mass objects, $\langle \Psi_{\mathrm{high}} \rangle$, and the other for low-mass systems, $\langle \Psi_{\mathrm{low}} \rangle$. Modifications of gravity endowed with mass-dependent screening mechanisms (e.g. chameleon screening) predict smaller $\langle \Psi_{\mathrm{high}} \rangle / \langle \Psi_{\mathrm{low}} \rangle$ ratios compared to GR, a fact that can be used to constrain these theories.[621] This methodology has the attractive advantage of dividing out both projection effects and theoretical inaccuracies, and forecasts for a survey like DESI (namely the Dark Energy Spectroscopic Instrument)[617] suggest that this probe can differentiate GR from $f(R)$ theories with $|f_{R0}| \approx 10^{-6}$ at the 95% confidence level.

Galaxy clusters form from large positive fluctuations in the primordial matter density field. The size of these overdensities initially inflates at an ever slower rate compared to the background expansion, until they reach a point when the self-gravitational pull completely decouples their evolution from the Hubble flow. At this stage the proto-clusters have reached their maximum size, the turn-around radius, and a phase of collapse and virialization follows. Idealizing galaxy clusters as spherical overdensities, in a flat ΛCDM cosmology an upper bound on the turn-around radius can be derived, which reads[622, 623]

$$r_{ta,max} = \left(\frac{3GM}{\Lambda c^2} \right)^{1/3}, \tag{37}$$

for a cluster of mass M. Equivalently, this radius can be interpreted as the maximum distance from the cluster center where the velocity of the infalling matter is equal and opposite to the Hubble speed, effectively remaining motionless in the cluster-centric rest frame. However, clusters are far from being perfectly spherical isolated objects, and the upper bound equation (37) should be really interpreted as a limit on the expectation value of the averaged turn-around radius. Occasionally this bound is violated by individual systems, and the probability of such occurrences can tell us something about the underlying theory of gravity. In fact, since the turn-around radius reflects how far from the cluster center the background acceleration can resist the gravitational attraction of the cluster, any additional fifth force would change the likelihood of bound violation.[624] Observationally, one needs to find filaments around galaxy clusters and measure the velocity profile along these structures using galaxies as tracers.[625, 626] For the purpose of constraining departures from the standard law of gravity, one should bear in mind that the nature of the screening mechanism determines the evolution of the scalar field in the different nonlinear structures of the cosmic web.[627, 628] More specifically, the Vainshtein screening is inactive in filaments regardless of their density, and tests based on the odds of bound violation can exploit this feature to distinguish it from other mechanisms (e.g. the chameleon screening) that are largely insensitive to the morphology of the environment.

5.3. *Internal properties of galaxy clusters*

Extensions to the laws of gravity can also affect the internal properties of galaxy clusters, such as their bulk rotation and ellipticity. For cosmologically viable and yet interesting modifications (e.g. $10^{-6} \lesssim |f_{R0}| \lesssim 10^{-5}$ and $1 \lesssim H_0 r_c \lesssim 10$, for $f(R)$ gravity and nDGP, respectively) the changes induced in these features are so minute that very large survey volumes are required to detect any signal with high enough statistical significance. Planned Stage IV experiments[235, 615, 616, 629, 630] will map wide areas of the sky with unprecedented depth allowing target statistical uncertainties of only a few percent. Assuming that systematics can be controlled at a comparable level, the internal properties of galaxy clusters can then provide a novel complementary probe of gravity on megaparsec scales.

The anisotropic shape of the host halos of galaxy clusters and groups can imprint a directional dependence on the efficiency of screening mechanisms. One should then expect systematic changes in the ellipticities of such halos induced by variations in the amplitude of the fifth force with direction. However, the extent and evolution of these modifications is complex and quite sensitive to the theory of gravity as well as to the nature of the screening mechanism,[285,398,631,632] so much so that for the Vainshtein screening no effect can indeed be observed.[632] The structural parameters describing the shape of galaxy clusters can be measured from gravitational lensing maps,[633,634] inferred from the X-ray emission of the intracluster medium,[635] or derived from a combination of imaging, X-ray, and SZ data capable of breaking key degeneracies.[636–638] Eventually, robust constraints on modified gravity theories from galaxy cluster ellipticities will only be possible with a solid quantification of the bias caused by, among other systematics, baryonic effects,[639] halo substructures,[640–643] and interlopers.[638]

The modified growth of structure in nonstandard gravity can also alter the halo concentration of galaxy clusters and groups, with changes strongly dependent on the details of the modification. Theories with linear scale-independent growth and screening mechanisms regulated by the local matter density (e.g. Vainshtein) preserve the standard power-law trend of the concentration–mass relation,[422] yet with different amplitudes and slopes.[253,344] This remains valid even in the absence of a screening mechanism as long as the linear growth is identical for all scales.[253] On the other hand, for theories of gravity characterized by a linear scale-dependent growth and a screening mechanism controlled by the local gravitational potential (e.g. chameleon, symmetron, dilaton) the concentration–mass relation reveals a more complex behavior, typically described by a broken power-law.[367] By employing lensing data for a small sample of galaxy clusters, a recent analysis of their halo concentrations found no evidence of deviations from GR.[344] In the future, thanks to redshift evolution information for large cluster samples becoming accessible with forthcoming surveys, the concentration–mass relation will be able to reach its full potential as a probe of gravity.

In some cases galaxy clusters have been observed to possess a coherent rotational velocity component (see e.g. Ref. 644 and references therein), which could result from the initial angular momentum of their primordial cloud, recent mergers, or interactions with close neighbors.[260,645] Observational techniques to detect this feature range from measuring the cluster-centric line-of-sight velocity of cluster galaxies to mapping distortions in the temperature and polarization of the CMB photons.[646] In this context, the effect of a scalar fifth force is that of shifting the overall rotational velocity distribution of a cluster sample to slightly higher values.[632,647] Given the smallness of the signal, the selection of fully unscreened systems (i.e. low-mass galaxy groups) from the next generation of cluster surveys[235,615,629] will be crucial to constrain departures from standard gravity at a competitive level. Furthermore, complementary information from high-quality spectra of cluster members[617] and of the intracluster gas,[648] together with high-resolution measurements of the thermal and kinematic SZ effects[630] will be necessary to reduce critical systematic uncertainties.

The radial distribution of dark matter and galaxies in clusters present a characteristic feature marking the physical scale where the accreted material is turning around for the first time after infall. This is known as the splashback radius.[649–651] Its location can be extracted from lensing and galaxy profiles,[652–654] with the two types of measurements showing interesting differences. In fact, contrary to dark matter particles, galaxies experience dynamical friction when moving through the cluster's halo.[655, 656] The effect is larger for more massive galaxies, and in turn reduces their splashback radius.[657] The higher infall velocities in modified gravity increase the splashback radius of dark matter particles while simultaneously weakening the impact of dynamical friction on the motion of galaxies, which effectively changes the relation between lensing and galaxy observations.[658] For viable theories, departures from splashback predictions in GR are of the order of a few percent, whereas current measurements in galaxy profiles are limited by systematic uncertainties at the 10% level.[653] Similar uncertainties, although statistical in nature, dominate lensing measurements at present.[654] With surveys like LSST,[615] Euclid[235] and WFIRST[616] both measurement techniques will reach enough statistical power to distinguish cosmologically interesting models, which upon careful control of systematics will make the splashback radius a valuable addition to the numerous tests of gravity on cluster scales.

Acknowledgments

We thank Adam Mantz and Steven Allen for helpful comments on this manuscript, and Yan Chuan Cai, Shadab Alam, and Tilman Tröster for useful discussions that greatly improved the clarity of this review. Matteo Cataneo acknowledges support from the European Research Council under the Grant No. 647112. David Rapetti was supported by a NASA Postdoctoral Program Senior Fellowship at the NASA Ames Research Center, administered by the Universities Space Research Association under contract with NASA.

Chapter 7

Towards Testing Gravity with Cosmic Voids

Yan-Chuan Cai

Institute for Astronomy, University of Edinburgh, Royal Observatory,
Blackford Hill, Edinburgh, EH9 3HJ, UK
cai@roe.ac.uk

Cosmic void as a cosmological probe is reviewed. We introduce the theoretical background and recent development in observations of cosmic voids, focusing on its potential to test theorists of gravity. Four observables about cosmic voids are introduced: void number counts, weak gravitational lensing, redshift-space distortions and ISW around voids. We discuss opportunities on their application to current and future galaxy surveys and highlight challenges for this subject.

Keywords: Cosmology; cosmological constant; dark energy; modified gravity; cosmological models.

1. Introduction

In light of the recent detection of gravitational waves from the binary neutron star merger GW170817 and simultaneous measurement of its optical counterpart GRB170817A, several popular classes of model of gravity beyond the ΛCDM are ruled out,[38, 41–44] but many other models remain viable and would affect the growth of large-scale structure of the Universe differently, such as Brans-Dicke type theories including $f(R)$ gravity,[86] the normal-branch Dvali-Gabadadze-Porrati (nDGP) model,[659] and more complex variants of dark energy within the standard gravity. Although these surviving models such as $f(R)$ gravity and nDGP are not completely satisfactory alternative to dark energy for cosmic acceleration, the key general physical properties, the screening mechanisms[628, 660] are likely to remain relevant and necessary for further development of alternative gravity models at cosmological scales. This is in part due to the fact that, by far, there is no strong evidence for a deviation from General Relativity (GR) from solar system tests and from conventional cosmological probes of the late-time large-scale structure which focus on high-density regions of the Universe. For example, weak gravitational lensing, galaxy clustering including redshift-space distortions and cluster abundances are, in general, weighted by the locations of galaxies, which are thought to form on peaks of matter distribution. Theoretical attempts trying to explain the cosmic acceleration by changing the law of gravity will have to pass tests of gravity delivered from these conventional observations. These stringent requirements leaves relatively little space for the development of alternative models. In essence, any viable model should converge to GR in high-density regions where most of these observations have set constraints on. This is why screening mechanism is usually needed to suppress any possible deviation from GR in those circumstances, which then leaves possible deviation from GR in the regime beyond the radar of conventional probes. For example, in large-scale cosmic voids or at the outskirt of galaxy clusters, behaviour of gravity may be allowed to be different.

At the mean time, as mentioned in Chapter 1, recent observations have indicated several low-level tensions between large-scale structure and the cosmic microwave background (CMB). For example, the constraints on Ω_m-σ_8 set by weak gravitational lensing from the late-time Universe is in 2-3σ tension with constraint from CMB temperature fluctuations from Planck.[10, 11] Likewise, the measurement from the brightest Sunyaev-Zel'dovich (SZ) cluster number counts seems to prefer lower ranges of $\Omega_m - \sigma_8$ than the best-fit Planck cosmology from the CMB temperature power spectrum. Perhaps the most prominent tension is the constraints for the local Hubble expansion rate H_0 delivered by CMB measurement versus that from the late-time supernovae (SNe) data, which is reported to be at the level of 3-4σ.[12, 661] Several anomalies in the CMB have also been reported, such as the Cold Spot in the southern Galactic plane,[662, 663] the alignment of quadrupole versus octopole in the CMB,[664, 665] the 'power deficit' at the low-ℓ range of the CMB power spectrum,[537] and the asymmetry of clustering between the north and the south Galactic plane[666, 667] – all these are confirmed by the Planck data.[668] While the reason and significance for these tensions and anomalies are disputed, i.e. it could be due to observational systematics, before they are verified or disproved by future observations or with new convincing analysis for the data, it remains possible that they indicate the imperfection of the standard model and thereby chances for new physics. There is yet not a satisfactory alternative theory of gravity which can convincingly explain those anomalies and tensions. Nonetheless, for the concern of cosmology, realising that GR being the undying governing law of gravity for the whole Universe is an assumption which comes from extrapolating the law extracted from the near-Earth environment to cosmological scales, it remains important to test the equivalence principle and GR at these scales. More generally speaking, scale is only one dimension of concern for the test of gravity. Other circumstances such as time evolution, density of local environment or the properties of local potentials are also relevant situations where the law of gravity needs to be tested.

A general feature for the surviving modified gravity models is that they often rely on screening mechanisms to suppress the fifth force in high-density regions. This is true for both the $f(R)$[87, 669] and nDGP models.[659] The former features a chameleon screening and the latter the Vainshtein screening mechanism.[58, 62] These non-linear behaviour of gravity inevitably alters structure formation in an environmentally dependent manner, i.e. in the regime where the fifth force is suppressed, gravity is back to GR and structure formation remains similar to that of the ΛCDM; in places where the fifth force is unscreened, such as in low-density regions in the

$f(R)$ model,[545, 628, 670] or outside the Vainshtein radius (see Chapter 1) in nDGP model,[545, 627] structure formation will be different due to the fifth force. This naturally leads to two branches of research in using cosmic voids to test theories of gravity: (1) on large scales, looking directly into the impact of the fifth force on the observable properties of cosmic voids; (2) at small scales, using cosmic voids as the environment for properties of dark matter haloes, galaxies or stars. This chapter focuses on recent research activities in the first area.

2. Cosmic Voids

Cosmic voids are under-dense regions of the Universe in terms of dark matter and galaxy distribution. In principle, the distribution of voids of different sizes is related to the N-point correlation function of the density field of the Universe, and is encoded with information beyond the variance of the field. Therefore, it can be used as an alternative to characterise the nature of large-scale structure beyond the galaxy/matter two point statistics.[671–675] Moreover, the density profiles of voids encode information about cosmological model and are shaped differently by different laws of gravity. These lay the basis for why it is useful to study voids, and in general, voids and clusters for cosmology. Therefore, the evolution of individual void and the statistical distribution of voids are important to be understood.

2.1. *Voids as individual objects*

The evolution of individual void is addressed by the pioneering work of Refs. 443, 676, 677, with the spherical model. In the GR ΛCDM universe, the acceleration of a spherical matter shell is solely governed by the total mass within the radius r, and the background dark energy density ρ_Λ.

$$\frac{\ddot{r}}{r} = -\frac{1}{6M_{\rm Pl}^2} \left[\rho_{\rm v} - 2\rho_\Lambda \right],\tag{1}$$

where $\rho_{\rm v}$ is the mean matter density within r; ρ_Λ is the background dark energy density; $M_{\rm Pl} = 1/\sqrt{8\pi G}$ is the reduced Planck mass with G being the gravitational constant. $\dot{r}$ represents the time derivative, or velocity, and so $\ddot{r}$ is the acceleration. *If the initial profile is a spherical top-hat under-density with radius $R_{\rm v}$, as illustrated in Fig. 1, the under-density will expand faster than the Hubble expansion. The matter shell right outside $R_{\rm v}$ will expand slightly slower than the matter shell at $R_{\rm v}$. When the inner shell overtakes the outer shell, shell crossing occurs. In an $\Omega_{\rm m} = 1$ universe, this happens when the radius of the under-density expands by a factor of 1.7, and the density contrast of the top-hat void is $\delta = -0.8$. These values have little dependence on cosmology.* The depth of the top-hat void at shell crossing has been taken as one of the conventional definitions for voids, and has been applied to predict the size distribution function of voids using the excursion set theory.[443] With more realistic initial conditions form N-body simulations, this model has been

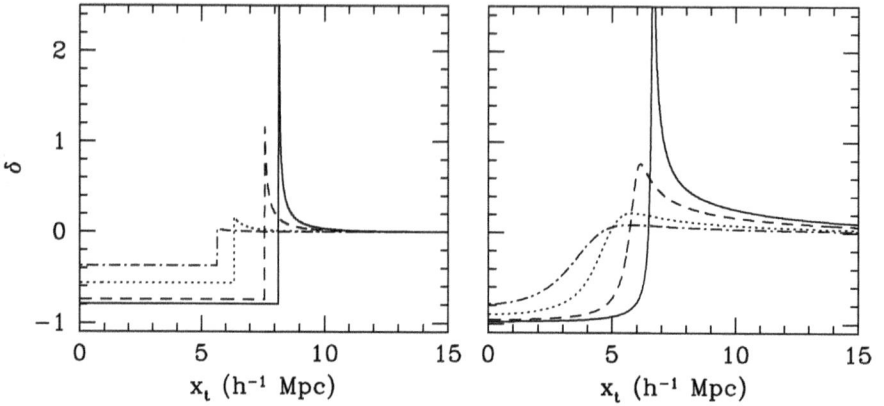

Fig. 1: Figure taken from Fig. 3 of Ref. 443. The evolution of the void density profile from its initial condition till shell-crossing, starting with a spherical top-hat under-density (left) and a void with an angular averaged SCDM profile (BBKS, Eq. (7.10)).[679] Different line styles represent different epochs: $a = 0.05, 0.1, 0.2$ and 0.3. See also Fig. 6 of the recent review article.[680]

shown to be successful in tracking the evolution of void density profiles for large voids.[678]

2.2. Void distribution function

Like the dark matter halo mass function, the statistical distribution for void population can be predicted with the excursion set theory[381] (see also Chapter 4 for more details). The basic idea of the excursion set theory is that for a Gaussian random field, the variance $S = \sigma^2(R_i)$ when smoothed by a filter of the size R_i, is specified given the linear matter power spectrum and the window function. A sequence of the density fluctuation $\delta(S)$ given by a series of increment in S, which is equivalent to decrease of the smoothing scale, follows a Brownian random walk, if the filtering of the density field is a k-space top-hat.[381, 443, 681] A random walk up-crossing the halo formation barrier δ_c corresponds to the formation of a halo, and one that down-crosses the void formation barrier δ_v represents a void. The probability density of a walk first crossing the barriers between $[\sigma, \sigma + d\sigma]$ is expressed as:

$$df(\sigma, \delta_{c,v}) = \frac{\delta_{c,v}}{\sigma}\sqrt{\frac{2}{\pi}}\exp\left(-\frac{\delta_{c,v}^2}{2\sigma^2}\right)\frac{d\sigma}{\sigma}, \tag{2}$$

where $\delta_c = 1.686$ for haloes and $\delta_v = -2.81$ for voids are the linearly extrapolated densities at shell-crossing, both coming from the spherical model assuming a top-hat initial density. The number density of haloes/voids as function of mass $n(M)$, or equivalently radius R_i, is then

$$\frac{dn(M)}{dM} = \frac{\bar{\rho}\,df(\sigma, \delta_{c,v})}{M dM} \tag{3}$$

where M is the mass within the spherical top-hat for either voids or haloes, and $\bar{\rho}$ is the mean matter density of the Universe. Once the linear matter power spectrum is specified, the abundance of haloes or voids can be predicted with the above equation. However, applying the above equation to void leads to violation of volume conservation. Individual voids expand as they evolve, by the time of shell-crossing, their radius would have increased by a factor of 1.7 and a factor of ~ 5 for their volume. The sum for the volumes of all voids therefore exceed the total volume of the Universe. An improved version of the above model was given by Jennings *et al.*[682] by imposing volume conservation, with which the agreement between N-body simulation and the new model is shown to improve.[682]

It is worth noting that the shell-crossing densities mentioned above are derived from the setting that the initial condition is a spherical top-hat, which is unrealistic in the real Universe. This assumption affects the shell-crossing density for voids more than for haloes, as the former is directly related to the derivative of the initial profile and the latter is governed by the total mass within a certain radius. Therefore, it is unsurprising that the predictions for the abundance of voids from this setting does not agree very well with simulations or observations. Nonetheless, this model provides a useful guidance for the general physical picture of voids in the following aspects: (i) voids are expanding relative to the background; (ii) as voids expand, they become emptier; (iii) the distribution of voids in terms of their sizes follows some Press-Schechter-type[380] function.

2.3. *Voids in simulation and observations*

In the real Universe, voids have complex shapes and different density profiles. Due to this nature, there is no good consensus for the definition of voids in simulations and observations. For example, there is no clear-cut answer for where the boarder of a void should be due to their irregular shapes. However, it is relatively unambiguous that they should be under-dense at their centres. For example, Fig. 2 compares the void regions identified in simulations using three different void finding algorithms, the spherical void finder,[683] the watershed void finder[684] and ZOBOV.[685] We can see that despite having different boundaries, the largest voids are found in all the three different algorithms. Broadly speaking, void finding algorithms can be classified into two categories: the first uses the distribution of mater or tracers of matter to define under-dense regions, such as the three void finding algorithms mentioned above. The second uses velocity divergence to identify outflow regions, e.g.[686] Like the example give in Fig. 2, the strongest characteristic of voids being under-dense usually converges among different ways of finding voids. But the quantitative measurement for the abundance of voids as function of radius may differ significantly. This is one aspect of voids one needs to keep in mind when applying it to cosmology or astrophysics. More detailed comparisons of different algorithms can be found in Ref. 445.

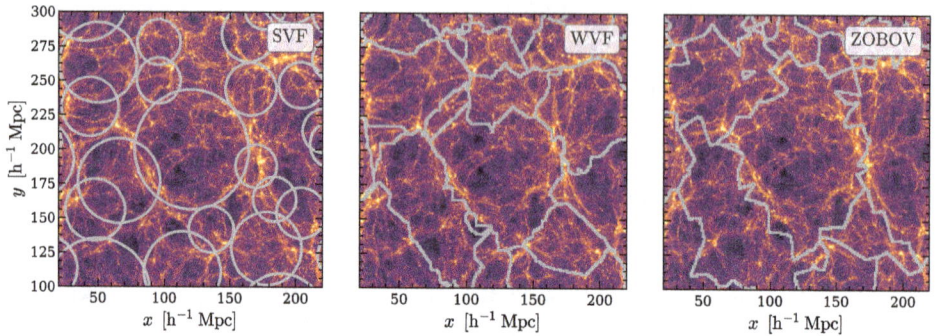

Fig. 2: Voids identified in simulations using dark matter particle distribution. Grey curves representing boundaries of voids are drawn on top of the projected density of dark matter. Three different void finding algorithms are shown for comparison: the spherical void finder,[683] the watershed void finder[684] and the zobov void finder.[685] Despite some detailed differences for the exact boarders of voids, the largest void region in the simulation are consistently found by all the three algorithms.

In observations, void catalogues have been generated from galaxy redshift surveys, and one of the most exploited dataset is in the SDSS area.[687–691] These have been used to study their imprints on the CMB via the integrated Sachs-Wolfe (ISW) effect,[689, 692–701] constrain cosmology via the Alcock-Paczyński Test (AP) test,[702–705] and measure the linear growth of structure[706–708] and weak gravitational lensing signal around voids.[709–713] On top of the above, there have been a number of astrophysical and cosmological applications of voids being proposed. These include: void ellipticity as a probe for the dark energy equation of state,[714–717] void abundances and profiles for testing theories of gravity[256, 257, 392, 670, 718–721] and constraining neutrino masses,[722] the nature of dark matter,[723] baryon acoustic oscillations in void clustering.[724, 725]

In this chapter, four observables of voids will be reviewed: void abundance, gravitational lensing of voids, redshift-space distortion around voids and ISW around voids, with a focus on their potentials to test theories of gravity.

3. The Fifth Force in Cosmic Voids

A general feature of modified gravity is that the presence of a fifth force, which changes the strength of gravity and violates the equivalence principle in certain circumstances. As shown in Refs. 58, 256, 726 for chameleon screening mechanism, violation of the equivalence principle for theories of gravity, often shown as a deviation of the inertial mass from the gravitational mass in under-dense regions, could be of the order of unity. This is also the case for the Cubic Galileon model,[721] which comes with the Vainshtein screening mechanism.[62]

In this section, we use a specific example of a coupled scalar field model to explain how the chameleon screening mechanism works in spherical structures, and illustrate how it affects large-scale structures, particularly voids. Starting from the

Lagrangian of this model

$$\mathcal{L} = \frac{1}{2}\left[M_{\mathrm{Pl}}^2 R - \nabla^a\phi\nabla_a\phi\right] + V(\phi) - C(\phi)\mathcal{L}_{\mathrm{M}}, \tag{4}$$

in which R is the Ricci scalar. $\mathcal{L}_{\mathrm{M}}$ lumps up the Lagrangian densities for dark matter and standard model species including radiation and neutrinos, except for dark energy, which in this model, is replaced by the scalar field ϕ, with $V(\phi)$ being its potential. The field is assumed to couple universally with dark matter and standard model species, characterised by the coupling function $C(\phi)$. Given the functional forms for $V(\phi)$ and $C(\phi)$, the coupled scalar field model is specified. To survive observational constraints mentioned in the introduction, the choice of these two functions for the model can be narrowed down. Qualitatively, we want the model to have the same expansion history as the ΛCDM, and to have the behaviour of gravity back to GR in high density regions. The model investigated by Refs. 294, 366 is designed to satisfy such requirements, with

$$C(\phi) = \exp(\gamma\phi/M_{\mathrm{Pl}}), \tag{5}$$

and

$$V(\phi) = \frac{\rho_\Lambda}{\left[1 - \exp\left(-\phi/M_{\mathrm{Pl}}\right)\right]^\alpha}. \tag{6}$$

In the above ρ_Λ is a parameter of order the present dark energy density, γ, α are dimensionless parameters controlling the strength of the coupling and the steepness of the potential respectively (see Fig. 1 of Ref. 294 for an illustration of the potential).

When having $\alpha \ll 1$ and $\gamma > 0$ as in Refs. 294, 366, which ensure that $V_{\mathrm{eff}}(\phi)$ has a global minimum close to $\phi = 0$ and that $\mathrm{d}^2 V_{\mathrm{eff}}(\phi)/\mathrm{d}\phi^2 \equiv m_\phi^2$ at this minimum is very large in high density regions, the above requirements are met in that: (1) ϕ is trapped close to zero throughout cosmic history so that $V(\phi) \sim \rho_\Lambda$ behaves as a cosmological constant; (2) the coupling is modulated by the local density ρ, i.e. the fifth force is strongly suppressed in high density regions where ϕ acquires a large mass, $m_\phi^2 \gg H^2$ (H is the Hubble expansion rate), and thus the fifth force cannot propagate far. The suppression of the fifth force also happens at early times when the density of the Universe is high, which naturally ensures that the model converges to GR in the early Universe. Therefore, the fifth force is mainly active at late times and in low density regions – voids. This model has similar environmentally dependent behaviour as the scalar field model first investigated by Ref. 58, and is an example of the chameleon models, which employs the chameleon mechanism (see Chapter 1) to suppress the fifth force.

The coupling of a scalar field with matter induces a fifth force, as shown by the geodesic equation for matter

$$\frac{d^2\vec{r}}{dt^2} = -\vec{\nabla}\Phi - \frac{C_\phi(\phi)}{C(\phi)}\vec{\nabla}\phi, \tag{7}$$

where $\vec{r}$ is the position vector, t the physical time, Φ is the Newtonian potential and $\vec{\nabla}$ is the spatial derivative; $C_\phi \equiv dC/d\phi$. The second term on the right hand side is the fifth force. The acceleration of a test particle is no longer solely provided by the gradient of the Newtonian potential, and part of it will be provided by the gradient of the scalar field, i.e., the fifth force.

To illustrate the physics of the fifth force in voids in the above model, one can set up an isolated spherical top-hat void, solve for its corresponding scalar field profile, and compute the profiles of the fifth-force. This has been been detailed in Ref. 256, and summarised here. The evolution equation of the spherical under-density (void) is the key relevant equation. In this model, it can be written (see Chapter 4) as

$$\frac{\ddot{r}}{r} = -\frac{1}{6M_{\rm Pl}^2}\left[\rho_{\rm v}(1+\eta) - 2\rho_\Lambda\right], \tag{8}$$

which differs from Eq. (1) by a factor of $(1+\eta)$, where η is the ratio between the fifth force and Newtonian gravity:

$$\eta = \frac{\sqrt{3a\Omega_\Lambda}\,\gamma\frac{d\psi}{d\tau}\Big|_{\tau=r/\lambda_{\rm out}}}{\frac{1}{2}\Omega_m\left(H_0 R_i\right)\left(ay_{\rm v}\right)^{-2}} \tag{9}$$

where $\psi = \phi/\phi_{out}$ and τ is the distance from the void centre normalised by the Compton wavelength of the scalar field at the background, $\lambda_{\rm out}$. The above equation can be solved iteratively with Eq. (8). This can be done for the entire density profile. Examples of the fifth force versus Newtonian force profile is shown in Fig. 3. A general feature for the chameleon model is that the fifth force is pointing outwards from the void centre, which will accelerate the expansion of voids, making them grow larger and deeper than their GR counterparts, as shown by the top and middle columns of Fig. 4. It is worth noting that for the same under-density, the expansion

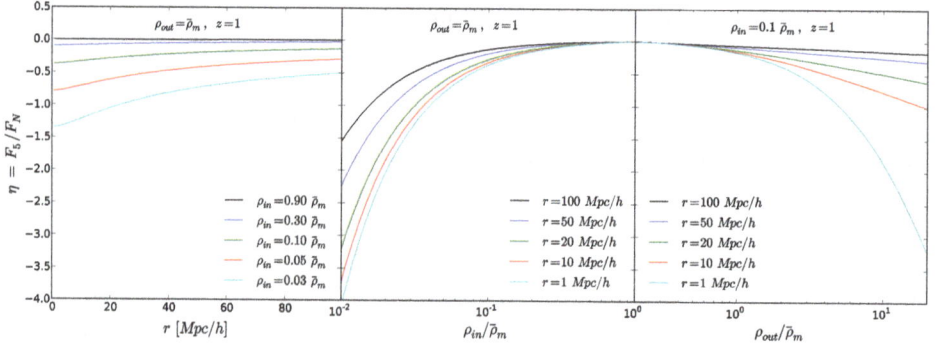

Fig. 3: Taken from Clampitt *et al.*[256] to illustrate the ratio between the fifth force versus Newtonian force η in the coupled scalar field model.[294, 366] *Left panel:* Variations of η with the spherical top-hat radius r. The exterior density is fixed to the cosmic mean today, $\rho_{\rm out} = \bar\rho_m$. Various values of interior density $\rho_{\rm in}$ are shown, with $\rho_{\rm in}$ decreasing from top to bottom. *Center panel:* The same, but for continuous variations of $\rho_{\rm in}$, fixed $\rho_{\rm out} = \bar\rho_m$ and various values of radius r, with r decreasing from top to bottom. *Right panel:* The same, but for continuous variations of $\rho_{\rm out}$, fixed $\rho_{\rm in} = 0.1\,\bar\rho_m$, and various values of radius r, with r decreasing from top to bottom.

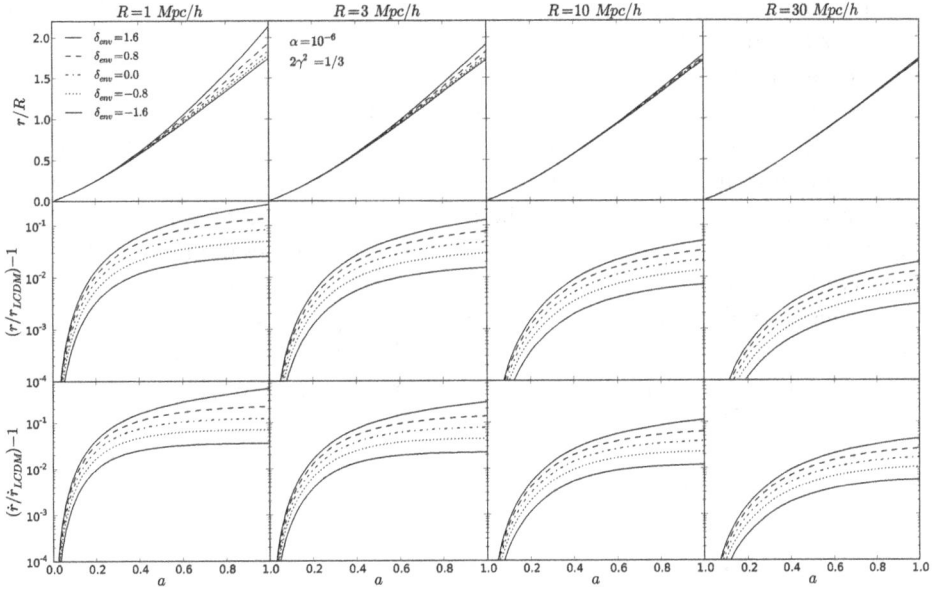

Fig. 4: Taken from Clampitt *et al.*[256] to illustrate the evolution of voids in the coupled scalar field model.[294, 366] *Top row:* void radius r in units of its initial comoving radius R, as a function of scale factor a. *Center row:* Fractional difference between the void radii in the coupled scalar field model versus their GR version of the same initial conditions. *Bottom row:* Fractional difference in the velocity. Columns show various values of initial comoving radius, $R = 1, 3, 10$ and $30\,\mathrm{Mpc}/h$, from left to right. Different values of the exterior density are shown, with δ_{env} decreasing from top to bottom. The largest deviations from GR occur for voids expanding within a larger over-dense region.

velocity of the matter shell is larger in the chameleon model than in GR, due to the larger acceleration it has been experiencing via Eq. (8).

From this physical picture, the following four observables are expected which can be used to test the effect of the fifth force.

- The fact that voids grow larger in chameleon models also means that they become emptier. The difference of matter distribution can be measured via gravitational lensing of voids. Perhaps more importantly, for models where there is order-unity difference between the lensing potential and the Newtonian potential, the gravitational lensing effect will respond to it directly, with potentially significant deviation from GR for the expected lensing signal. These will be addressed in Sec. 4.2.
- The fact the voids expand faster in chameleon models suggests that the velocity field around voids may provide a smoking gun to test these models. Moreover, since the velocity is the first time integral of acceleration, it responses more sensitively to the onset of the fifth force than the density field does. The information about the velocity field is contained naturally in the void-galaxy correlation func-

tion in redshift space, i.e., redshift-space distortion around voids. This will be discussed in Sec. 4.3.

- The difference for the evolution of individual voids also lead to statistical differences for the void population. One may expect small voids to merge and form larger voids, which will change the distribution function of the size of voids. The rate it occurs will be affected by the presence of a fifth force and lead to changes of the void population. Counting the number of voids as a function of their size is therefore another probe that can be used to test theories of gravity, which will be detailed in Sec. 4.1.

- The fact that the fifth force is suppressed in high-density regions and active in under-dense regions opens up a series of tests by comparing astrophysical phenomena in over-dense and under-dense regions and looking for the differences.[68, 69, 726–728] For example, the properties of dark matter haloes or galaxy clusters and groups, galaxies, gas and stars components in galaxies may all be different in different environment due to the fifth force.

4. Observables for Voids

4.1. *The size distribution function of voids*

Using the well-developed excursion set theory mentioned in Sec. 2.2 and Chapter 4, one can predict the number of voids passing a density threshold for GR and modified gravity models. Note however, if one takes shell-crossing as the void formation barrier, the corresponding linearly extrapolated density δ_v in modified gravity models may not necessarily be a constant. For example, because of the screening mechanism, δ_v can depend on scale and environmental density,[256, 729] and in some specific cases, shell-crossing does not occur at all due to the varying strength of the fifth force along the radial direction of a void, i.e. for top-hat voids, the outer matter shell may have more rapid total acceleration than its inner shell and so the latter will not be able to catch up with the former.

Nonetheless, one can take the default GR value of $\delta_v = -2.81$ as the common void formation barrier for different models, and compare the resulting abundances to the GR version. The example for the coupled scalar field model is presented in Fig. 5. Indeed, we expect the abundance of voids to be higher than in the GR model for the range of void radius shown in the figure. It is worth noting that the difference between the chameleon model and GR in terms of void abundance is nearly a factor of 10 greater that of the difference for the halo mass function. This suggests that the abundance of voids may be more sensitive than haloes for constraining theories of gravity. Predictions for the $f(R)$ and symmetron models from N-body simulations by[729] qualitatively confirms this. Although, it was also shown that when (mock) galaxies are used to define voids, it becomes more difficult to distinguish different models.[729]

A major challenge in using void abundance to test gravity is the ambiguity of void definition. To some extent, the conclusion may depend on the sample of

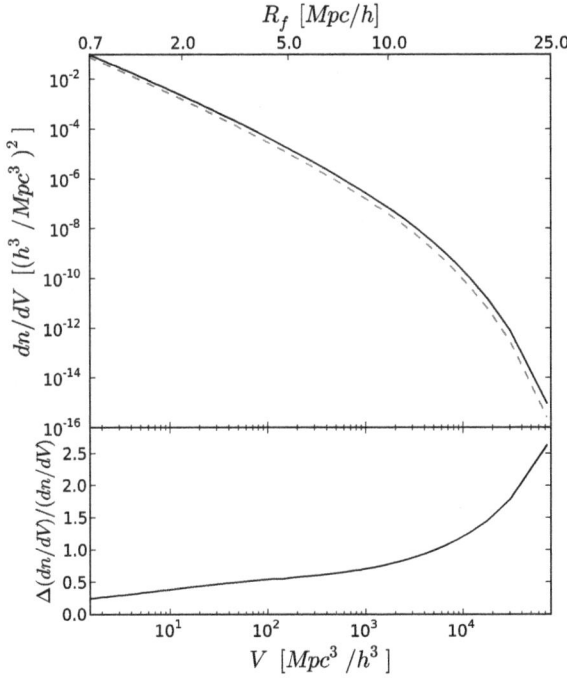

Fig. 5: From Clampitt *et al.*,[256] top-panel, comparing the void number counts as a function of their radius/volume between the coupled scalar field model (solid line) and the GR ΛCDM model (blue-dashed line). The bottom panels shows their fractional differences.

galaxies and the specific void finding algorithm used to find voids. This leaves space for comparing different methods to find the best signal-to-noise for distinguishing different models. To overcome this challenge, mock galaxy catalogues have been employed to compare with data. As a consistency test, agreements have been found between the simulated void abundance with a fixed ΛCDM cosmology versus data from the SDSS CMASS galaxy catalogue,[690, 701] although, a $\sim 3\sigma$ tension between data and simulation was noted in.[690] This may turn out to be another statistical fluke or due to unknown observational systematics. Future survey such as DESI[730] with its much larger survey volume will be able to tell more decisively if the tension remains.

The abundance of voids and clusters have also been combined to study cosmology. Using the extreme values statistics on the largest void from the 2MASS-WISE galaxy catalogue and cluster from the Atacama Cosmology Telescope (ACT), Sahlén *et al.*[731] were able to set constraints on Ω_m-σ_8 for a flat ΛCDM universe, as shown in Fig. 6. It has also been demonstrated that combining void abundance with cluster abundance is a powerful way to break degeneracies among cosmological parameters and potentially offers tighter constraints on modified gravity via the linear growth rate index γ from $f = \Omega_m^\gamma$.[732] γ is shown to be confined within a narrow range of $\gamma \approx 0.55$ in GR[733] (see also Chapter 2).

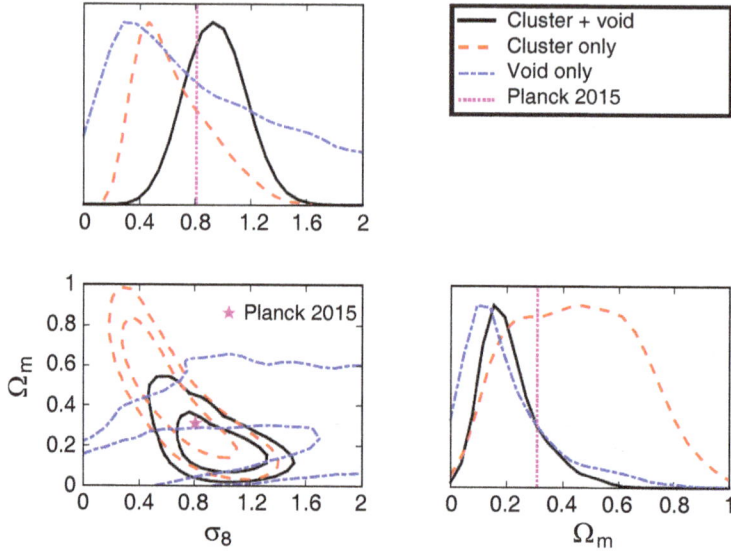

Fig. 6: From,[731] constraints on Ω_m-σ_8 for a flat ΛCDM universe using extreme value statistics with the largest void from the 2MASS-WISE galaxy catalogue and the largest cluster from the Atacama Cosmology Telescope (ACT). 68% and 95% confidence levels are shown for void (blue) and cluster (red) individually as well as for the combined case (black).

4.2. Lensing around voids

For individual voids, the density profile of voids will be altered by the presence of a fifth force. This can be measured using gravitational lensing around these objects. Weak gravitational lensing by large-scale structure of the Universe has been well studied via cosmic shear, γ, the distortions of background galaxy images by the foreground structures. This signal can be thought of being contributed by the lensing signal from peaks and troughs of the density field, and is sourced by the lensing potential

$$\psi_{\text{lens}} = \frac{1}{c^2} \int \frac{D_{ls}}{D_s D_l} (\Phi + \Psi) dz, \tag{10}$$

with Φ and Ψ being respectively the time and space parts of the metric potentials in the perturbed FRLW metric, and z being the redshift. D_l, D_{ls} and D_s are the line-of-sight angular diameter distances of the lens, the source and that between the lens and the source respectively. The 2D Laplacian of the lensing potential is related to the lensing convergence κ via $\kappa = \frac{1}{2}\nabla^2 \psi_{\text{lens}}$. In the GR limit where $\Phi = \Psi$, it can be expressed as

$$\kappa = \frac{3H_0^2 \Omega_m}{2c^2} \int \frac{D_l D_{ls}}{D_s} \frac{\delta}{a} dz. \tag{11}$$

The lensing convergence is therefore related to the projected mass density along the line of sight. It is related to the tangential component of the shear γ_t via

$$\gamma_t(R_p) = \kappa(< R_p) - \kappa(R_p) = [\Sigma(< R_p) - \Sigma(R_p)]/\Sigma_{\text{crit}}, \qquad (12)$$

where $\kappa(R_p)$ and $\Sigma(R_p)$ are the convergence and projected mass density at the projected radius R_p; $\kappa(< R_p)$ and $\Sigma(< R_p)$ are the averaged convergence and projected mass density within R_p (see Figs. 7 and 8 for examples). $\Sigma_{\text{crit}} = \frac{c^2}{4\pi G} \frac{D_s}{D_{ls} D_l}$ is the geometry factor. The density profiles of voids (or clusters) can then be measured by stacking the shear signal of the background galaxies around them. Like galaxy-galaxy lensing, this usually requires identifying voids using 3D positions of galaxies and having lensing source galaxies behind the voids – a lensing survey overlapping with a spectroscopic redshift survey. Attempts have been made to identify voids using lensing photo-z surveys alone,[713, 734] or to measure the projected under densities on the sky plane.[735–739] The first forecast for the detectability of void lensing via stacking was made by Krause *et al.*[709] and Higuchi *et al.*[710] Detections of void lensing signal has been achieved by Ref. 711, 712 in the SDSS area and the DES science verification data.[713] CMB lensing of voids has also been detected by Refs. 701, 740. These detections paves the way for using the signal to test theories of gravity using future galaxy redshift surveys and lensing surveys.

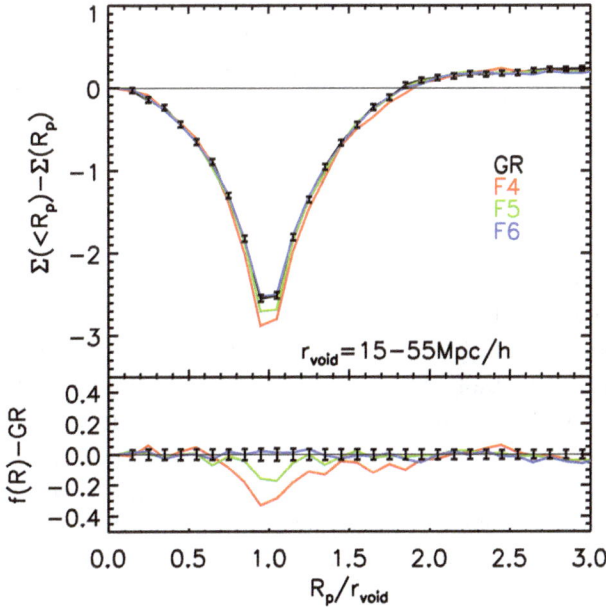

Fig. 7: From Cai *et al.*,[670] the lensing tangential shear profiles around voids identified in N-body simulations of ΛCDM model and $f(R)$ models, with increasing strength deviation from GR labels as F6, F5 and F4. The bottom panels shows the differences from GR. $\Sigma(< R_p) - \Sigma(R_p)$ is proportional to the surface mass density within the projected radius of R_p to which we subtract the surface mass density at R_p.

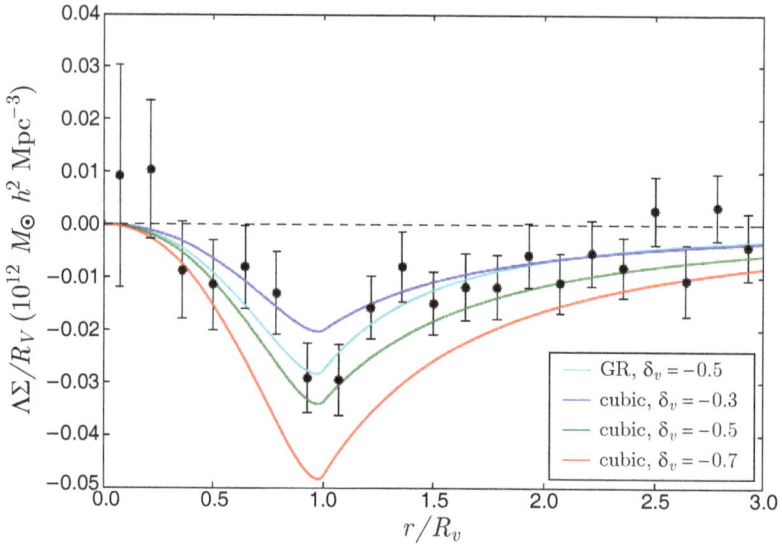

Fig. 8: From Baker *et al.*,[741] the observed lensing tangential shear profiles (data points with errors) around SDSS voids and SDSS LRG galaxy image data. Colour lines are model predictions from GR and the Cubic Galileon model with different void central density.

For chameleon models which have not been ruled out such as $f(R)$ gravity,[72] the effect of the scalar field on the lensing potential is minor and so the lensing signal is unchanged if the density profile of voids are the same. In these models, the fifth force affects the lensing signal indirectly via the change of the dark matter density profile (Fig. 7). For Galileon models (or the massive gravity model), which features the Vainshtein screening mechanism, the relation between the lensing potential and the density perturbation can be strongly modified and the lensing signal around voids can be very different even the mass density profile is the same as in GR. For example, in the Cubic Galileon model, the lensing effects around voids of the same profiles as in GR can be a factor of two larger,[721, 741] see Fig. 8. The difference is dominated by the direct effects of the fifth force on the lensing signal.[721] For this kind of model, lensing of voids is potentially one of the most powerful probes to distinguish them from GR. Even the Galileon model has been ruled out,[742] it remains interesting for the community to explore this possibility for compelling models of this kind. It is also worth noting that for the part of the parameter space in the cubic and quartic Galileon models, no solution for the scalar field can be found in deep voids at the late time,[300] which is a caveat for the model. This problem persist even when the the quasi-static approximation is dropped.[343]

In observations, lensing signal around voids is usually sample-variance limited,[446, 709, 710] i.e. the signal-to-noise can be improved by having larger survey areas. The measurement is best conducted with a lensing survey over the same area as a spectroscopic survey, within which voids can be identified using the 3D positions of galaxies (although 2D voids can also be identified with a lensing survey alone with photo-z, see Ref. 713). Near future surveys such as LSST and Euclid will meet such a requirement for at least part of their areas. A recent forecast for the detectability of the $f(R)$ model has been made in,[446] finding that the distinguishing power with LSST and Euclid is still limited to $F5$, which is compatible to other cosmological tests over similar scales[446] (see Table 1 of Ref. 239 and references therein). With a similar survey setting and using 3D voids, the constraining power for the nDGP model is found to be somewhat weaker.[743]

A recent development on lensing around voids has generalised the idea to troughs, which is to measure the lensing signal around under densities in the projected galaxy field.[735–738] This has the advantage over the 3D void method for the following reasons: (1) identifying troughs in projected galaxy density field does not require 3D positions of galaxies, therefore, a lensing survey alone allows the measurement to be conducted. (2) An analytical model to predict the signal has been developed in Refs. 735, 738, which allows the constraints on cosmology, and possible extensions to modified gravity. (3) The lensing signal around troughs can be much stronger than 3D voids if one chooses a relatively small size for the troughs (or aperture), although the price to pay is that the difference between GR and modified gravity may be weaker due to the projection effect. In a recent study by Cautun *et al.*,[446] it has been shown that in their specific setting, the distinguishing power for $f(R)$ vs GR is indeed greater using troughs than using 3D voids. Note that the conclusion of Cautun *et al.* may not necessarily apply to other settings of galaxy surveys and lensing surveys. It remains important to investigate the optimal distinguishing power for testing gravity for specific surveys.

4.3. *Redshift-space distortions around voids*

While gravitational lensing probes the sum of the two metric potentials, the motion of galaxies usually responses only to the Newtonian potential. The combination of lensing and kinematic measurement such as redshift-space distortions can be used to test the gravitational slip.[150, 252, 464, 465, 467, 744] This applies to voids, where the gravitational slip may be the strongest.[745]

Redshift-space distortions (RSD) arise from the fact that the observed distance of a galaxy s is perturbed by its peculiar velocity along the line of sight $v_{\rm pec}$:

$$s = r + v_{\rm pec}/(aH), \tag{13}$$

where r is the real space comoving distances of a galaxy, a is the scale factor of the Universe and H is the Hubble constant. On large scales, infall motion of galaxies towards high density regions causes the amplitudes of galaxy clustering to increase along the line of sight. The galaxy two point correlation function appears flattened along the line of sight. For voids, we expect to have coherent outflow velocities around void centres, but the expected distortion pattern is not simply the opposite of the case in high density regions. Several studies have developed and tested models for RSD around voids in the standard ΛCDM model[706, 746–748] in order to extract the growth rate. At the linear order, the redshift-space void-galaxy correlation function is

$$\xi^s(\mathbf{r}) = \xi^r(r) + \frac{1}{3}\beta\bar{\xi}^r(r) + \beta\mu^2[\xi^r(r) - \bar{\xi}^r(r)], \tag{14}$$

where $\mu = \cos\theta$ and θ is the subtended angle from the line of sight; $\beta = f/b$ and $f = d\ln D/d\ln a$ is the linear growth rate with D being the linear growth factor and b is the linear galaxy bias. ξ^r is the real space void-galaxy correlation function.

In this model, the void-galaxy correlation function may be flattened or elongated depending on the real space density profile of voids, and its corresponding velocity profile[706] (see Fig. 9 for an example). Linear theory applies to regions where the density fluctuation is small, which may not be the case near the centre of voids. As shown in Nadathur et al.,[747] the distortion pattern with a second order term included may differ from the linear model (see Ref. 747 for an expression where some higher order terms are kept). This is also hinted by the study of Achitouv and Cai.[749] They have shown that in regions where the local density is significantly lower than the mean density, the growth rate is non-linear. This suggests that more sophisticated model may be needed to fully describe the distortion patterns near the void centre. This has been a topic of active investigation in recent years, with

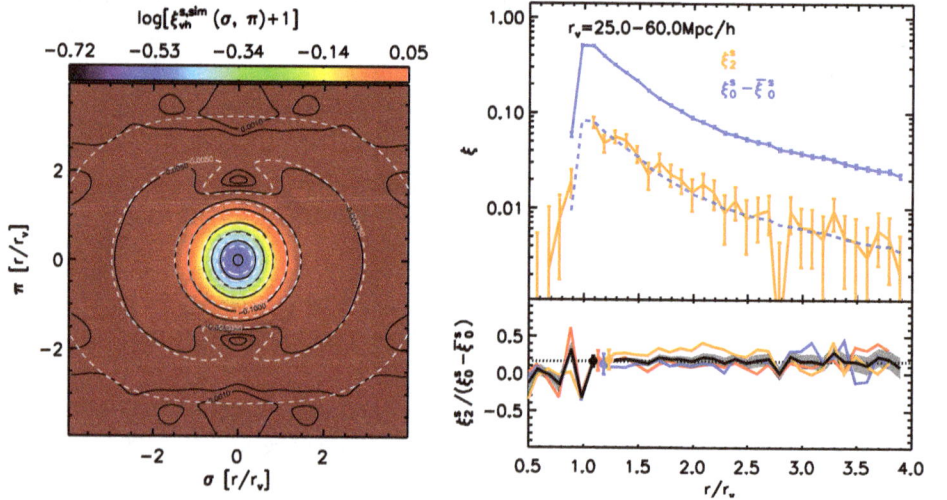

Fig. 9: From Cai et al.,[706] Left: comparing the the void-halo correlation function in redshift space measured from simulations with the linear theory. The black contours give results from the simulation and the white contours are the best-fit linear model. Right: the upper panel shows the monopole and quadrupole moments of the correlation function. The black curve is from taking all measurements of voids along three different major axes of the simulation box, with the shaded region showing the error on the mean. The other three curves represent results from viewing the simulation along three different major axes. The black filled circle with error bars is the best-fit value from viewing voids along three different directions. The red, blue and orange filled circles and errors are from individual viewing directions. They are slightly offset from each other to aid visibility.

some noticeable theoretical problems and debates not fully resolved in GR. I listed a few of them here: (1) to what accuracy the galaxy bias around voids remains linear; (2) to what accuracy the mapping between density and velocity around voids remains linear. Nonetheless, an accurate modelling for the distortion pattens induced by peculiar velocities can be used to constrain the growth of structure.

RSD around voids encodes information about the density and velocity. This is ideal for capturing the difference between GR and non-standard gravity models, where violation of the equivalence principle may occur most evidently in voids. The fact that voids become deeper in chameleon models will show as a higher amplitude of monopole in the void-galaxy correlation function in real space. This effect will be enhanced further when observing voids in redshift space because of the larger amplitudes of expansion velocity around voids. Therefore, RSD around voids is a promising observable which can capture the combination of these two effects.[670, 706] The linear model can be used as a consistency check for deviations from GR. It has been demonstrated in Ref. 706 that the linear growth rate can be recovered using RSD around voids.

Observational constrains for the growth rate parameter using RSD around voids have been reported by Achitouv *et al.*[750] from the 6dF Galaxy Survey, Hawken *et al.*[751] from the VIPERS survey and Hamaus *et al.*[707, 708] using SDSS void catalogues. Among them, a somewhat lower growth rate was found at $z \sim 0.7$ from the

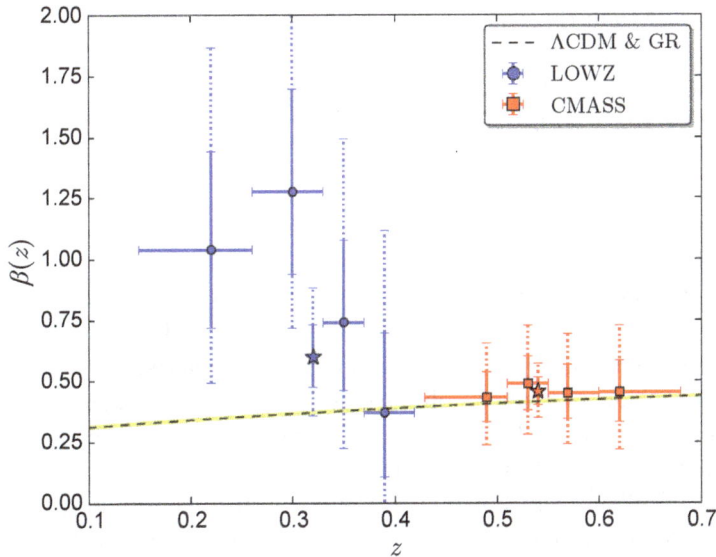

Fig. 10: From Hamaus *et al.*,[708] constraints on the growth rate parameter $\beta(z) = f(z)/b(z)$ using void-galaxy correlation functions from the SDSS LOWZ (blue circles) and CMASS (red squares) galaxy samples. Stars represent the joint constraint from voids of all redshifts in each sample. Vertical solid lines indicate one and two σ errors. The dashed line with yellow shading shows fiducial GR ΛCDM model with the linear bias $b = 1.85$. The tow data points at the lowest redshifts are not fully consistent with the model line.

VIPERS results than from other constraints from other measurements (Fig. 12 in the paper), though not significant. Also, at lower redshift, there is a $2 - 3\sigma$ tension with GR from the SDSS low-z sample,[708] as shown in Fig. 10. These may indicate an imperfection of current linear model, or a tension with GR. One also needs to caution the possible imperfect calibration of survey systematics in the SDSS area.[752]

4.4. ISW signal from voids and superclusters

The integrated Sachs-Wolfe effect[692] is a CMB secondary effect which arises from the evolution of large-scale gravitational potentials. In a ΛCDM universe, dark energy stretches cosmic voids and superclusters, causing their gravitational potentials to decay. Photons from the CMB will lose/gain energy when traversing a void/supercluster, and so the CMB temperature is expected to be colder/hotter when a void/supercluster sits along the line of sight. The induced temperature fluctuations are

$$\frac{\Delta T}{T_{\text{CMB}}} = -\frac{1}{c^2} \int (\dot{\Phi} + \dot{\Psi}) dt, \tag{15}$$

where the $\dot{\Phi}$ and $\dot{\Psi}$) are the time derivative of the metric potentials and T_{CMB} is the CMB temperature. The signal is sensitive to the time variation of the metric potentials, and its detection would give direct evidence for dark energy and can be used to constrain theories of gravity.[252, 753] In the standard model, the signal is expected to be much smaller than the primordial CMB temperature fluctuation, posing a challenge for its direct detection. Therefore, cross-correlation between galaxy samples with the CMB is usually employed to extract the signal. A $\sim 4\sigma$ detection of the signal has been reported by combining multiple galaxy samples and cross-correlate them with the CMB,[754, 755] and these have placed constraints for the chameleon models.[252, 753]

Another way to detect the signal is by stacking the CMB with voids and superclusters found in the late-time large-scale structure. This should yield a cold and hot spot respectively for voids and superclusters. It was first conducted by Granett et al.[693] where a somewhat unexpected high significant ($\sim 4\sigma$) result was reported by stacking superstructures found form the photo-z galaxy catalogue in the SDSS area. However, the amplitudes of the signal were higher than expected from the standard model. Follow up analysis using similar method with spectroscopic redshift catalogue (but still in the SDSS area) confirms the results qualitatively,[689, 694–699, 701, 756] which may indicate a tension with the standard model, but the study by Nadathur and Crittenden[700] has found an agreement with the ΛCDM model when using a (different) matched filter approach.

While the debate over the amplitudes and significance of the stacked ISW signal around voids and superclusters may continue, all the above studies have found that the sign of the ISW signal to be consistent with expectation from the standard model. These analysis, together with the positive cross-correlation between the entire galaxy sample with the CMB, strongly indicate that the late-time large-scale potentials are indeed decaying. This has placed one of the strongest constraints for the Galileon model.[742] Combining superstructures from galaxy surveys which cover a different area from the SDSS should be able to beat down the statistical error and hence help to settle the debate and provide better constraints for cosmology.

5. Summary and Notes for the Future

Observables of cosmic voids focus on low density regions of the Universe. They are potentially powerful for testing theories of gravity where violation of the equivalence principle is expected to occur most prominently in low density regions. A common reason behind this is that void statistics such as abundance, void lensing and RSD are sensitive to the non-Gaussianity of the density field, which is expected to be different in modified gravity models than in GR.

Using the four observables about voids mentioned above, some level of tension or inconsistency for cosmological constraints have been reported, in contrast to the constrains from conventional methods such as two-point statistics. For example, both the void abundance and void RSD analysis using SDSS data have indicated some level of tension with the standard ΛCDM model,[690, 708] which is not seen in most of the SDSS analysis using galaxy clustering. We know that there are observational systematics such as star contamination and galactic extinction in the SDSS data.[752, 757, 758] Although they have been calibrated to meet the precision requirements needed for galaxy clustering, it is unclear if such systematic calibration is sufficient for the relevant void statistics.

Therefore, while these new analysis using voids may open up windows for new physics, one needs to be cautious about possible effect from uncharted observational systematics. Next generation galaxy surveys such as DESI, LSST and Euclid promises to increase the current survey volume and number of galaxies by an order of magnitude. The statistical errors of these measurements will go down substantially, and so the importance of controlling systematic errors for galaxy surveys will be paramount. Calibration of observational systematics will likely to remain the major challenge for constraining theories of gravity using large-scale structure.

Facing this challenge, it is perhaps more pressing to employ novel cosmological probes and new methods of data analysis. Given that different probes may suffer from different systematics, using multiple probes will allow them to be cross-checked with each other and deliver more stringent results.

Acknowledgments

YC was supported by supported by the European Research Council under grant numbers 670193. YC thank Baojiu Li and Vasiliy Demchenko for useful comments.

Chapter 8

Astrophysical Tests of Screened Modified Gravity

Jeremy Sakstein

Center for Particle Cosmology, Department of Physics and Astronomy,
University of Pennsylvania, 209 S. 33rd St., Philadelphia, PA 19104, USA
sakstein@sas.upenn.edu

Screened modified gravity models evade solar system tests of relativistic gravitation but exhibit novel and interesting effects on scales between the solar system and the Hubble flow: astrophysical scales. In this article we review how astrophysical tests using stars, galaxies, and clusters can be used to constrain these theories. We classify screening into three categories: thin-shell (chameleon, symmetron, and dilaton models), Vainshtein screening (e.g. galileons and Horndeski), and Vainshtein breaking (e.g. beyond Horndeski and DHOST) and discuss the optimal strategy for probing each. In many cases, this is driven by whether a specific category violates the equivalence principles (strong or weak). We summarize the general model-independent bounds on each screening category that have been derived in the literature.

1. Introduction

Screened modified gravity theories evade the solar system tests that have proved prohibitive for classical alternative gravity theories such as Brans-Dicke. In many cases, they do not fit into the PPN formalism. The environmental dependence of the screening (see Chapter 1) has motivated a concerted effort to find new and novel probes of gravity using objects that are well-studied but have hitherto not been used to test gravity. Astrophysical objects—stars, galaxies, clusters—have proved competitive tools for this purpose since they occupy the partially-screened regime between solar system and the Hubble flow. In this chapter, we review the current astrophysical tests of screened modified gravity theories. We begin by introducing the theories we will study and outline the strategy typically employed to identify astrophysical probes.

Throughout this chapter we will use the mostly positive metric signature so that the Minkowski metric $\eta_{\mu\nu} = (-1, 1, 1, 1)$. With this convention, the trace of the energy momentum tensor is $T = T^\mu_\mu = -\rho + 3P$ with $T^0_0 = -\rho$. The Ricci scalar $R > 0$. We will consider the scalar fields to be universally coupled to all particle species (including dark matter) with the same coupling strength.

1.1. *Searching for screening mechanisms*

We will focus on three distinct categories of theories with screening mechanisms, with theories within each exhibiting similar effects on astrophysical objects. This allows us to identify the optimum strategy for testing each theory.

Thin-Shell Theories: Chameleon,[57, 58] symmetron,[60] and dilaton[59] models all screen using the thin-shell effect. For this reason we will refer to them as *thin-shell* theories; the theoretical aspects of these models are covered in detail in Chapter 1. The specific details of each model are not important for astrophysical tests and one can completely parameterise them using the effective coupling $\beta(\phi_{\rm BG})$ where $\phi_{\rm BG}$ is the asymptotic (background) field value and the self-screening parameter is given by (see Sec. 4.4.1 of Chapter 1)

$$\chi_{\rm BG} = \frac{\phi_{\rm BG}}{2\beta(\phi_{\rm BG})M_{\rm pl}}. \tag{1}$$

For $f(R)$ models one has $f_{R0} = 2\chi_0/3$ where χ_0 is the value of $\chi_{\rm BG}$ evaluated at cosmic densities. If the self-screening parameter is larger than an object's Newtonian potential $\Psi = GM/R$ then this object will be self-screening. If not, then an object will be partially unscreened. This implies that the best objects for testing these theories are non-relativistic ones. In particular, main-sequence stars have $\Psi \sim 10^{-6}$ whereas post-main-sequence stars have $\Psi \sim 10^{-7}$–10^{-8} (owing to their larger radii) and are therefore more constraining probes. Similarly, rotationally-supported galaxies have

$$\Psi \sim \frac{GM}{R} = v_{\rm circ}^2, \tag{2}$$

where $v_{\rm circ}$ is the circular velocity. The most unscreened galaxies are therefore dwarf galaxies with $v_{\rm circ} \sim 50$ km/s so that $\Psi \sim 10^{-8}$. (Spirals like the Milky Way have $v_{\rm circ} \sim 200$ km/s implying $\Psi \sim 10^{-6}$.) There is the added complication of environmental screening whereby a potentially unscreened dwarf could be screened by its cluster companions. Therefore, one needs to use void dwarfs as laboratories for testing thin-shell screening theories. Reference 69 has complied a *screening map* of the nearby universe using criteria developed by[68] and calibrating on N-body simulations. Recently, this has been revisited by.[759]

Vainshtein Screening Theories: Theories that screen using the Vainshtein mechanism so that the ratio of the scalar to Newtonian force outside a spherical object is[a]

$$\frac{F_\phi}{F_{\rm N}} = 2\alpha^2 \left(\frac{r}{r_{\rm V}}\right)^n \tag{3}$$

will be referred to as *Vainshtein screening theories*. We parameterise the coupling strength α and Vainshtein radius $r_{\rm V}$ (below which the force is screened) using a crossover scale $r_c (= \Lambda_c^{-1})$ akin to the DGP model so that $r_{\rm V}^3 \sim \alpha GM r_c^2$. These

---

[a]Note that in Chapter 1 the α variable is replaced with β. In this chapter we use α to distinguish the Vainshtein screening theories from the other screened models.

theories include very general theories including Horndeski[103] but here we will only focus on cubic and quartic galileon models[104] for which $n = 3/2$ and $n = 2$ respectively. In the case of Vainshtein screening, the Vainshtein radius is typically larger than the radius of stars and galaxies, making astrophysical tests difficult but not impossible.

Vainshtein Breaking Theories: Theories such as beyond Horndeski[121, 122] and degenerate higher-order scalar-tensor theories (DHOST) (see Ref. 760 and references therein) exhibit a "breaking of the Vainshtein mechanism" such that the Newtonian potential and lensing potential ($g_{ij} = (1 - 2\Phi)\delta_{ij}$) are corrected inside extended objects to[128, 129, 131]

$$\frac{d\Psi}{dr} = \frac{GM(r)}{r^2} + \frac{\Upsilon_1 G}{4} \frac{d^2 M(r)}{dr^2} \tag{4}$$

$$\frac{d\Phi}{dr} = \frac{GM(r)}{r^2} - \frac{5\Upsilon_2 G}{4r} \frac{dM(r)}{dr} + \Upsilon_3 G \frac{d^2 M(r)}{dr^2} \tag{5}$$

where the three dimensionless parameters Υ_i are related to the cosmological values of the functions and parameters appearing in a specific theory and also the effective description of dark energy[42, 129, 130, 133] (introduced by Ref. 111). The form of the corrections do not suggest the best objects for testing these theories and one must calculate on an object by object basis. See Chapter 1 for a detailed discussion of Vainshtein breaking and the parameters Υ_i.

1.2. *Roadmap of astrophysical tests*

We will begin by discussing the most important difference between Vainshtein and thin-shell screened theories in Sec. 2: equivalence principle violations. Next, we will introduce the theory of stellar structure in modified gravity in Sec. 3. In the subsequent sections we review current astrophysical bounds by object. Non-relativistic stars in Sec. 4, galactic tests in Sec. 5, galaxy cluster tests in Sec. 6, and tests using relativistic stars in Sec. 7.

2. Equivalence Principle Violations

2.1. *Weak equivalence principle*

One important difference between thin-shell and Vainshtein screening is the presence of weak equivalence principle (WEP) violations.[b] It was pointed out in the original chameleon and symmetron papers that thin-shell screening violates the WEP[57, 58, 60] because the thin-shell factors for each body, which determines their motion in an external field, are composition and structurally dependent. This issue was studied in more detail by Ref. 726 who also studied equivalence principle violations in galileon theories. Consider an extended object of inertial mass M_i and gravitational

---

[b]We define the WEP as the statement that the motion of a test body in an external gravitational field depends only on its mass and is independent of its composition and internal structure.

mass M_g in an applied external Newtonian potential Ψ^{ext} and scalar field ϕ^{ext} (chameleon, symmetron, or galileon). The equation of motion for this object in the non-relativistic limit is

$$M_i \ddot{\vec{x}} = -M_g \vec{\nabla} \Psi^{\text{ext}} - Q \frac{M}{M_{\text{pl}}} \vec{\nabla} \phi^{\text{ext}}, \tag{6}$$

where M is the baryonic mass.[c] The gravitational mass can be thought of as a gravitational charge that parameterises the response of the object to an externally applied Newtonian potential and so we have defined an analogous scalar charge Q that quantifies the response of an object to an externally applied scalar field.[d] In theories without a scalar field, the WEP is obeyed if $M_i = M_g$. This is not generically the case in scalar field theories because Q can depend on the structure and composition of the object. If one chooses to couple the scalar to different particle species (including dark matter) with different coupling strengths then there is an additional explicit breaking of the WEP by construction. We will not consider such theories in this chapter.

Thin-shell screening: For theories that screen using the thin-shell effect (chameleon and symmetron theories) one has (see Secs. 4.4.1–4.4.3 of Chapter 1)

$$Q = \beta(\phi_{\text{BG}}) \left(1 - \frac{M(r_{\text{scr}})}{M} \right), \tag{8}$$

where $M(r_{\text{scr}})$ is the mass enclosed inside the screening radius r_{scr}. We note that this expression applies no matter the thickness of the shell. In particular, Eq. (8) is derived by separating the object into two regions separated by the screening radius: the central screened region and the outer unscreened region. In the screened region the field is assumed to minimise its effective potential so there is no contribution to the scalar charge and in the unscreened region the effective mass of the field is ignored so that $\nabla^2 \phi = \beta(\phi_{\text{BG}})\rho/M_{\text{pl}}$ i.e. $\phi = 2\beta(\phi_{\text{BG}})M_{\text{pl}}\Psi_{\text{N}}$ (Ψ_{N} is the Newtonian potential). The effects of including the mass and small rolling of the field in the unscreened region are negligible for astrophysical scenarios where precision tests are not possible. Note also that the derivation assumes that the object is sufficiently far away from the source of the background field so that the monopole dominates but close enough that the variation of the background fields can be ignored. (See Refs. 398 for the effects of ellipticity and higher-order multipoles; ellipticity tends to

---

[c]We define the baryonic mass M as the mass calculated by integrating over the 00-component of the energy-momentum tensor i.e.

$$M = -\int d^3 \vec{x} \, T^\mu_\mu = \int d^3 \vec{x} \rho(x), \tag{7}$$

where the second equality holds in the non-relativistic limit. Note that this may include the mass of dark matter but not the self-energy of the gravitational field, which is found by integrating the Landau-Lifshitz energy-momentum pseudo-tensor.

[d]The factor of M_{pl} is needed because ϕ^{ext} has different units to Ψ^{ext}. It is chosen so that Q is dimensionless.

weaken chameleon screening.) We refer the reader to Ref. 726 for the full technical derivation.

The force between two bodies with masses M_1 and M_2 is[761, 762]

$$F_{1,2} = 2Q_1 Q_2 \frac{GM_1 M_2}{r^2}, \tag{9}$$

where Q_i is given by Eq. (8) with $M \to M_i$. Thus the WEP is violated unless either $Q = 0$ or $Q = \beta(\phi_{\mathrm{BG}})$ i.e. the objects are fully screened or fully unscreened.

Vainshtein screening: In the case of Vainshtein screening, there is no thin shell suppression. Furthermore, the equation of motion can be written in the form of a current conservation law $\nabla_\mu J^\mu = 8\pi\alpha G\rho$, which ensures that

$$Q = \alpha \tag{10}$$

i.e. independent of the object's internal structure. The WEP is therefore satisfied in galileon theories. One possible caveat to this is many-body effects. The equation of motion for galileon theories is non-linear in second-derivatives, which leads to severe violations of the superposition principle.[e] The above argument circumvents this by assuming that the external galileon field is only slowly varying so that the galilean shift symmetry can be used to superimpose the fields, and it is not clear what happens away from this approximation. The full two-body problem has been studied by Ref. 81 for an Earth-Moon like system; they found a mass-dependent reduction of the galileon force of $\sim 4\%$ indicating that the WEP may be broken by non-linear many-body effects. The non-linear nature of the equations makes modeling of such systems difficult. Indeed, departures from spherical symmetry do not have analytic solutions except in highly symmetric cases.[763, 764] See Refs. 146, 765 for some detailed studies of this issue.

2.2. *Strong equivalence principle*

The strong equivalence principle (SEP) is the statement that an object's motion is independent of its self-gravity. Unlike the WEP, the SEP is violated by all of the theories considered in this section.[f] This is because the scalar field is sourced only by the baryonic mass (defined in Eq. (7)) and not the curvature so that the no-hair theorems hold and strongly gravitating objects have no scalar charge.[g] A no-hair theorem for the galileon was proved by Ref. 771 for static, spherically symmetric black holes and subsequently generalized by Ref. 767 to the case of slow rotation. Thus, in a system composed of baryonic matter (including dark matter) and black holes, the baryonic component will have $Q = \alpha$ while the black holes will have $Q = 0$. The baryons will therefore fall at a faster rate than the black holes in an

---

[e]The equations of motion for chameleon and symmetron theories are quasi-linear and so there is always some regime in which superposition approximately holds.
[f]In fact, scalar-tensor theories generically violate the SEP, the statements made in this subsection have nothing to do with screening.
[g]One exception to this is scalar couplings to the Gauss-Bonnet scalar.[766–770]

externally applied gravitational field, violating the SEP. In the case of chameleon theories, the presence of an accretion disk around black holes may source secondary scalar hair.[772, 773]

3. Stellar Structure in Modified Gravity

Stars are complicated objects whose lives, existence, and stability are a result of the interface between diverse and disparate areas of physics, including gravitational physics, atomic physics, nuclear physics, hydrodynamics, and particle physics.[774] Modern theoretical modeling of stellar structure and evolution therefore utilizes sophisticated numerical simulations that solve a large number (often in the thousands depending on the type of star) of coupled differential equations simultaneously. Fortuitously, the effects of gravitational physics appears in a single equation, the momentum equation, which describes the Lagrangian velocity $\vec{v} = \dot{\vec{r}}$ of a fluid element located at Lagrangian position $\vec{r}$ due to some external force (per unit mass) $\vec{f}$ and the hydrodynamic (Eulerian) pressure P:

$$\dot{\vec{v}} = -\frac{1}{\rho}\vec{\nabla}P + \vec{f}, \tag{11}$$

where $\rho(\vec{r})$ is the Eulerian density. In the case of general relativity, the force per unit mass is simply the gradient of the Newtonian potential

$$\vec{f} = -\vec{\nabla}\Psi. \tag{12}$$

For alternative theories, one must solve for the force per unit mass within the new framework. Typically this involves solving for additional scalar (or other spin) field profiles sourced by the star's mass. Note that we will only discuss non-relativistic objects here, postponing relativistic stars for a later section.

3.1. *Equilibrium structure*

The velocity of each fluid element is constant for a static, spherically symmetric object in equilibrium and so the left hand side of Eq. (11) is zero. In GR, the force per unit mass is simply the Newtonian force and one has the well-known hydrostatic equilibrium equation (HSEE)

$$\frac{dP(r)}{dr} = -\frac{GM(r)\rho(r)}{r^2}, \tag{13}$$

where $M(r)$ is the mass enclosed inside r and therefore satisfies the continuity equation

$$\frac{dM(r)}{dr} = 4\pi r^2 \rho(r). \tag{14}$$

For thin-shell screening theories, the HSEE is modified to[h]

$$\frac{dP(r)}{dr} = -\frac{GM(r)\rho(r)}{r^2}\left[1 + 2\beta^2(\phi_{\text{BG}})\left(1 - \frac{M(r_{\text{scr}})}{M(r)}\right)\Theta\left(r - r_{\text{scr}}\right)\right], \qquad (15)$$

where $M(r_{\text{scr}})$ is the mass enclosed within the screening radius r_{scr}, $\Theta(x)$ is the Heaviside step function, and the new factor arises from the fifth-force $F_\phi = -\beta(\phi_{\text{BG}})M_{\text{pl}}d\phi/dr$ with ϕ_{BG} being the background (asymptotic) value of the scalar. If the star's host galaxy is self-screened then this is the field value that minimizes the effective potential at mean galactic density, if the host galaxy is unscreened then the relevant density is the mean cosmic density. In theories that exhibit Vainshtein breaking the corresponding HSEE is[131, 134, 780, 781]

$$\frac{dP(r)}{dr} = -\frac{GM(r)\rho(r)}{r^2} - \frac{\Upsilon_1 G\rho(r)}{4}\frac{d^2M(r)}{dr^2}, \qquad (16)$$

which can be expressed in alternate forms by taking derivatives of Eq. (14) to find

$$\frac{d^2M(r)}{dr^2} = 8\pi r\rho(r) + 4\pi r^2\frac{d\rho(r)}{dr}. \qquad (17)$$

The Vainshtein radius of stars is several orders of magnitude larger than their radius (if not, the theory would be ruled out using traditional solar system tests of GR) and therefore stars do not exhibit observational deviations from GR. For this reason we do not give the HSEE for Vainshtein screened theories unless they also exhibit Vainshtein breaking.

These equations presented thus far do not form a closed set because the equation of state $P(\rho)$ is not known. One must either couple these equations to microphysical and macrophysical processes such as radiative transfer, nuclear burning, opacity, and convection to calculate the equation of state (EOS), or provide a known (or approximate) equation of state. Two important equations that will arise at various points in this chapter are the equation of radiative transfer, which describes the temperature gradient of the star due to photon transport:

$$\frac{dT}{dr} = -\frac{3}{4a}\frac{\kappa}{T^3}\frac{\rho L}{4\pi r^2}, \qquad (18)$$

where κ is the opacity, and the energy generation equation

$$\frac{dL}{dr} = 4\pi r^2\sum_i \rho(r)\varepsilon_i(T, \rho). \qquad (19)$$

This equation describes the photon luminosity gradient produced by the interaction process i with rate ε_i per unit mass.

---

[h]Note that we have ignored the mass of the field in the region exterior to the screening radius. This is a commonly used and very good approximation.[775–779]

3.1.1. *Polytropic models*

One simple and well-studied equation of state is the polytropic equation of state[782]

$$P = K\rho^{\frac{n+1}{n}}, \tag{20}$$

which are good approximations for many stars, or at least some region of them. In the context of modified gravity, polytropic equations of state allow one to decouple to gravitational and non-gravitational physics. This means one can discern the effects of changing the theory parameters without the need to account for possible degeneracies with non-gravitational processes. The stellar structure equations are self-similar for polytropic equations of state, which means one can work with dimensionless variables to extract the structure of the star independently of the central conditions. In particular, it is useful to work with the dimensionless radial coordinate

$$r = r_{\rm LE} y, \quad \text{where } r_{\rm LE}^2 \equiv \frac{(n+1)P_{\rm LE}}{4\pi G \rho_{\rm LE}^2}, \tag{21}$$

and $P_{\rm LE}$ and $\rho_{\rm LE}$ are the central pressure and density respectively. Note that these quantities are often denoted by subscript c's in the literature but we use 'LE' here to avoid confusion with the DGP crossover scale r_c. One can define the dimensionless function $\theta(y)$ via

$$\rho = \rho_{\rm LE}\theta(y)^n \quad \text{and} \quad P = P_{\rm LE}\theta(y)^{n+1} = K\rho_{\rm LE}^{\frac{n+1}{n}}\theta^n, \tag{22}$$

which encodes the structure of the star. In GR, one can take a derivative of Eq. (13) and apply Eq. (14) to find the Lane-Emden equation (LEE)

$$\frac{1}{y^2}\frac{d}{dy}\left[y^2\frac{d\theta(y)}{dy}\right] = -\theta(y)^n. \tag{23}$$

The equivalent equation for thin-shell screening theories is[776, 778, 779]

$$\frac{1}{y^2}\frac{d}{dy}\left[y^2\frac{d\theta(y)}{dy}\right] = -\theta(y)^n \begin{cases} 1 & y \le y_s \\ (1 + 2\beta^2(\phi_{\rm BG})) & y > y_s, \end{cases} \tag{24}$$

where $r_{\rm scr} = r_{\rm LE}y_s$ (i.e. y_s is the dimensionless radius where screening begins) and the factor of $(1 + 2\beta^2(\phi_{\rm BG}))$ assumes that the star is fully unscreened outside the screening radius.[i] In Vainshtein breaking theories the LEE is[131, 134]

$$\frac{1}{y^2}\frac{d}{dy}\left[\left(1 + \frac{n\Upsilon_1}{4}y^2\theta(y)^{n-1}\right)y^2\frac{d\theta}{dy} + \frac{\Upsilon_1}{2}y^2\theta(y)^n\right] = -\theta(y)^n, \tag{25}$$

which has been derived using Eq. (17) and using the relations in Eq. (22). The boundary conditions for the LEE are $\theta(0) = 1$ ($P(r = 0) = P_{\rm LE}$) and $\theta'(0) = 0$ ($dP(r)/dr = 0$ at the origin, which is a consequence of spherical symmetry). (See Ref. 784 for a detailed study of the LEE in GR.) One can find analytic solutions

---

[i]If one were to attempt to go beyond this approximation and include the thin-shell factor $(1 - M(r_{\rm scr})/M(r))$ the self-similarity would be lost and, with it, the simplicity of the LEE.[776, 783]

for specific values of n but these are typically not relevant for astrophysics and so one must solve the LEE numerically.

The radius of the star is defined as the radial coordinate where the pressure falls to zero, which defines y_R such that $\theta(y_R) = 0$. One then has

$$R = r_{LE} y_R. \tag{26}$$

The stellar mass can be found by integrating Eq. (14) to find

$$M = 4\pi r_{LE}^3 \int_0^{y_R} y^2 \theta(y)^n dy = \begin{cases} \omega_R & \text{GR and Vainshtein breaking} \\ \frac{\omega_R + 2\beta^2(\phi_{BG})\omega_s}{1 + 2\beta^2(\phi_{BG})} & \text{Thin-shell} \end{cases}, \tag{27}$$

where we have replaced $\theta(y)^n$ using the appropriate Lane-Emden equation and defined

$$\omega_Y = -y^2 \frac{d\theta(y)}{dy}\bigg|_{y=Y} \tag{28}$$

with $Y = s$ being short for $Y = y_s$.

Two important properties of polytropes that will be useful later, are the mass-radius relation

$$R = \gamma \left(\frac{K}{G}\right)^{\frac{n}{3-n}} M^{\frac{n-1}{n-3}}; \quad \gamma \equiv (4\pi)^{\frac{1}{n-3}} (n+1)^{\frac{3}{3-n}} \omega_R^{\frac{n-1}{3-n}} y_R \tag{29}$$

and the central density in terms of the mass and radius

$$\rho_c = \delta \left(\frac{3M}{4\pi R^3}\right); \quad \delta = -\frac{y_R}{3 d\theta/dy|_{y=y_R}}. \tag{30}$$

These relations are derived in Refs. 782, 785 (and other similar textbooks). They apply to GR and Vainshtein breaking theories but not chameleon theories. We do not give the chameleon equivalents here since they will not be necessary.[j]

3.1.2. *Numerical models: MESA*

In order to model more complicated stars that do not have simple polytropic equations of state, one needs sophisticated numerical codes. One publicly available code that has proven invaluable for stellar structure in modified gravity is MESA.[786] MESA solves the stellar structure equations coupled to the equations describing micro and macrophysical processes. The reader is referred to the instrumentation papers[786–788] for a comprehensive review of MESA's capabilities.

In the context of modified gravity, MESA has been modified to solve the modified HSEE for both thin-shell screening (Eq. (15))[775, 776, 778, 779] and Vainshtein breaking theories (Eq. (16)).[131] MESA is a one-dimensional code (meaning that it assumes spherical symmetry) that splits each star into cells of varying lengths (the number of

---

[j]In fact, they have never been formally derived in the literature, although such a derivation is simple and straight forward.

cells depends on the complexity of the star) and assigns relevant quantities (radius, density, temperature etc.) to each cell. The set of cells and these quantities then defines a stellar model at a specific time-step. Given a specific stellar model, the stellar structure equations are discretized on each cell solved to produce a new stellar model at a later time. Thus, the star is simulated over its entire lifetime. The publicly available version of MESA solves the GR HSEE, Eq. (13). The modified versions of MESA solve either Eq. (15) or (16). We will briefly describe how these modifications work below.

Thin-shell: There are two independent chameleon modifications of MESA. The first[775] solves the full scalar differential equation using a Gauss-Seidel relaxation algorithm. The second,[776, 778, 779] uses the thin-shell approximation. Both codes agree very well but here we will only describe the latter implementation since it is more commonly used in the literature. Given a starting stellar model, the screening radius is computed by solving[775, 776, 778, 779]

$$\chi_{\rm BG} \equiv \frac{\phi_{\rm BG}}{2\beta(\phi_{\rm BG})M_{\rm pl}} = 4\pi G \int_{r_{\rm scr}}^{R} r\rho(r)dr. \tag{31}$$

The code numerically integrates $r\rho(r)$ from the first cell until the cell where Eq. (31) is satisfied. If the central cell is reached before this happens the screening radius is set to zero. In the latter case, the code simply rescales $G \to G(1 + 2\beta^2(\phi_{\rm BG}))$. In the former case, the mass inside the screening radius is found and used as an input for Eq. (15). The next stellar model is then found by solving Eq. (15). The screening radius is recomputed at every time-step to account for the changes in the star's structure.

Vainshtein breaking: MESA was first updated to include Vainshtein breaking by Refs. 131, 780. In this case, the default (GR) HSEE is replaced by Eq. (16). A numerical derivative of the density is taken by differencing across adjacent cells so that $d^2M(r)/dr^2$ can be computed in each cell using Eq. (17). The code then evolves to the next time-step using the modified HSEE for any input value of Υ_1, allowing the stellar evolution to be computed.

3.2. *Radial perturbations*

Moving away from equilibrium, one can consider Lagrangian perturbations so that

$$\vec{r} = r\hat{\vec{r}} + \vec{\delta r} \tag{32}$$

and the velocity is $\vec{v} = \dot{\vec{\delta r}}$. The dynamics of $\vec{\delta r}$ describe perturbations of the star about its equilibrium configuration and, specializing to linear time-dependent radial perturbations[k]

$$\delta r = |\vec{\delta r}| = \frac{\xi(r)}{r}e^{i\omega t} \tag{33}$$

---

[k]Non-radial modes are not important for thin-shell screened theories because they cannot be observed in galaxies other than our own (which is screened) and their governing equations have yet to be derived in modified gravity theories.

one can linearize the other quantities (pressure, density, etc.) and combine their governing equations to find, assuming GR for now,

$$\frac{d}{dr}\left[r^4\Gamma_{1,0}P_0(r)\frac{d\xi(r)}{dr}\right] + r^3\frac{d}{dr}\left[(3\Gamma_{1,0}-4)P_0(r)\right]\xi(r) + r^4\rho_0(r)\omega^2\xi(r) = 0, \quad (34)$$

where subscript zeros refer to equilibrium quantities (found by solving the HSEE and other stellar structure equations) and $\Gamma_{1,0} = d\ln P_0/d\ln\rho_0$ is the first adiabatic index ($\Gamma_{1,0} = (n+1)/n$ for polytropic equations of state). Equation (34) is referred to as the *linear adiabatic wave equation* (LAWE). It is a Sturm-Liouville eigenvalue problem that must be solved given certain boundary conditions[789] defined at the center and surface of the star. The eigenfrequencies ω_n give the period of oscillation about the minimum $\Pi_n = 2\pi/\omega_n$. Just like the equilibrium equations, the LAWE is self-similar and one can scale all of the dimensionful quantities out of the equation to find a dimensionless form in terms of a dimensionless frequency $\Omega^2 = \omega^2 R^3/(GM)$ so that the frequencies scale as

$$\omega_n^2 \sim \Omega_n\frac{GM}{R^3} \quad (35)$$

or $\Pi \propto G^{-1/2}$. Theories where gravity is stronger therefore make stars of fixed mass and composition pulsate faster (or with a shorter period).

Reference 778 has derived the equivalent wave equation for thin-shell screened theories

$$\frac{d}{dr}\left[r^4\Gamma_{1,0}P_0(r)\frac{d\xi(r)}{dr}\right] + r^3\frac{d}{dr}\left[(3\Gamma_{1,0}-4)P_0(r)\right]\xi(r)$$
$$+ r^4\rho_0(r)\left[\omega^2 - 8\pi\beta^2(\phi_{\mathrm{BG}})\rho_0(r)\Theta(r-r_{\mathrm{scr}})\right]\xi(r) = 0, \quad (36)$$

which is typically referred to as the *modified linear adiabatic wave equation* (MLAWE). The boundary conditions are the same as in GR. One can see that the effect of the scalar field is to add a density-dependent mass term for $\xi(r)$ that increases ω (makes the period shorter) at fixed mass and composition, in line with our scaling arguments above. This is borne out by numerical simulations of polytropic and MESA models.[778] Another possible effect of stellar oscillations is that they may source scalar radiation, although detailed work for both non-relativistic[790, 791] and relativistic stars[792] (who studied the orbital-decay of binary pulsars) have found this to be negligible.

For Vainshtein breaking theories, the derivation of the MLAWE is incredibly complicated but follows the relativistic derivation of Ref. 793 starting from perturbations of a relativistic gas sphere in a de Sitter background and taking the weak-field sub-horizon limit. The result is[794]

$$\frac{d}{dr}\left[r^4\left(\Gamma_{1,0}P_0(r) + \pi\Upsilon_1 Gr^2\rho_0(r)^2\right)\frac{d\xi(r)}{dr}\right] + r^3\frac{d}{dr}\left[(3\Gamma_{1,0}-4)P_0(r)\right]\xi(r)$$
$$+ r^4\rho_0(r)\omega^2\left(1 - \frac{\pi\Upsilon_1 r^3\rho_0(r)M(r)}{2[M(r)+\pi r^3\rho_0(r)]^2}\right)\xi(r) = 0, \quad (37)$$

with modified boundary condition at the center (see Ref. 794).

3.2.1. *Stellar stability*

In GR, and thin-shell and Vainshtein breaking theories, the wave equation is a Sturm-Liouville eigenvalue equation of the form of a differential operator $\hat{\mathcal{L}}$ acting on a function $\xi(r)$ with weight function $W(r)$ i.e. $\hat{\mathcal{L}}\xi = W(r)\omega^2\xi$. This means we can bound the lowest eigenfrequency using the variational method by constructing the functional

$$\omega_0^2 < F[\chi] \equiv \frac{\int_0^R \chi^*(r)\hat{\mathcal{L}}\chi(r)dr}{\int_0^R W(r)\chi^*(r)\chi(r)dr}, \tag{38}$$

for some trial function χ. Taking this to be constant, we find

$$\omega_0^2 < \frac{\int_0^R 3r^2(3\Gamma_{1,0} - 4)P_0(r)dr}{\int_0^R r^4\rho_0(r)dr}, \tag{39}$$

using the GR wave equation. When $\Gamma_{1,0} < 4/3$ the lowest frequency is necessarily complex, signaling a tachyonic instability. In thin-shell screening theories, the equivalent of Eq. (39) is[778]

$$\omega_0^2 < \frac{\int_0^R 3r^2(3\Gamma_{1,0} - 4)P_0(r)dr + \int_{r_{\rm scr}}^R 8\pi\beta^2(\phi_{\rm BG})Gr^4\rho_0(r)dr}{\int_0^R r^4\rho_0(r)dr}, \tag{40}$$

so that the instability is mitigated in a screening-dependent manner. Objects that are more unscreened can have $\Gamma_{1,0} < 4/3$ and still be stable due to the compensating effect of the (positive) new term. This is borne out in the numerical computations of Ref. 778. Finally, in Vainshtein breaking theories the expression is[794]

$$\omega_0^2 < \frac{\int_0^R 3r^2(3\Gamma_{1,0} - 4)P_0(r)dr}{\int_0^R r^4\rho_0(r)\left(1 - \frac{\pi\Upsilon_1 r^3\rho_0(r)M(r)}{2[M(r)+\pi r^3\rho_0(r)]^2}\right)dr}. \tag{41}$$

When $\Upsilon_1 < 0$ the instability is the same as in GR but when $\Upsilon_1 > 0$ there is a second potential instability. For a star of constant density this always occurs when $\Upsilon_1 > 49/6$. For more general models, one needs to integrate over the equilibrium structure to determine the presence of the instability, although, given the large value for constant density stars, it is unlikely that the instability is realized in practice for sensible choices of Υ_1.

4. Stellar Structure Tests

In this section we review different objects that the theory developed in the last section has been applied to and the resulting bounds on screened modified gravity theories.

4.1. *Main-sequence stars*

4.1.1. *The Eddington standard model*

One of the simplest treatment of main-sequence stars which works well for low-mass objects is the *Eddington standard model*, which makes the assumption that the star is supported by a combination of radiation pressure from photons generated by nuclear burning in the core and hydrodynamic gas pressure (ideal gas law):

$$P_{\text{rad}} = \frac{1}{3}aT^4 \text{ and } P_{\text{gas}} = \frac{k_{\text{B}}\rho T}{\mu m_{\text{H}}}, \tag{42}$$

where m_{H} is the mass of a hydrogen atom, a is the radiation constant ($a = 8\pi^5 k_{\text{B}}^4/15c^3h^3$), and μ is the mean molecular weight (number of particles per atomic unit). Introducing $b = P_{\text{gas}}/P$, where $P = p_{\text{rad}} + P_{\text{gas}}$, Eq. (42) implies that

$$\frac{b}{1-b} = \frac{3k_{\text{B}}\rho}{a\mu m_{\text{H}}T^3}. \tag{43}$$

This implies b is a constant if one makes the approximation that the specific entropy ($s \propto \rho/T^3$) is constant. The total pressure is then

$$P = K(b)\rho^{\frac{4}{3}} \text{ with } K(b) = \left(\frac{3}{a}\right)^{\frac{1}{3}} \left(\frac{k_{\text{B}}}{\mu m_{\text{H}}}\right)^{\frac{4}{3}} \left(\frac{1-b}{b^4}\right)^{\frac{1}{3}} \tag{44}$$

so that the star is polytropic with $n = 3$ and its structure can be found by solving the Lane-Emden equation for the theory of gravity in question.

For modified gravity, the most important quantity for main-sequence stars is the luminosity, which must be determined from the radiative transfer equation. (In this section we assume that the opacity is constant, which is a good approximation for main-sequence stars where the dominant contribution comes from electron scattering.) Differentiating Eq. (42), one can find an expression for the surface luminosity using the appropriate HSEE (Eq. (13) for GR, Eq. (15) for thin-shell models, and Eq. (16) for Vainshtein breaking)

$$L = \frac{4\pi GM(1-b)}{\kappa} \begin{cases} 1 & \text{GR and Vainshtein breaking} \\ 1 + 2\beta^2(\phi_{\text{BG}})\left(1 - \frac{M(r_{\text{scr}})}{M}\right) & \text{thin-shell,} \end{cases} \tag{45}$$

where $M = M(R)$ is the stellar mass. Thus, in order to determine the luminosity (at fixed mass) we must calculate b. This is accomplished by inserting the definition of r_{LE} (Eq. (21)) into Eq. (27) to find a quartic equation (Eddington's quartic equation):[131, 776]

$$\frac{1-b}{b^4} = \left(\frac{M}{M_{\text{Edd}}}\right)^2 \begin{cases} 1 & \text{GR} \\ \left(\frac{\bar{\omega}_R}{\omega_R}\right)^2 & \text{Vainshtein Breaking} \\ \left[(1 + 2\beta^2(\phi_{\text{BG}}))\frac{\bar{\omega}_R}{\omega_R + 2\beta^2(\phi_{\text{BG}})\omega_s}\right]^{\frac{2}{3}} & \text{Thin-shell} \end{cases} \tag{46}$$

where $\bar{\omega} \approx 2.018$ is the GR value and the Eddington mass is

$$M_{\text{Edd}} = \frac{4\bar{\omega}_R}{\sqrt{\pi}G^{\frac{3}{2}}} \left(\frac{k_{\text{B}}}{\mu m_{\text{H}}}\right)^2 \left(\frac{3}{a}\right)^{\frac{1}{2}} \approx 18.3 M_{\odot}\mu^{-2}. \tag{47}$$

Note that the GR and Vainshtein breaking luminosities are not identical despite having the same expression since b is determined from different equations.

At this point, one can discern the gross effects of modified gravity on the stellar luminosity. First, note from Eq. (45) that when $b = 1$ the luminosity is zero. This is because this extreme value corresponds to no radiation pressure and hence no photons. When $b \ll 1$ the star is dominated by radiation pressure and one has $L \propto GM$. Conversely, when b is close to unity (so that the star is gas-pressure supported) one can write $b = 1 - \delta$ for $\delta \ll 1$ and Eq. (46) shows that $\delta \propto (M/M_{\text{Edd}})^2 \propto G^4 M^2$. One then has $L \propto G^4 M^3$. This means that the effects of modified gravity are more pronounced in pressure supported stars. Equation (46) requires $b \approx 1$ for $M < M_{\text{Edd}}$ whereas $b \ll 1$ for $M > M_{\text{Edd}}$ so that low-mass stars are gas-supported and high-mass stars are pressure supported. We therefore expect the effects of modified gravity to be more pronounced in low-mass stars.

The procedure for calculating the luminosity in any given gravity theory is as follows: first, one numerically solves the relevant $n = 3$ Lane-Emden equation for a given set of parameters (there are no free parameters in GR) to find ω_R ($= \bar{\omega}_R$ in GR). For thin-shell models, one must also find the screening radius and ω_s using Eq. (31) (see Refs. 776, 779 for the details). Once ω_R (and ω_s for chameleons) have been obtained, Eq. (46) can be solved numerically to find b. This can then be put into Eq. (45) to find the luminosity.

Plots of the ratio L/L_{GR} are shown in Fig. 1 for both thin-shell and Vainshtein breaking theories. In both cases $\mu = 1/2$, appropriate for hydrogen stars. Evidently, the effects of modified gravity are indeed more pronounced in low-mass objects due to their gas pressure support. We have chosen $\beta(\phi_{\text{BG}}) = 1/\sqrt{6}$ for thin-shell models corresponding to the (constant) value predicted by $f(R)$ models. When $\chi_{\text{BG}} \geq 10^{-5}$ the enhancements plateau at low masses because the stars are fully unscreened. The asymptotic value is precisely $(1 + 2\beta(\phi_{\text{BG}})^2)^4 = (4/3)^4 \approx 3.16$, in agreement with our prediction above that $L \propto G^4$ for fully unscreened gas pressure supported stars. We chose $\Upsilon_1 > 0$ for the Vainshtein breaking models, which, evidently, lowers the

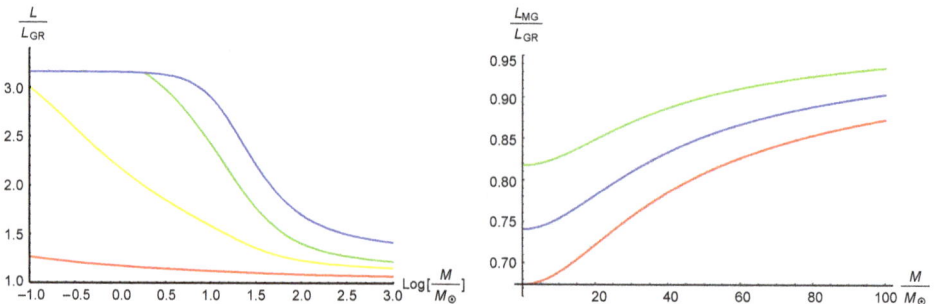

Fig. 1: The luminosity enhancement for main-sequence stars assuming the Eddington standard model. *Left panel*: Thin-shell screening theories (taken from Ref. 776). The plot shows the enhancement for $\beta(\phi_{\text{BG}}) = 1/\sqrt{6}$ (corresponding to $f(R)$ theories) with (from top to bottom) $\chi_{\text{BG}} = 10^{-4}$ (blue), 10^{-5} (green), 5×10^{-6} (yellow), and 10^{-6} (red). *Right panel*: Vainshtein breaking theories (taken from Ref. 131). From top to bottom, $\Upsilon_1 = 0.4$ (green), $\Upsilon_1 = 0.6$ (blue), and $\Upsilon_1 = 0.8$ (red).

luminosity compared with GR. A good rule of thumb (but by no means a concrete feature) is that positive values of Υ_1 weaken gravity (compared with GR) in the Newtonian limit.[1] Had we chosen $\Upsilon_1 < 0$ we would have found the converse behavior i.e. the luminosity would have been enhanced, a consequence of strengthened gravity.

4.2. *MESA models*

4.2.1. *Thin-shell stars*

For thin-shell screening theories, Refs. 775, 776 have used a modified version of MESA to compute the color-magnitude or Hertzprung-Russell (HR) tracks for solar mass stars. An example of this is shown in Fig. 2. The curves show the evolution of a solar mass and metallicity star from the zero-age main-sequence to the tip of the red giant branch in $f(R)$ chameleon theories $(2\beta^2(\phi_{BG}) = 1/3)$. Also shown are the radius and ages of the star when the central hydrogen mass fraction $X = 0.5$, 0.1, and 10^{-5} so that one can compare stars at similar points in their evolution. The parameters are chosen so that the stars are progressively more unscreened from bottom to top. The curve at $\chi_{BG} = 10^{-7}$ mimics GR on the main-sequence because the star is fully screened (recall main-sequence stars have $\Psi \sim 10^{-6}$) but becomes unscreened on the red giant branch when the radius of the star increases

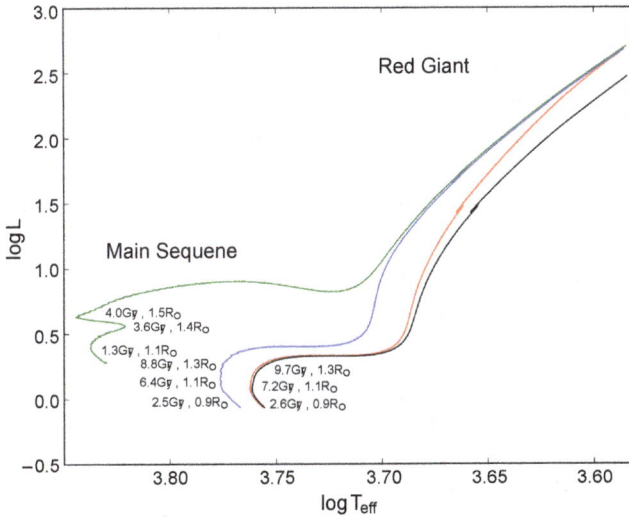

Fig. 2: The HR tracks of a solar mass star of solar metallicity $(Z = 0.02)$. From bottom to top: GR (black), 10^{-7} (red), 10^{-6} (blue), 5×10^{-6} (green). In each case, the value of $\beta(\phi_{BG}) = 1/\sqrt{6}$ so that the chameleon theory can be re-written as an $f(R)$ theory. The radius and age at the points where the central hydrogen mass fraction has fallen to 0.5, 0.1 and 10^{-5} are indicated in the figure (from bottom to top) with the exception of the red curve, which mimics GR on the main-sequence. Figure taken from Ref. 776.

---

[1] This is because $d^2 M/dr < 0$ for low mass homogeneous stars.

by about a factor of 10, lowering its Newtonian potential. The blue curve has a comparable shape to GR but is shifted to higher temperatures and luminosities, indicating that the star is brighter and hotter than its GR counterpart. The green curve corresponds to a star that is fully unscreened, and looks like the HR track for a $2M_\odot$ star. In all cases, at fixed X more unscreened stars are younger, indicating that stellar evolution has proceeded at a faster rate. This is because the amount of nuclear fuel is fixed (at fixed mass) but more unscreened stars need to consume it at a faster rate in order to combat the increased gravity. Thin-shell screening stars are therefore hotter, brighter, and more ephemeral the more unscreened they are. Unfortunately, these predictions have yet to be utilized as a test of chameleon theories. The main reason for this is that one requires unscreened galaxies—dwarf galaxies in cosmic voids—in order for the stars to become sufficiently unscreened. Main-sequence and post-main-sequence stars are typically not resolvable in such galaxies.

4.2.2. *Vainshtein breaking stars*

The HSEE for Vainshtein breaking theories was implemented into MESA by[131] using the method outlined in Sec. 3.1.2. The HR tracks for solar mass stars and two solar mass stars are shown in the left and right hand panels of Fig. 3 respectively. One can see that, at fixed metallicity the effects of increasingly positive Υ_1 is to make the star dimmer and cooler. This is because positive values of Υ_1 act in an equivalent manner to weakening gravity and therefore the star needs to burn nuclear fuel at a slower rate to stave off gravitational collapse. Another consequence of this is that stars evolve more slowly when Υ_1 is more positive, as evidenced by the location of the filled circles in the left panel. Negative value of Υ_1 have the opposite effect (i.e. gravity is strengthened); fuel is consumed at a faster rate and the star is hotter, brighter, and more ephemeral. On the main-sequence, these effects are degenerate with metallicity; it is evident from the figures that a GR $Z = 0.03$ star

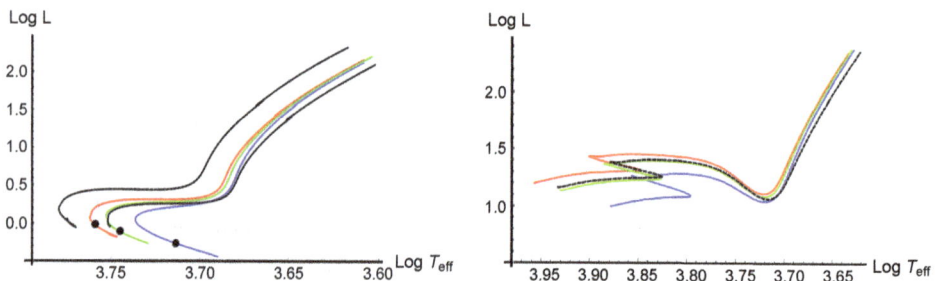

Fig. 3: HR tracks in GR and Vainshtein breaking theories. From top to bottom: GR, $Z = 0.02$ (red, solid), GR, $Z = 0.03$ (black, dotted), $\Upsilon_1 = 0.1$, $Z = 0.02$ (green, solid), $\Upsilon_1 = 0.3$, $Z = 0.02$ (blue, solid). *Left panel:* One solar mass. The solid circle shows the star when its age is 4.6×10^9 yr. *Right panel:* Two solar masses. Figures taken from Ref. 131.

has a similar main-sequence track to a Vainshtein breaking star with $\Upsilon_1 = 0.1$ and $Z = 0.02$. (If $\Upsilon_1 < 0$ the effects of Vainshtein breaking are degenerate with decreasing the metallicity.) This degeneracy vanishes on the red giant branch. In theory, the effects of Vainshtein breaking should be present in all stars in our local neighborhood. In practice, to date there have been no local tests, either proposed or performed. This is due partly to the degeneracy with metallicity, although this can either be corrected for with other measurements or avoided by using post-main-sequence stars.

4.3. *A stellar bound for Vainshtein breaking theories*

One important requirement for the stability of stars is that $P''(r) < 0$.[795] If this is violated then the outwards pressure gradient would be an increasing function of the radius and would therefore overcompensate for the gravitational force and drive an outward expansion. At the center of the star, the pressure, density, and mass can be expanded as

$$P(r) = P_{LE} - \frac{P_2}{2}r^2 + \cdots , \quad \rho(r) = \rho_{LE} - \frac{\rho_2}{2}r^2 + \cdots , \quad \text{and } M(r) = \frac{M_3}{3}r^3 + \cdots , \quad (48)$$

where the linear terms are absent in the expansions of $P(r)$ and $\rho(r)$ because one needs $P'(0) = \rho'(0) = 0$; the expansion for $M(r)$ begins at cubic order in order to be consistent with Eq. (14). Plugging these expansions into the HSEE, Eq. (16), one finds

$$P_2 = \frac{G\rho_c M_3}{3}\left(1 + \frac{3\Upsilon_1}{2}\right) < 0, \quad (49)$$

implying the bound $\Upsilon_1 > -2/3$. This bound was first derived by Ref. 134 using similar arguments.

4.4. *Dwarf stars*

Dwarf stars are those that populate the mass range between Jupiter mass planets $(M_J \sim 10^{-3}M_\odot)$ and main-sequence stars with masses $M \sim \mathcal{O}(0.1M_\odot)$. When first formed, a star will contract under its own self-gravity liberating energy and increasing the temperature and density. The contraction must be halted by the onset of pressure support either due to electron degeneracy pressure or thermonuclear fusion. In the former case, the star is inert and is referred to as a *brown dwarf*. In the latter case, it is a *red dwarf*. Only stars that are sufficiently heavy can achieve the requisite core density and temperature for hydrogen burning to proceed efficiently. Thus, low-mass stars are brown dwarfs and higher mass stars are red dwarfs. The transition mass, the minimum mass for hydrogen burning (MMHB), is $M_{MMHB} \approx 0.08M_\odot$ in GR. A detailed account of low-mass stars can be found in Ref. 796. In the context of modiied gravity, dwarf stars are good probes of Vainshtein breaking theories,[135, 136] and so we focus exclusively on these in this subsection.

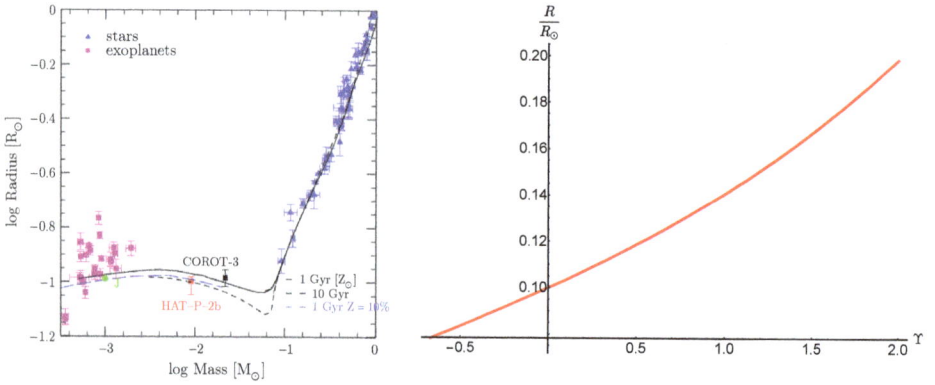

Fig. 4: *Left panel*: The mass-radius relation for low-mass stars. The radius plateau is visible for stars with masses less than the MMHB, which is the the sudden upturn at $\log_{10}(M/M_\odot) = -1.2$. Figure taken from Ref. 797. *Right panel*: The radius of brown dwarf stars in Vainshtein breaking theories as a function of Υ_1. Taken from Ref. 136.

4.4.1. *Brown dwarf stars: The radius plateau*

Brown dwarfs are inert (non-hydrogen burning) stars[m] composed primarily of molecular hydrogen and helium in the liquid metallic phase with the exception of a thin layer near the surface, which is composed of a weakly coupled plasma that is well-described by the ideal gas law. They are fully convective and therefore contract along the Hyashi track with a polytropic $n = 1.5$ EOS.[774] In fact, Coulomb corrections to the electron scattering processes shift the EOS of lower mass brown dwarfs ($M \lesssim 4M_J$) to lower values $n \approx 1$.[796, 798] For $n = 1$, one has $P_{LE} = K\rho_{LE}^2$ (c.f., Eq. (22) and recall $\theta(0) = 1$) so that Eq. (21) gives $r_{LE}^2 = K/(2\pi G)$ and the radius, $R = r_{LE}y_R$ is, is independent of the mass. This leads to a *radius plateau* in the mass-radius relation for stars with masses $M_J < M < M_{MMHB}$. An example of this is shown in the left panel of Fig. 4, which plots the mass-radius relation for low-mass stars. (The figure shows small differences due to small degeneracies with the stellar metallicity and age are not important here.) One can see that the radius of low mass stars (with masses lighter than the MMHB) is approximately constant[n] (note that the scale is logarithmic). In GR, the plateau lies at $R \approx 0.1R_\odot$ but in Vainshtein breaking theories y_R depends on Υ_1 and therefore so does the plateau radius. This is shown in the right panel of Fig. 4. One can see that the changes in the radius are significant for $|\Upsilon_1| \sim \mathcal{O}(1)$, although whether this can be used to place new bounds is not clear since the data pertaining to the radius plateau is

---

[m]Higher mass brown dwarfs may burn deuterium or lithium (or both) for a short time (until the reserve is depleted). In fact there are minimum masses for deuterium and lithium burning analogous to the MMHB.
[n]The radius is not exactly constant because the EOS is not exactly polytropic. One typically fits more precise numerical equations of state to an $n = 1$ polytrope in order to extract the polytropic constant K and make analytic predictions.

currently sparse.[797] Future data releases from Gaia, Kepler, or their successors may be able to populate the brown dwarf mass-radius diagram sufficiently.

4.4.2. *Red dwarf stars: The minimum mass for hydrogen burning*

The central conditions in low-mass stars are not sufficient for efficient burning on the PP-chains. In particular, the Coulomb barrier for the ^{3}He-^{3}He and ^{3}He-^{4}He cannot be overcome at the relevant central temperatures and densities (10^6 K and 10^3 g/cm^3). Instead, proton burning proceeds via deuterium burning with the end point being Helium-3. The MMHB is the smallest mass where the luminosity generated by this reaction process can balance the luminosity lost from the star's surface.

A simple model of red dwarf stars first presented in Ref. 796 for GR was adapted for Vainshtein breaking theories by[135, 136] who showed that the MMHB is sensitive to Υ_1. In this model, the star is supported by a combination of degeneracy pressure and the ideal gas law, which are both described by $n = 1.5$ polytropic equations of state. Stable hydrogen burning is achieved when

$$3.76 M_{-1} = \left[\frac{\left(1 + \frac{\Upsilon_1}{2}\right)}{\kappa_{-2}} \right]^{0.11} \left(1 + \frac{3\Upsilon_1}{2}\right)^{0.14} \frac{\gamma^{1.32} \omega_R^{0.09}}{\delta^{0.51}} I(\eta); \quad I(\eta) = \frac{(\alpha + \eta)^{1.509}}{\eta^{1.325}}, \tag{50}$$

where $M_{-1} = M/(0.1 M_\odot)$, $\kappa_{-2} = \kappa_R/10^{-2}$ with κ_R being the Rosseland mean opacity, $\alpha = 4.82$, and γ and δ are defined in Eq. (29) and Eq. (30), respectively. The degeneracy parameter η is the ratio of the Fermi energy to $k_B T$ and measures the relative contribution of each type of pressure, degeneracy pressure being more important for larger η.

The function $I(\eta)$ has a minimum value of 2.34 at $\eta = 34.7$ and so there is a minimum value of M for which Eq. (50) can be satisfied, the MMHB. Assuming $\kappa_{-2} = 1$ (we will discuss this later), the MMHB in GR is $M_{\text{MMHB}}^{\text{GR}} \approx 0.08 M_\odot$ whereas in Vainshtein breaking theories it depends on Υ_1 as shown in Fig. 5. One

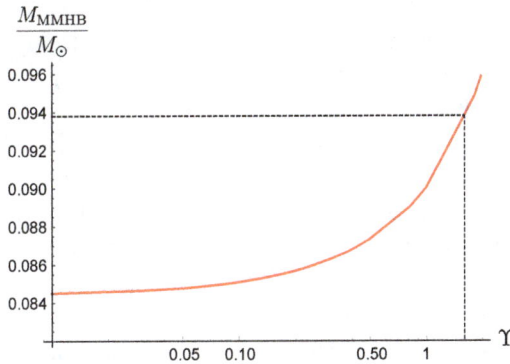

Fig. 5: The MMHB as a function of Υ_1. The black dashed line shows the upper limit on the mass of the lightest red dwarf presently observed and the corresponding value of Υ_1. Figure taken from Ref. 135.

can see that, for positive values of Υ_1, the MMHB is larger than the GR value. This is because the weakened gravity results in lower central densities and temperatures at fixed mass so that heavier objects are needed to reach the requisite conditions for hydrogen burning. One cannot take theories with Υ_1 too large because the theory would predict that observed red dwarf stars should be brown dwarfs. Indeed, the lightest red dwarf (M-dwarf) is Gl 866 C with a mass $M = 0.0930 \pm 0.0008 M_\odot$.[799] Vainshtein breaking theories are only compatible with this observation if the bound $\Upsilon_1 < 1.6$ is satisfied.

This bound is incredibly robust. Indeed, there are few degeneracies with other astrophysical effects. There is a degeneracy with the opacity but, as is evident in Eq. (50), this is very mild and is not strong enough to impart any uncertainty onto this bound. Similarly, variations in the chemical composition between different dwarf stars are small and the compositions themselves do not evolve significantly over the life-time of the star. Another possible degeneracy is rotation, but this acts to increase the MMHB[800, 801] and can therefore only make the bound stronger. Finally, the method used to infer the star's mass is insensitive to the theory of gravity. The mass is either inferred from empirical relations, which do not assume any gravitational physics, or from the orbital dynamics of binaries,[802] which occurs in a regime where there is no Vainshtein breaking (i.e. outside the objects) so that the equations are identical to GR. See Refs. 135, 136 for an extended discussion on this.

4.4.3. *White dwarf stars: The Chandrasekhar mass and mass-radius relation*

White dwarf stars are the remnants of low-mass stars ($M \lesssim 8 M_\odot$) that have gone off the main-sequence to become giant stars and have subsequently had their outer layers blown away by stellar winds leaving only the core. In the absence of any thermonuclear fusion, electron degeneracy pressure provides the counter-gravitational support. Low mass white dwarf stars are well described by $n = 1.5$ polytropic equations of state ($P \propto \rho^{\frac{5}{3}}$) corresponding to a non-relativistic gas whereas high-mass white dwarfs are best described by $n = 3$ ($P \propto \rho^{\frac{4}{3}}$) corresponding to a fully relativistic gas. Following Eq. (29), this means that low-mass white dwarfs follow the mass-radius relation $R \propto M^{-\frac{1}{3}}$ whereas fully relativistic white dwarfs have a fixed mass (the Chandrasekhar mass). If one tries to go to higher masses, the star is unstable and a thermonuclear explosion occurs, resulting in a type Ia supernova. This is the same instability found using perturbation theory (see Eq. (39)).

The majority of white dwarf stars are composed primarily of ^{12}C, for which an equation of state can easily be found. We will follow the method of Ref. 803, which[781] have adapted to Vainshtein breaking theories. Defining $x = p_F/m_e$ where p_F is the Fermi momentum, the number density of degenerate electrons is

$$n_e = \frac{m_e^3 x^3}{3\pi^2} \tag{51}$$

while the electron pressure and energy density are $P_e = m_e^4 \Psi_1(x)$ and $\epsilon_e = m_e^4 \Psi_2(x)$ with $\Psi_i(x)$ given in Ref. 803. The density receives contributions from both the carbon atoms and the electrons but the former are far heavier than the latter and so the density is $\rho \approx \rho_C$. On the other hand, the pressure comes primarily from the electrons and so one has $P \approx P_e$. One can use these approximations with the appropriate HSEE to construct white dwarf models. In this case, the unknown functions are $m(r)$ and $x(r)$ which satisfy $M(0) = 0$ and $x(0) = x_0$ so that their is one free parameter defined at the center for the star. The radius is defined by $P(R) = 0$ (which implies $x = 0$) so that $M = M(R)$. Varying x_0 allows one to build up the mass-radius relation.

Reference 781 have studied white dwarfs in Vainshtein breaking theories using the above equation of state by solving both the GR (Eq. (13)) and Vainshtein breaking (Eq. (16)) HSEEs. The mass-radius relation that they obtained is shown in the left panel of Fig. 6. A χ^2 test was performed using the observed masses and radii of 12 carbon-oxygen white dwarfs taken from Ref. 804 treating Υ_1 as a fitting parameter. The resultant bounds were $-0.18 \leq \Upsilon_1 \leq 0.27$ at 1σ and $-0.48 \leq \Upsilon_1 \leq 0.54$ at 5σ. For $\Upsilon_1 < 0$ the effects of Vainshtein breaking are equivalent to strengthening gravity, which has the effect of lowering the Chandrasekhar mass as shown in the central panel of Fig. 6. The mass of the heaviest observed white dwarf therefore places a bound on negative values of Υ_1 since values too negative would predict that this object should have gone supernova. The heaviest observed white dwarf has a mass $M = 1.37 \pm 0.01 M_\odot$,[805] which places the bound $\Upsilon_1 \geq -0.22$.[781]

Recently, Ref. 806 have studied white dwarfs in Vainshtein breaking theories using more sophisticated modeling that accounts for degeneracies, includes finite-temperature effects, and implements a detailed modeling of the various layers of the star. They studied both carbon-oxygen and helium white dwarfs and were able to

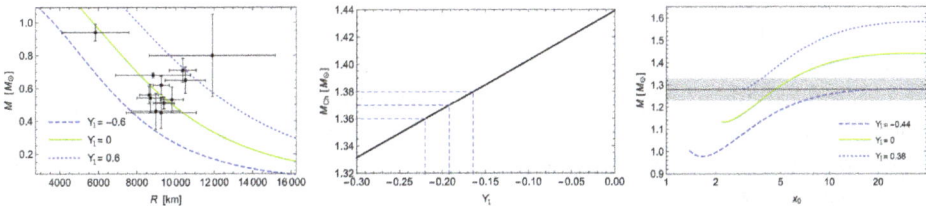

Fig. 6: *Left panel*: The mass-radius relation for white dwarf stars in GR (green, center, solid) and Vainshtein breaking theories with $\Upsilon_1 = -0.6$ (blue, dotted, upper) and $\Upsilon_1 = 0.6$ (blue, dashed, lower). Also shown are the masses and radii of 12 observed white dwarfs from Ref. 804 that were used by Ref. 781 for their analysis. *Central panel*: The Chandrasekhar mass as a function of Υ_1 for $\Upsilon_1 < 0$ (black, solid). The blue dashed lines show the values of Υ_1 that correspond to the central value and upper and lower limits of the mass of the heaviest observed white dwarf ($M = 1.37 \pm 0.01 M_\odot$). *Right panel*: The mass-central density (parameterized by x_0) relation for GR (green, center, solid), $\Upsilon_1 = -0.44$ (blue, upper, dotted), and $\Upsilon_1 = 0.38$ (blue, lower, dashed). The black solid line represents the mass of RX J0648.0-4418 ($M = 1.28 \pm 0.05 M_\odot$) and the grey region shows the 1σ error bars. These figures were adapted from Ref. 781.

place the tighter bound $\Upsilon_1 < 0.14$ at 2σ ($\Upsilon_1 < 0.19$ at 5σ). This was partly due to their improved modeling and partly because the 26 stars they used came from a dataset where the masses and radii were determined using the eclipsing binary technique so that no modeling of the white dwarf interior was necessary. This helped to eliminate some degeneracies.

A final bound can be found by considering rotating white dwarfs. If the white dwarf is rotating with angular frequency ω then the HSEE must be augmented by a centrifugal force

$$\frac{dP}{dr} = -\frac{GM(r)\rho(r)}{r^2}\left[1 + \frac{\pi\Upsilon_1 r^3}{M(r)}\left(2\rho(r) + r\frac{d\rho(r)}{dr}\right)\right] + \rho(r)\omega^2 r. \qquad (52)$$

If at any point the pressure gradient is outward i.e. $dP/dr > 0$ then the star is unstable and so we must require[781]

$$\Upsilon_1 > \left(\frac{\omega^2}{\pi G\rho} - \frac{M(r)}{\pi r^3 \rho(r)}\right)\left(2 + \frac{d\ln\rho}{dr}\right)^{-1} \qquad (53)$$

at every r. Note that the equality changes to an upper bound if $d\ln\rho/dr < -2$. For the simple case of constant density one recovers the bound of Ref. 134, $\Upsilon_1 > -2/3$. The positive pressure contribution implies that there is a minimum stellar mass for given values of ω and Υ_1, and that the strongest bounds should come from the most rapidly rotating objects. The majority of white dwarfs are slowly rotating but some rapidly rotating objects have been observed, in particular, RX J0648.0-4418, which has a mass $M = 1.28 \pm 0.05 M_\odot$ which rotates with a period of 13.2 s.[807] Fixing ω to this value,[781] have scanned a range of x_0 for different values of Υ_1 to find the range of parameters where such a star can be stable. Their results are illustrated in the right hand panel of Fig. 6. Accounting for the error bars, only values of Υ_1 in the range $-0.59 \le \Upsilon_1 \le 0.50$ can successfully model this object.

Before leaving white dwarf stars, we note that they may be used to constrain the time-variation of G. Reference 808 have shown that type Ia supernovae (whose progenitors are white dwarf stars that have accreted matter and exceeded their Chandrasekhar mass) are no longer standarizable distance indicators in theories where G varies on cosmological time-scales. This is because their peak luminosity depends on their Chandrasekhar mass, which decreases in theories where gravity is stronger than GR. An independent measurement of the luminosity distance to a type Ia supernova using gravitational wave standard sirens (such as coalescing neutron stars)[809] could constrain the time-variation of G since the absolute magnitude of the supernova can now be determined.

4.5. *Distance indicator tests*

Distance indicators have proved a highly constraining novel probe of theories that screen using the thin-shell mechanism. Distance indicators are a method of inferring the distance to a galaxy based on some proxy, for example, by measuring the apparent magnitude of a standard candle such as a type Ia supernova. Typically, the

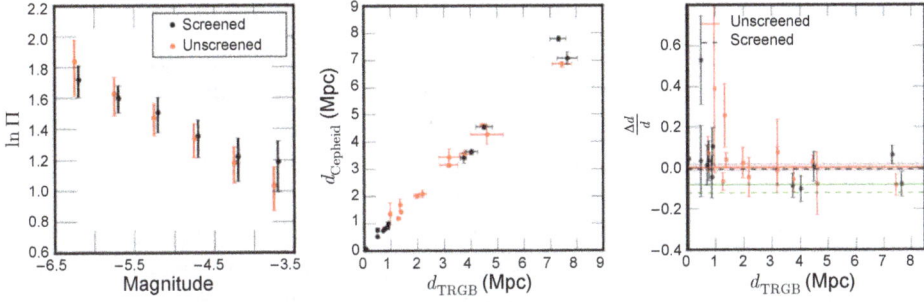

Fig. 7: Comparison of Cepheid and TRGB distances to a sample of screened (black dots) and unscreened galaxies (red dots). Figures adapted from Ref. 777. *Left panel*: Comparison of the PL relation. *Center panel*: Comparison of Cepheid and TRGB distances. *Right panel*: Comparison of the difference between the two distance estimates as a function of the TRGB distance. The black dashed and red solid lines are the best-fit for the screened and unscreened samples respectively. The solid and dashed green lines show predictions for models with $2\beta^2(\phi_{\rm BG}) = 1$ and $1/3$ ($f(R)$ models) respectively.

formula used to infer the distance is based upon empirical calibrations made locally or theoretical calculations. In the former case, the calibration has been performed in a screened environment and in the latter the calculations assume GR. Therefore, if one compares two distance estimates to the same galaxy, one sensitive to the theory of gravity and the other not, then the two will not agree if the galaxy is unscreened. The amount by which they agree therefore constrains the model parameters. In what follows, we will summarize how Ref. 777 used two different distance indicators, Cepheids and tip of the red giant branch (TRGB) stars, to constrain thin-shell models.

4.5.1. *Screened distance indicators: Tip of the red giant branch*

When stars of 1–$2M_\odot$ leave the main-sequence and ascend the red-giant branch the stellar luminosity is due to a thin shell of hydrogen burning outside the helium core. As the star ascends, the core temperature increases until it is hot enough for the triple-α process to proceed efficiently. At this point, known as the *helium flash*, the star moves rapidly onto the asymptotic giant branch (AGB), leaving a visible discontinuity in the I-band at $I = 4.0 \pm 0.1$ with the spread being due to a slight dependence on metallicity. This discontinuity can be used as a distance indicator since the luminosity is known. The details of the helium flash depend on nuclear physics and not the theory of gravity so the TRGB is a screened distance indicator.°

°In fact, if $\chi_{\rm BG} \gtrsim 10^{-6}$ MESA simulations reveal that the tip luminosity can decrease by 20%. This is because the core is unscreened in these cases and the temperature is increased. For this reason, the temperature needed for the helium flash is reached faster and therefore the discontinuity occurs lower on the red giant branch. In what follows we will only consider $\chi_{\rm BG} \lesssim 10^{-6}$.

4.5.2. *Unscreened distance indicators: Cepheid stars*

Stars with masses 3.5–$10 M_\odot$ execute semi-convection-driven *blue loops* in the color-magnitude diagram where the temperature increases at roughly fixed luminosity. During this phase, the stars can cross the instability strip where they are unstable to pulsations driven by the κ-mechanism (see Ref. 789 for details of this process). In this phase, a layer of doubly ionized helium acts as a dam for energy so that small compressions of the star go towards increasing the temperature in the ionization zone and not into increasing the outward pressure. This energy dam drives pulsations which result in a periodic variation of the luminosity and gives rise to a period-luminosity (PL) relation.[810] These stars are known as *Cepheid variable stars* and are used as distance indicators. In thin-shell screened theories, the inferred distance depends on the level of screening because the period of pulsation Π is faster with stronger gravity. This can be calculated either by solving the MLAWE, Eq. (36), or by using the fact that $\Pi \propto G^{-1/2}$ to find[777]

$$\frac{\Delta d}{d} = -0.3\frac{\Delta G}{G} \approx -0.6\beta^2(\phi_{\mathrm{BG}})\left(1 - \frac{M(r_{\mathrm{scr}})}{M}\right). \tag{54}$$

Thus, in thin-shell screened theories Cepheid distance indicators are unscreened and under-estimate the true distance.

4.5.3. *Comparisons and constraints*

Using the screening map, Ref. 728 compared the TRGB and Cepheid for a sample of screened galaxies as well as a control sample of unscreened galaxies. The TRGB distance was taken as the true (screened) distance and the theoretical value of $\Delta d/d$ was computed by using MESA Cepheid profiles at the blue edge of the instability strip[p] to calculate $\Delta G/G$ in Eq. (54). An example is shown in Fig. 8. One can see that the two samples are consistent and a statistical analysis yielded the constraints shown in Fig. 8. In this case χ_{BG} probes the cosmological value of χ (or, equivalently f_{R0} for $f(R)$ models) since the galaxies are unscreened. The bounds are the strongest astrophysical ones to date and $|f_{R0}| > 4 \times 10^{-7}$ is ruled out for $f(R)$ models.

4.6. *Astroseisemology*

The use of radial stellar oscillations in Vainshtein breaking theories has been studied by Ref. 794 who solved the MLAWE, Eq. (37), for some simple polytropic stellar models with the results shown in Fig. 9. The effects are small with the exception of brown dwarfs where $\Delta\Pi/\Pi \sim \mathcal{O}(1)$. The authors also investigated MESA models and found large changes in the period of Cepheid pulsations, although this was primarily driven by the altered equilibrium structure, which changed the intersection of the Hertzprung-Russell track with the instability strip.

---

[p]The location of the instability strip may change in modified gravity models but, to date, this has never been investigated.

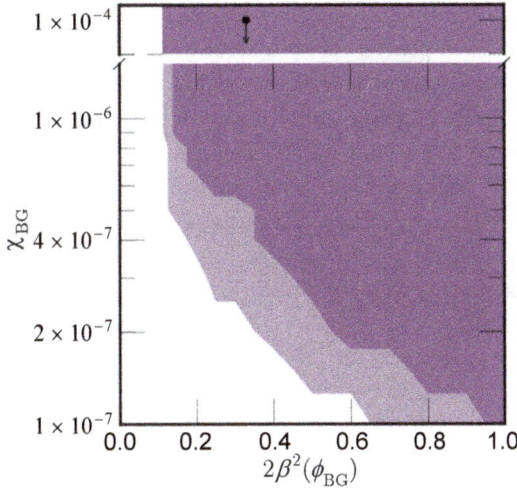

Fig. 8: The 68% (dark purple) and 95% excluded regions in the χ_{BG}–$\beta(\phi_{BG})$ plane by comparing Cepheid and TRGB distance indicators. The black arrow shows an older constraint coming from galaxy cluster statistics. Figure adapted from Ref. 777.

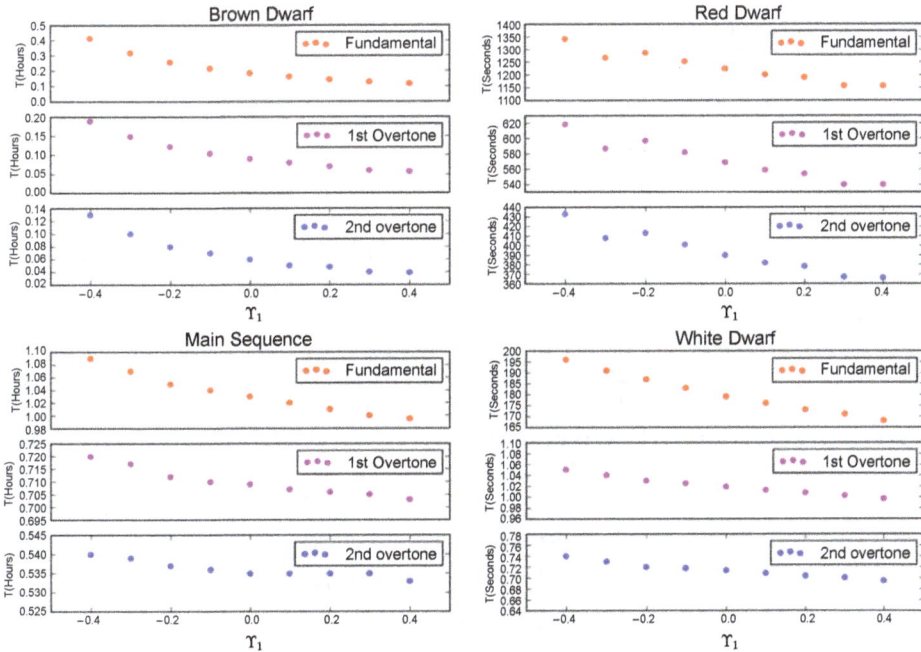

Fig. 9: The fundamental oscillation mode and first two overtones for representative polytropic stellar models in Vainshtein breaking theories. Figure adapted from Ref. 794.

5. Galactic Tests

The morphology and dynamics of galaxies, in particular dwarf or low surface brightness (LSB) galaxies, have proved to be a strong tool for testing screened modified
gravity theories, especially those that screen using the thin-shell mechanism. This is
partly because they have multiple components—dark matter, stars, gas—that can
be screened to different levels and partly because they themselves have Newtonian
potentials of $\mathcal{O}(10^{-8})$ making them some of the most unscreened objects in the
universe.[q] In this section, we discuss several novel tests that can, and in some cases
have, been used to constrain thin-shell screening theories. We will use two common
models for the dark matter density profile to aid in our computations: the cored
isothermal sphere (CSIS)

$$\rho(r) = \frac{\rho_0}{1 + \left(\frac{r}{r_0}\right)^2} \tag{55}$$

and the Navarro-Frenk-White (NFW) profile[547]

$$\rho(r) = \frac{\rho_s}{\frac{r}{r_s}\left(1 + \frac{r}{r_s}\right)^2} \tag{56}$$

where ρ_0 and r_0 are the core density and radius (CSIS) and ρ_s and r_s are the
scale density and radius (NFW). The former profile is typically a good fit to dwarf
galaxies with core radii of order 1–4 kpc[811] whilst the latter are well-motivated both
theoretically and observationally.

5.1. *Rotation curves*

Theories that violate the equivalence principle i.e. those that screen using the thin-
shell effect allow for a novel test of gravity using the rotation curves of different
galactic components.[728] In particular, a galaxy is composed of stars (with Newtonian
potentials $\Psi \sim \mathcal{O}(10^{-7}\text{--}10^{-6})$) and diffuse gas (with Newtonian potential $\mathcal{O}(10^{-11}\text{--}$
$10^{-12})$[726]) that rotate around the center with a radially-dependent circular velocity
given by

$$\frac{v_{\text{circ}}^2}{r} = \frac{GM_{\text{gal}}(r)}{r^2} + \frac{Q_{\text{obj}}}{M_{\text{pl}}}\frac{d\phi_{\text{gal}}}{dr}, \tag{57}$$

where a subscript 'gal' refers to fields sourced by the galaxy and Q_{obj} is the scalar
charge of the object (see Sec. 2). Let us make two simplifying assumptions: that
the galaxy is unscreened so that $d\phi_{\text{gal}}/dr = 2\beta(\phi_{\text{BG}})M_{\text{gal}}(r)/M_{\text{pl}}$ and that stars
are fully screened ($Q_{\text{obj}} = 0$) whilst the gas is fully unscreened $Q = \beta(\phi_{\text{BG}})$. In this
case, the circular velocity for the stars, $v_\star$ and gas v_{gas} satisfy[728]

$$\frac{v_{\text{gas}}}{v_\star} = \sqrt{1 + 2\beta^2(\phi_{\text{BG}})}. \tag{58}$$

---

[q]Of course, one must use dwarf galaxies that are sufficiently isolated so as to avoid environmental screening by their neighbors. In practice, this means using dwarf galaxies in voids. See the
discussion in Sec. 1.1 for more information on this matter.

Thus, a comparison of the rotation curves of stars and gas can provide a novel probe of thin-shell screening theories.

In practice, performing this test is not so simple because traditional probes of the galactic rotation curves use either $H\alpha$ or 21 cm lines, both of which probe the gaseous, unscreened component. Another useful line is the OIII line that results from a forbidden transition in doubly ionized oxygen. This is particularly useful for thin-shell screening theories since the line is only present at very low densities. The stellar component can be probed independently using absorption lines for metals found in stellar atmospheres, for example, the MgIb triplet or the CaII lines, found in the atmosphere of K- and G-dwarfs (main-sequence stars). These stars have Newtonian potentials of order 10^{-6} and hence values of χ_{BG} smaller than this (where they are screened) can be probed provided that their host galaxies are unscreened for the same parameters.

The screening map contains six galaxies that have both OIII and MgIb information available that Ref. 812 have used to perform this test. Their method is as follows: first, the gaseous rotation curve is used to fit a density profile for the galaxy accounting for systematic errors and astrophysical scatter. (Note that the gaseous curve is measured at more finely-spaced radial intervals so this provides a more accurate fit.) Next, this model is used to predict the stellar rotation curve and deviations from the measured curve are quantified to determine the statistical significance with which any deviation can be rejected. The results for each individual galaxy are then combined to obtain the constraints in Fig. 10. (Note that these constraints probe the self-screening parameter ($\chi_{BG} = \chi_0 = 3|f_{R0}|/2$) at cosmic densities since the galaxies are unscreened.) Also shown are the distance indicator

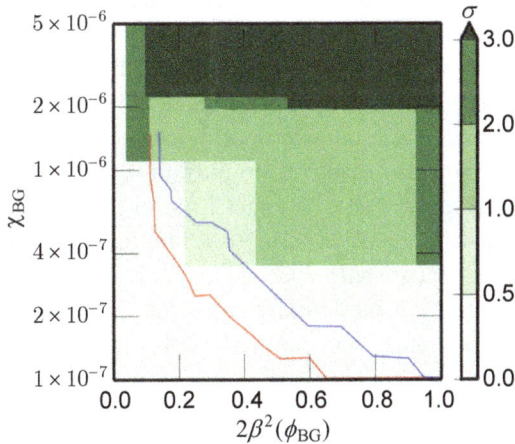

Fig. 10: Constraints on the value of χ_{BG} and $\beta(\phi_{BG})$ resulting from comparing the stellar and gaseous rotation curves of six unscreened galaxies in the screening map. The shade of green indicates the significance (σ) with which the models can be rejected; the exact scale is indicated in the figure. The 1 and 2σ bounds from distance indicator tests described in Sec. 4.5 are shown in blue and red respectively. Figure adapted from Ref. 812.

constraints for comparison. One can see that distance indicators are more constraining for large couplings but rotation curves can push into the regime $2\beta^2(\phi_{\rm BG}) < 0.1$. (Effects on distance indicator tests are subdominant to GR in this range.) The jaggedness of the contours is a result of the small sample size. A larger sample size with better kinematical data from both gaseous and stellar emission lines would greatly improve the constraints. It is possible that data from SDSS IV-MaNGA could provide such a sample although, to date, no analysis has been performed.

5.2. *Morphological and kinematical distortions*

Another consequence of the WEP violations discussed in Sec. 2 is that when $\chi_{\rm BG} \lesssim 10^{-6}$ (the Newtonian potential of main-sequence stars) it is possible for the stellar component of a dwarf galaxy to be self-screening whilst the surrounding dark matter halo and gaseous component is unscreened. This leads to several novel morphological and kinematical tests of thin-shell screened theories.[728] If a galaxy of mass M_1 is falling edge-on towards another larger (but unscreened) galaxy of mass M_2 a distance d away then the gas and dark matter will feel a larger external force than the stars and will hence fall at a faster rate. The stellar disk will then lag behind the gas and dark matter and become offset from the center. In the case of face-on infall the stars are displaced from the equatorial plane by a height[728]

$$z = \frac{2\beta^2(\phi_{\rm BG})M_1 R_0^3}{GM_2(R_0)d^2}, \tag{59}$$

where R_0 is the equilibrium distance from the galaxy's center. (z and R_0 can be taken to define cylindrical coordinates centered on the falling galaxy.) Since $M_2 \propto R_0^2$ with $n < 3$ for any sensible density profile this is an increasing function of distance from the center and one hence expects the stellar disk to be warped into a U-shape that curves away from the direction of in-fall.

   Reference 728 have simulated these scenarios by solving for the orbits of galaxies composed of 4000 stars for dark matter halos described using both NFW and CSIS profiles. The halo and gas are taken to be fully unscreened with $\beta(\phi_{\rm BG}) = 1/\sqrt{2}$ (corresponding to a fifth-force that is equal in strength to the Newtonian force). The halo falls from a distance of 240 kpc to a final distance of 100 kpc in 3 Gyr. The orbits are initially circular with a Gaussian scatter of 1 km/s. They considered two simple scenarios: edge-on infall and face-on infall. They identify the following three observational consequences of the WEP violation:

- **Offset stellar disk**: The reduced force on the stellar disk causes it to lag behind the HI gaseous component of the galaxy. An example of this is shown in Fig. 11 where $\mathcal{O}(\rm kpc)$ offsets are evident for CSIS galaxies. The offset is smaller for NFW profiles owing to the larger slope near the center and therefore larger restoring force.
- **Morphological warping**: The face-on infall cases exhibit a warping of the galactic disks whereby the stars were displaced from the principal axis by an amount

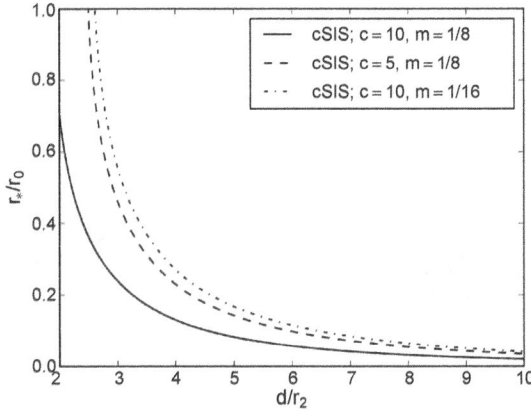

Fig. 11: The offset of the stellar disk as a function of d/r_2 (r_2 is the Virial radius of the falling galaxy) for CSIS profiles with parameters indicated in the figure. The figure uses concentrations $c = r_0/r_{\rm vir}$ with $r_{\rm vir}$ being the Virial radius and $m = M_2/M_1$. Figure taken from Ref. 728.

that increases with distance from the galactic center. An example of this is shown in Fig. 12.

- **Asymmetries in the rotation curves**: For edge-on in-falling galaxies, the stellar rotation curve becomes asymmetric compared with the HI curve. An example of this is shown in Fig. 13. One can see that the zero-velocity point of the stellar rotation curve is off-axis (in the opposite direction to the galaxy's motion) whilst the HI curve is symmetric and sits on-axis. Note that the effect discussed in the previous section—faster HI circular velocities than stellar circular velocities due to self-screening of the stars and unscreening of the dwarf galaxy—is also evident in the plot.

All of the effects found above are observable and the first attempt to use them to place constraints was made by Ref. 813 who analyzed data circa 2013. They searched for potential offsets between the HI and optical centroids using SDSS r-band optical measurements to trace the stellar centroid and ALFALFA radio observations of the 21 cm line to trace the HI centroid. In both cases they used a sample of unscreened galaxies taken from the screening map as well as a control sample of screened galaxies. A similar test was performed by looking for offsets between the optical centroid and kinematic HI centroid measured using the rotation curve. Both samples were consistent and a statistical analysis accounting for both astrophysical and modified gravity scatter did not allow the authors to place any meaningful constraints. The same authors searched for U-shaped warpings of nearly edge-on galaxies by aligning each galaxy image so that the principal axis lies along the horizontal direction then finding the centroids in each vertical column; no constraints could be placed due to the large error bars. The authors estimate that 8000 dwarf galaxies would be needed to test down to $\chi_{\rm BG} \sim 10^{-6}$ and 20000 to reach 10^{-7}. Finally, the authors

Initial Disk $t = 0$ Gyr   cSIS$_{2\text{kpc}}$ $t = 3$ Gyr

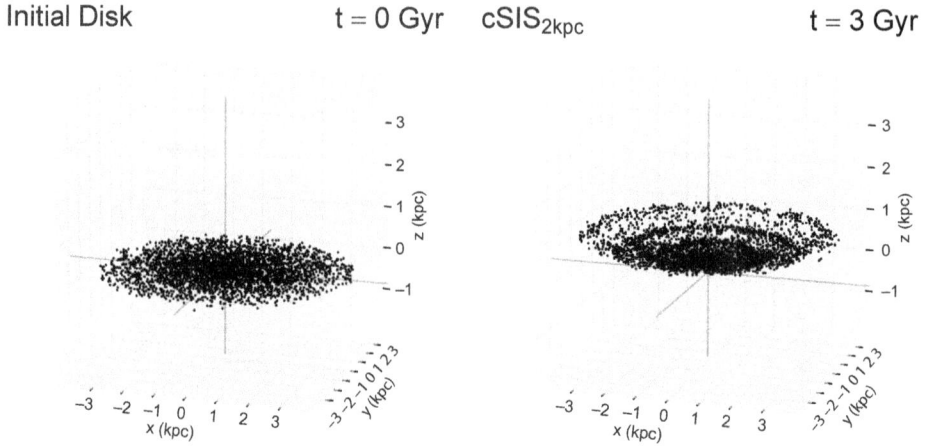

Fig. 12: Face-on infall, the neighboring galaxy lies in the negative z direction. *Left panel*: The initial stellar distribution in the galaxy. *Right panel*: Final distribution for a CSIS profile with $r_0 = 2$ kpc and $\rho_0 = 1.2 \times 10^7 M_\odot/\text{kpc}^3$. Figure adapted from Ref. 728 where more examples can be found.

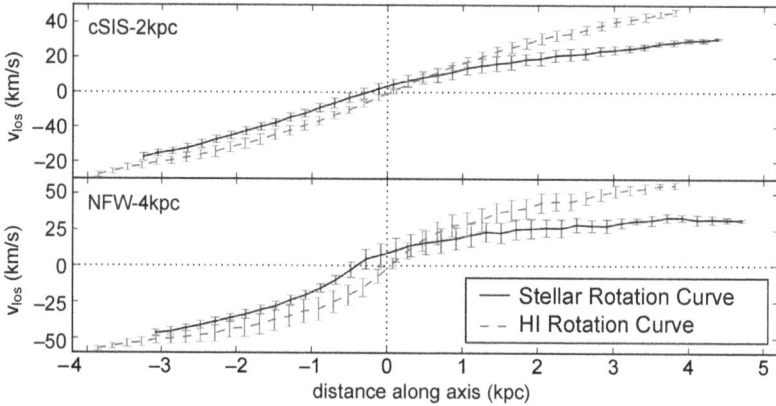

Fig. 13: The rotation curves (v_{los} is the line of sight velocity) for the stellar component (black, solid) and gaseous HI component (blue, dashed) after edge-on infall. The upper panel shows the same galaxy as Fig. 12 and the lower panel shows an NFW profile with $r_s = 4$ kpc and $\rho_s = 10^7 M_\odot/\text{kpc}^3$. Figure taken from Ref. 728.

tested the prediction of asymmetric rotation curves by using a weighted average of the difference in the velocity Δv of the approaching and receding sides of the rotation curves for Hα about the optical (stellar centroid) normalized to the maximum rotation velocity v_{max}. The GHASP Hα survey was used for this purpose. No constraints could be placed due to large uncertainties in the modeling of the inner halo as well as systematic uncertainties due to asymmetric drift and non-circular motion.

Very recently, Ref. 70 have used ALFALFA observations of a sample of 10,822 galaxies taken from an updated screening map[759] to constrain thin-shell theories by searching for offsets between the optical and HI centroids. Using a forward-modeling Bayesian likelihood method, they were able to obtain a new bound $\chi_0 = 3|f_{R0}|/2 < 1.5 \times 10^{-6}$. Improved measurements and larger samples from future surveys such as VLA or SKA could markedly improve these constraints. In particular, one could constrain $\beta(\phi_{BG}) \lesssim 10^{-3}$.[70]

6. Galaxy Cluster Tests

Galaxy clusters are another useful probe of modified gravity models. One reason for this is that they can be probed using both non-relativistic (dynamical and kine-matic) and relativistic (weak lensing) tracers and many modified gravity theories predict that the dynamical and lensing masses differ. Another is that they are some of the most massive objects in the universe and may enhance small fifth-forces (although they are also likely to be highly self-screening). The comparison of dynamical and lensing masses as a probe of modified gravity is covered in Chapter 6 and so here we will only focus on tests using fifth-force enhancements.

6.1. *Strong equivalence principle violations: Black hole offsets*

Galileon theories are difficult to test on small scales (unless they include Vainshtein breaking) due to the efficiency of the Vainshtein mechanism and, until recently, the strongest constraints came from the lack of deviations in the inverse-square law found using lunar laser ranging (LLR).[80] (Laser ranging to Mars could improve these by several orders of magnitude.[814]) One way the Vainshtein mechanism has been successfully constrained is using the SEP violations discussed in Sec. 2.2. The principle of this test, as first pointed out by Ref. 815, is the following: consider a galaxy falling in an external Newtonian and galileon field. The baryons (stars and gas) and dark matter all have scalar charge $Q = \alpha$ but the central black hole (in fact, any black hole) has zero scalar charge. The stars, gas, and dark matter therefore fall faster than the central black hole, causing it to lag behind and become offset from the center. Eventually, the restoring force from the remaining baryons at the center will compensate for the lack of the galileon force, leading to a visible offset.[r] This offset would be correlated with the direction of the galaxy's acceleration, thereby providing a smoking-gun signal.

One scenario for testing this (inspired by the proposal of Ref. 815) was proposed by Ref. 816. Satellite galaxies orbiting inside massive clusters are accelerating towards the center. When they are far away they can be outside the Vainshtein radius and see an unscreened galileon field but even inside the Virial radius there can be a large galileon contribution to the acceleration. The reason for this twofold. First, the Vainshtein mechanism is not as efficient for extended objects[545] as it is for

---

[r]In fact, it is possible for the black hole to escape the galaxy all together in some circumstances.[816]

point masses because the equations of motion are akin to Gauss' law where only the mass inside the radial coordinate contributes to the field profile. For this reason, the fifth-force profile of a point mass will be more suppressed than that of an extended object of radius R at distances $r < R$ due to some of the cluster's mass being at distances greater than r. When $r > R$ the force profiles will be identical, again due to Gauss' law. Second, 2-halo corrections boost the cluster mass at large radii[817] so that there is a larger source for the fifth-force outside the Vainshtein radius.

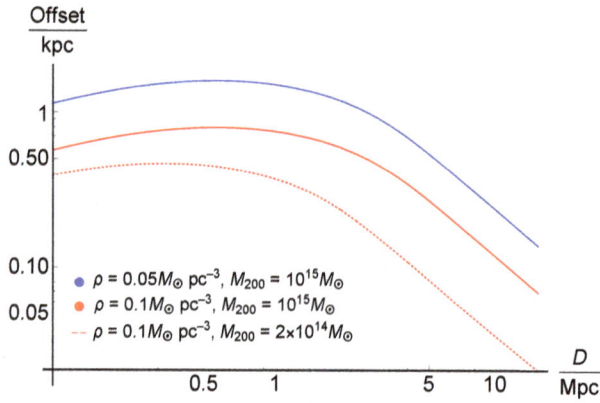

Fig. 14: The predicted offset for a cubic galileon model with $r_c = 500$ Mpc as a function of distance from the Virgo cluster center for different satellite central densities and cluster masses (M_{200} is the mass enclosed inside R_{200}) indicated in the figure. Figure taken from Ref. 816.

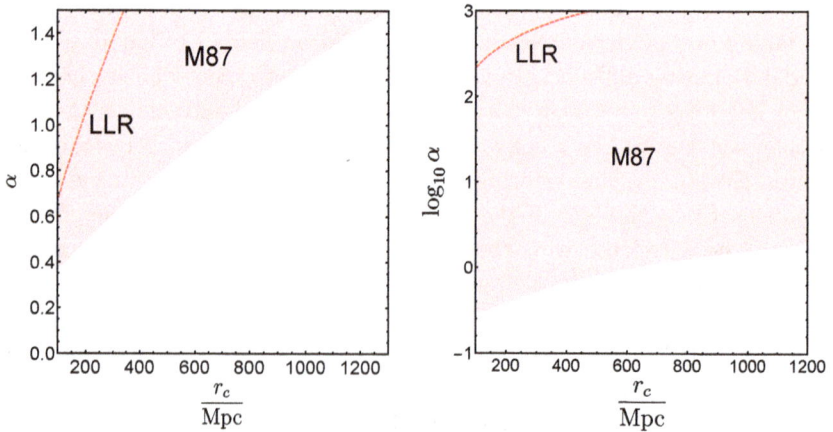

Fig. 15: Constraints on the galileon model parameters from the lack of a black hole offset in M87 in the Virgo cluster (pink shaded region). The red dashed line shows the older constraints from LLR. *Left panel*: Cubic galileon *Right panel*: Quartic galileon. Note that the scale for α is logarithmic. Figure in left panel adapted from Ref. 816.

Figure 14 shows the predicted offsets for the Virgo cluster (modeled using a concentration $c = 5$ NFW profile) for satellite galaxies with constant density profiles. One can see that offsets of $\mathcal{O}(\text{kpc})$ are predicted. Using a dynamical model of M87,[818] which is falling towards the center of the Virgo cluster, one finds that the central black hole is offset by no more than 0.03 arcseconds so that the galileon force $\lesssim 1000(\text{km/s})^2/\text{kpc}$. Combining this with the model for the Virgo cluster above,[816] obtained the constraints on cubic and quartic galileon models shown in Fig. 15. One can see that self-accelerating models ($r_c \sim 6\text{–}10 \times 10^3$ Mpc) are currently unconstrained but smaller values of r_c are excluded. Of course, this is just one system and[816] discuss how future X-ray and optical surveys could improve these bounds.

It is worth mentioning here that the black hole offset test is not unique to Vainshtein screened theories or, indeed, screened modified gravity. Any scalar-tensor theory will predict similar SEP violations. What is novel is the screening mechanism. In the absence of Vainshtein screening, scalar-tensor theories are best tested in the laboratory or solar system[64, 819] (or with the other astrophysical probes discussed here in the case of thin-shell screening). Vainshtein screening is so efficient that this difficult test is the most competitive.

7. Relativistic Stars

Relativistic stars are a good probe of alternative gravity theories but many of the classic tests (absence of dipole radiation for example) are not competitive for screened modified gravity. In the case of thin-shell screening, the screening is more efficient for objects with larger Newtonian potentials (i.e. relativistic objects) and any effects are highly degenerate with the EOS.[820] In the case of Vainshtein screening, the Vainshtein radius is several orders of magnitude larger than the radius of neutron stars and any deviations from GR are highly suppressed.[821] The exception is theories with Vainshtein breaking since the deviations inside astrophysical bodies can be important for the structure of compact objects. Given the above considerations, the entirety of this section will focus on Vainshtein breaking theories. See Refs. 822, 823 for reviews of compact objects in modified gravity theories.

One generic feature of scalar tensor theories with coupling strength α ($= 2\beta^2(\phi_{\text{BG}})$ for chameleons) is that there is a tachyonic instability for the scalar when the quantity[s]

$$1 - 3\frac{\tilde{P}}{\tilde{\rho}} \geq -\frac{1}{12\alpha^2}, \tag{60}$$

where tildes refer to Jordan frame quantities. This instability is never realized for screened modified gravity theories when sensible equations of state are used.[824, 825]

---

[s]This condition is equivalent to demanding that the trace of the energy-momentum tensor, which sources the scalar, becomes negative.

7.1. *Vainshtein breaking*

7.1.1. *Static spherically symmetric stars*

Unlike non-relativistic objects, for which the HSEE depends universally on Υ_1 independently of the specific theory, the Tolman-Oppenheimer-Volkoff (TOV) equation for Vainshtein breaking theories depends on both the theory, and the asymptotics. Indeed, since Υ_1 is a function of the cosmological time-derivative of the scalar (as well as H and second time-derivatives) one has $\Upsilon_1 = 0$ in an asymptotically Minkowski spacetime whereas in an asymptotically FRW spacetime Υ_1 may be non-zero. The situation is complicated further by the fact that there are three branches of solution for the scalar field (the equation of motion reduces to a cubic after manipulation) and the correct branch (the one which gives the correct asymptotics) requires a fully relativistic calculation to determine. For this reason, current works have used specific models that admit exact de Sitter (dS) solutions so that one can determine the correct branch of solution (and therefore Υ_1) in a controlled and systematic manner. References 794, 826, 827 have identified several models that have exact dS solutions and exhibit Vainshtein breaking. The derivation of the TOV equation for these models is long and complicated so we refer the reader to Refs. 794, 826, 827 for the full details. Here we will only sketch the derivation. One first solves the equations of motion to find an exact de Sitter solution. Next, the metric potentials and scalar are perturbed by introducing a perfect fluid source. One finds an exact Schwarzschild-de Sitter metric in the exterior of the star and the correct branch of solution for the scalar inside the star is the one that matches onto this solution. Taking the sub-Horizon limit, one can eliminate the scalar completely leaving a system of three equations that must be solved, two for the metric potentials and one for the pressure. These are the Vainshtein breaking counterparts to the TOV equation found in GR. Given an equation of state, these can be solved with appropriate boundary conditions (spherical symmetry at the center and vanishing pressure at the radius) to find the structure of the star. In the simplest models, one finds a universal parameter $\Upsilon_1 = \Upsilon_2 = \Upsilon$ ($\Upsilon_3 = 0$).

Reference 823 has solved the TOV equations using both $n = 2$ relativistic polytropic and two realistic (BSK20 and SLy4) neutron star equations of state. Furthermore, Ref. 827 have solved using 32 equations of state, including some that include hyperons, kaons, and strange quark matter. An example of the neutron star mass-radius relation found using the SLy4 EOS is shown in Fig. 16. Also shown is the mass of the heaviest neutron star observe to date (PSR J0348+0432 with mass $M = 2.01 \pm 0.04 M_\odot$).[828] One can see that positive values of Υ make the stars less compact i.e. they have lower masses at fixed radii. In the case of SLy4, even reasonably small values of $\Upsilon > 0$ result in maximum masses that are not compatible with the mass of PSR J0348+0432 but since the EOS of dense neutron matter is not known this just implies that the SLY4 EOS is not compatible with Vainshtein breaking theories with these values of Υ and, indeed, changing the EOS one can find masses in excess of $2M_\odot$. Negative values of Υ_1 produce stars that are more

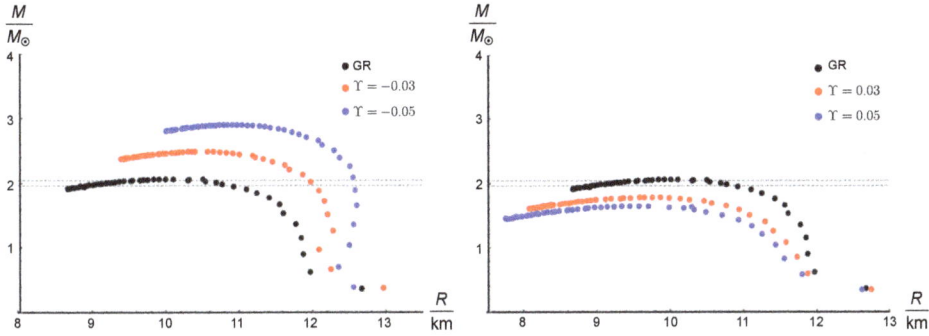

Fig. 16: The mass-radius relation for neutron stars in Vainshtein breaking theories for the values of Υ given in the figures. The GR prediction corresponds to $\Upsilon = 0$ and the grey dashed line shows the upper and lower bounds on the mass of the heaviest neutron star presently observed.[828] Figures taken from Ref. 826. *Left panel*: $\Upsilon < 0$. From top to bottom: $\Upsilon = -0.05$, $\Upsilon = -0.03$, GR. *Right panel*: $\Upsilon > 0$. From top to bottom: *GR*, $\Upsilon = 0.03$, $\Upsilon = 0.05$.

compact so that they have larger masses and fixed radii. Although Fig. 16 only shows one EOS, the features exhibited there are generic for all the equations of state studied by Refs. 827. Another interesting prediction evident from the figure is that the maximum mass can be far in excess of the GR prediction. While an exact mass is dependent on the EOS, Ref. 827 has noted that, for some equations of state, the radius and mass of higher-mass stars can violate the GR causality bound (i.e. these stars would host internal pressure waves that would travel superluminally in GR) and so the observation of stars with these properties would be in tension with GR. The equivalent causality bound in Vainshtein breaking theories is unknown at present since calculating it is a more difficult task. In particular, the kinetic mixing of the scalar and metric would require one to find the sound speeds of both the scalar and pressure modes simultaneously.

7.1.2. *Slowly rotating stars*

A more robust method of testing gravity with relativistic stars is to use relations that are independent of the equation of state. In particular, it is well known in GR that there is a relation between the dimensionless moment of inertia $\bar{I} = I/(G^2 M^3)$, where I is the moment of inertia, and the compactness $\mathcal{C} = GM/R$.[829,830] The compactness of a given star can be computed by solving the appropriate TOV equation but in order to compute the moment of inertia one needs to solve the equations for a slowly rotating star to first order in its angular velocity ω. Given the complexity of the equations, we will once again sketch the procedure for calculating these quantities and refer the reader to Ref. 794 for the full details. The method essentially follows the method of Hartle[831] applied to scalar-tensor theories.

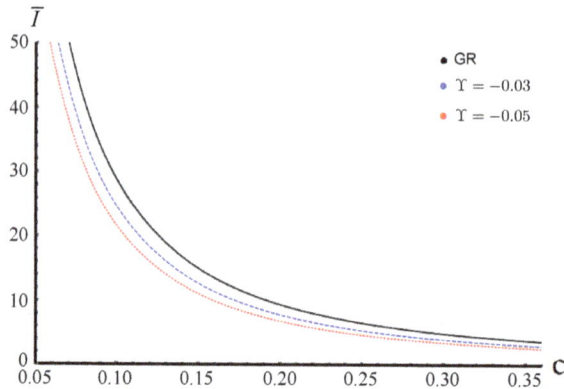

Fig. 17: The $\bar{I}$–$\mathcal{C}$ relation in GR and Vainshtein breaking theories with parameters indicated in the figure. From top to bottom: GR (black, solid), $\Upsilon = -0.03$ (blue, dashed), $\Upsilon = -0.05$ (red, dotted). Figure taken from Ref. 794.

The first step is to perturb the Schwarzschild-de Sitter metric to include the star's rotation with an angular velocity Ω.[t] For slowly rotating objects, this plays the rôle of the small perturbation parameter. The quantity that must be calculated is $\omega(r)$, the coordinate angular velocity of the star as measured by a freely-falling observer. One finds that the scalar is only perturbed at $\mathcal{O}(\Omega^2)$ whereas the $\mathcal{O}(\Omega)$ contribution to the perturbed tensor equations yield and equation of motion for ω of the form

$$\omega'' = K_1(\delta\nu, \delta\lambda, \rho, \rho, P, P', \Upsilon)\omega' + K_0(\delta\nu, \delta\lambda, \rho, \rho, P, P', \Upsilon)\omega, \qquad (61)$$

where K_1 and K_0 are given in Ref. 794 and reduce to their GR values of $K_1 = 4/r$ and $K_0 = 0$ when $\Upsilon = 0$. The moment of inertia is found using the relation

$$\omega(R) = \Omega\left(1 - \frac{GI}{R^3}\right). \qquad (62)$$

Reference 829 found that a function of the form

$$\bar{I} = a_1\mathcal{C}^{-1} + a_2\mathcal{C}^{-2} + a_3\mathcal{C}^{-3} + a_1\mathcal{C}^{-4} \qquad (63)$$

fitted the GR $\bar{I}$–$\mathcal{C}$ relation well and so Ref. 794 fit the relation found in Vainshtein breaking theories to the same function (the reader is referred to the original reference for the numerical coefficients). Their results are shown in Fig. 17. One can see that Vainshtein breaking theories also predict an $\bar{I}$–$\mathcal{C}$ that depends on Υ and furthermore that it is distinct from the GR prediction. Therefore, in principle, measuring the $\bar{I}$–$\mathcal{C}$ relation could place new bounds on Vainshtein breaking theories. In practice, this measurement is a while away since one needs to find highly relativistic systems where the (post-Newtonian) spin-orbit contribution to the precession can be measured. There are few known systems at the present time, although the next generation of

---

[t]The perturbation decays at infinity so that there is no change to the asymptotic spacetime.

radio surveys should be able to find more, making this measurement possible on a time-scale of a decade or so. Since Vainshtein breaking theories screen outside bodies, the measurement of the spin-orbit coupling, and therefore I itself is the same as in GR.

8. Astrophysical Tests of Couplings to Photons

There have been many studies of chameleon theories that couple to photons via a term

$$\Delta \mathcal{L} = \beta_\gamma \frac{\phi}{M_{pl}} F_{\mu}\nu F^{\mu}\nu = \frac{\phi}{M_\gamma} F_{\mu}\nu F^{\mu}\nu \tag{64}$$

in the Lagrangian.[64,819,832] Mixing between the scalar and photons can induce both linear and circular polarizations into the (nominally unpolarized) starlight in the inter-galactic medium (IGM).[833] The lack of any observed polarization places the bound $M_\gamma > 1.1 \times 10^9$ GeV provided $m_{eff}(\phi_{IGM}) < 1.1 \times 10^{-11}$ eV. The coupling (64) can also act as a loss mechanism whereby photons are converted to chameleons. This can result in deviations in the X-ray luminosity functions of active galactic nuclei (AGN), the lack of which places imposes $M_\gamma > 10^{11}$ GeV for $m_{eff}(\phi_{AGN}) < 1.1 \times 10^{-12}$ eV. Similarly, the lack of any observed depletion of CMB photons in the COMA cluster constrains $M_\gamma > 1.1 \times 10^9$ GeV.[834] Finally, the depletion of CMB photons increases the opacity of the universe and alters the distance-duality relation,[835] although current constraints are not competitive with those discussed above. This may change with data releases from current and next generation cosmological surveys.

Acknowledgment

J.S. is supported by funds provided to the Center for Particle Cosmology by the University of Pennsylvania.

Chapter 9

Laboratory Constraints

Philippe Brax

Institut de Physique Théorique, Université Paris-Saclay, CEA, CNRS,
F-91191 Gif sur Yvette, France

Clare Burrage

School of Physics and Astronomy, University of Nottingham,
Nottingham, NG7 2RD, UK

Anne-Christine Davis

DAMTP, Centre for Mathematical Sciences,
University of Cambridge, CB3 0WA, UK

1. Introduction

Dark energy is a cosmological phenomenon *per se*.[26,836] In this chapter we will describe attempts to detect effects of the physics of modified gravity,[18] motivated by dark energy and the cosmological constant problem,[837] in the laboratory. Classical effects of modified gravity can be tested by fifth force searches where new classical interactions could influence the motion of test masses.[838] The quantum nature of the modifications can also be probed using the Casimir effect,[839] atom[840] and neutron[841] interferometry, and the neutron energy levels in vacuum,[842] see Ref. 843 and references therein. Theoretically we have restricted the models to scalar-tensor theories with a coupling between a scalar field and matter.[19] The coupling to photons and its quantum-mechanical origin is also recalled.[832]

Laboratory tests of gravity have a long history and, as discussed in Sec. 2, the need to make cosmological theories of dark energy and modified gravity compatible with laboratory and solar system tests was a key motivation for the introduction of screening mechanisms.[660] In broad terms the goal of screening mechanisms is to allow the additional scalar field to give rise to modifications of the standard cosmology on the largest scales in the Universe, whilst being un-observable on the shorter distances and at the higher densities present on Earth. However this is not the end of the story, as we will see in this section; carefully designed experiments can allow the effects of the scalar field to be unscreened. The additional level of precision and control that we have in the laboratory then means that these measurements tend to be extremely constraining for cosmological modified gravity theories.

In the last ten years, the variety of experimental techniques which have been introduced to test modified gravity is quite astonishing.[819] From the classic Casimir and fifth force experiments to levitating microspheres and atomic interferometry, the effects of the scalar fields are all in the non-linear regime of the theories. This is different from most of the tests on cosmological scales where the linear regime is the easiest to probe. So far no sign of modified gravity with screening has been unravelled in the linear regime motivating studies of the non-linear properties of these theories. Hence the laboratory experiments are a useful complement to future large cosmological surveys. On the other hand, and as the non-linear regime is what will be the subject of this chapter, the analysis has to be mostly dealt with in a case by case basis. No model independent parameterised description is yet available for laboratory tests, and the non-linear regime they probe can only be connected to the parameterised descriptions of the linear perturbation theory relevant on the largest cosmological scales on a case by case basis. As an example and as there are already similar reviews in the literature[19, 64, 142, 819] we have decided to concentrate some of the technical aspects to less developed models such as the environmentally dependent dilaton[59] which can be treated almost completely analytically and provides a nice template for more complex models.

The types of laboratory tests that will be presented here really only probe the screening properties of modified gravity theories, i.e. the fact that very light scalar fields involved in the late time acceleration of the expansion rate of the Universe would induce far too large deviations from General Relativity in the solar system and therefore must be shielded from matter locally. Types of screening can be classified in two different ways: Firstly by the highest order of the derivative terms which appear in the non-linear terms. From zero to two derivatives these are: chameleon[57, 58] and dilaton-symmetron[59, 60, 844–846, a], K-mouflage[61, 847, 848] and Vainshtein.[62] Secondly by the class of term in the scalar Lagrangian in which the non-linear terms are present. For the chameleon, symmetron (or dilaton) and kinetic models (Vainshtein and Galileon) the non-linearities appear in the scalar potential, matter coupling and kinetic terms respectively. Unfortunately, so far, it does not appear to be possible to probe screening which relies on derivative self interactions with laboratory experiments. This is because the nature of the screening means that variations in the field occur only over longer distance scales than can be probed terrestrially. Only chameleon and symmetron or dilaton models are sufficiently local to respond to variations of matter densities on laboratory scales. They will be the main focus of this chapter although the techniques presented here can be applied to other models too.

---

[a]Chameleons screen with an increase of the scalar mass in a dense environment whereas dilaton-symmetrons have the Damour-Polyakov property of decreasing couplings to matter in dense regions.

2. Experimental Constraints

Laboratory experiments are most effective at constraining theories which screen through a chameleon-like mechanism, i.e. *thin shell* theories. Thin shell theories have the advantage that the scalar field responds rapidly to changes in density, meaning that even if the effects of the scalar are screened in the solar system they can be unscreened by a laboratory vacuum chamber. In this section we will first describe how a thin shell scalar behaves in a laboratory vacuum, and then go on to detail the laboratory experiments which currently are the most constraining for chameleon models. We will discuss in the following section the case of the dilaton which corresponds to the Damour-Polyakov screening,[844] and we will see that laboratory experiments are less effective. For convenience sake, we recall that the standard chameleons have an inverse power law potential[25, 849]

$$V(\phi) = \frac{\Lambda^{4+n}}{\phi^n} + \dots \tag{1}$$

where the ... include a cosmological constant piece as the chameleons do not lead to self-acceleration. The coupling of ϕ to matter is constant $\beta = \frac{M_{pl}}{M}$ corresponding to an exponential coupling function $A(\phi) = e^{\beta\phi/M_{pl}}$ acting as a rescaling between the Jordan frame, where particles couple to the metric independently of the scalar field, and the Einstein frame where Newton's constant is scalar field independent but particle masses are rescaled by the coupling function. Chameleons can be embedded in supersymmetric models for instance where the two functions $V(\phi)$ and $A(\phi)$ are more complex, with no changes to the chameleon mechanism itself.[850]

2.1. *Chameleons in laboratory vacuums*

The chameleon scalar field changes its mass as a function of the local density. We consider here an idealized vacuum chamber, which is spherical with internal radius L, internal density ρ_{vac} and walls of density ρ_{wall} and thickness T. If $T > 1/m(\rho_{wall})$, where $m(\rho_{wall})$ is the mass of the scalar field in a uniform density ρ_{wall}, then we know that within the walls of the vacuum chamber the chameleon reaches the field value which minimises its effective potential. This greatly simplifies our calculations, as it means that we can ignore the behaviour of the chameleon in the exterior of the vacuum chamber, and just focus on the interior, as long as we impose the boundary condition that the chameleon minimises its effective potential within the walls. Whilst the condition $T > 1/m(\rho_{wall})$, needs to be checked experiment by experiment, and chameleon model by chameleon model, in general we find that this is satisfied for chameleon models of interest if $T \gtrsim 1$ mm.

In the interior of the vacuum chamber the density is much less than within the walls, therefore the chameleon will try to adjust its value to reach the value which minimises the effective potential for this lower density. If $L > 1/m(\rho_{vac})$ then at the centre of the vacuum chamber the chameleon will have reached this minimising value. For smaller vacuum chambers the chameleon field will still evolve in the interior, it just will not have enough room to reach the value which minimises the effective potential. In this case we find that at the centre of the vacuum chamber the chameleon takes a value so that its Compton wavelength is of order the size of the vacuum chamber $1/m(\rho_{\text{vac}}) \sim L$. The order one constant of proportionality varies depending on the choice of chameleon model.[64, 819] It is not generally possible to solve analytically for the full form of the chameleon profile inside the vacuum chamber, but it is possible to compute it numerically.[851, 852] In Sec. 3, we present analytical approximation for the field profile inside a cylinder where the different regimes can be easily seen.

2.2. *Unscreening inside the vacuum chamber*

The conditions for a source mass to be screened will be reviewed in the next section in Eqs. (4), (45). Whether or not an object is screened depends on the value of the scalar field in the interior of the source mass, and the background value that the scalar would take if the source were absent. The advantage of performing experiments in laboratory vacuua is that, if the field is able to respond to the lower density of the vacuum as described in the previous subsection, then the difference between the background value of the scalar field, and the value that minimises the effective potential in the interior is increased. In a cavity the value of the field is in general lower, for chameleons, than the value which minimises the effective potential[b] implying that $|\phi_{\text{BG}} - \phi_A|$, see Eq. (4), decreases therefore implying that objects A are more screened; here BG refers to the background in which a given object is embedded, where ϕ_{BG} is the background value of the scalar field. In particular dense objects embedded in a cavity with $m_{\text{vac}}L \lesssim 1$ are such that $\phi_{\text{BG}} \sim \phi_A \sim \phi_{\text{walls}}$ and are therefore screened when the coupling β does not vanish. For symmetrons, the situation is exactly the opposite as when $\phi_{\text{BG}} \sim \phi_A \sim \phi_{\text{walls}} = 0$ for $m_{\text{vac}}L \lesssim 1$, the coupling β vanishes and dense objects are unscreened. This is a major difference between chameleon-like and symmetron-like models.

For instance it has been shown that, at least in parts of the chameleon parameter space neutrons, atomic nuclei and silica microspheres can be unscreened in vacuua with $L \sim 10$ cm and $\rho_{\text{vac}} \sim 10^{-17}$ g/cm^3.

In the following subsections we review the most constraining experiments for chameleon models. These constraints are summarized in Fig. 1.

2.3. *A comparison with astrophysical tests*

In the astrophysical tests of screened and unscreened models of dark energy and modified gravity it is often useful to use

$$\chi_{\text{BG}} = \frac{\phi_{\text{BG}}}{2\beta_{\text{BG}} M_{\text{pl}}} \qquad (2)$$

---

[b]The profile of the scalar field inside a cavity is bubble-like with a low value inside the walls and a higher value in the middle of the cavity.[853] This large value is always lower than the value in infinite vacuum as the field does not have enough space to reach this maximal value. This is a finite size effect which disappears when the size of the cavity is pushed to infinity.

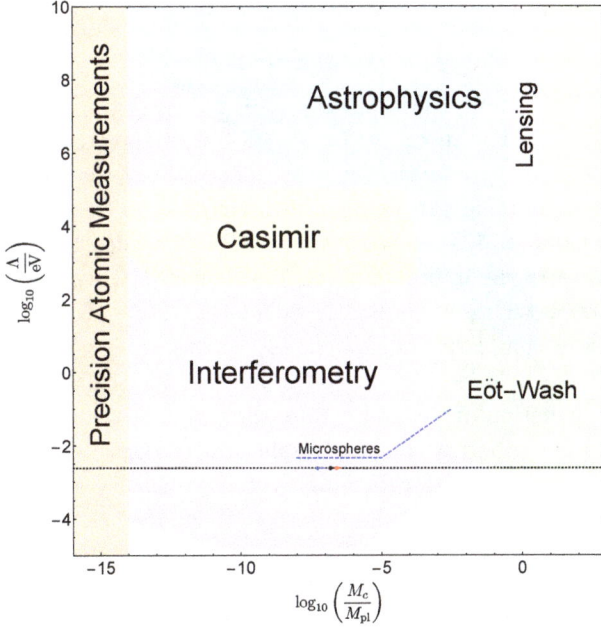

Fig. 1: Experimental constraints on the chameleon self-coupling Λ and coupling to matter $M_c = \frac{M_{\rm pl}}{\beta}$, for the chameleon potential $V(\phi) = \Lambda^5/\phi$.[819]

in order to discriminate screened and unscreened objects. Here $\beta_{\rm BG} = \beta(\phi_{\rm BG})$ refers to the background value of the scalar field coupling to matter. A given object is screened in this environment when

$$\Psi_A \geq \chi_{\rm BG} \tag{3}$$

where Ψ_A is the Newtonian potential at the surface of the object A. This is particularly useful in astrophysics where a host of phenomena happen in the same environment, e.g. within a galaxy where for instance one may be interested in the screening of stars or the galactic gas. Hence a uniform criterion depending uniquely on the Newtonian potential makes sense and is practical. For laboratory experiments, the background is far from being standardisable, i.e. it could be a cylindrical cavity, the two plates of Casimir experiments or the single mirror for neutron energy levels. As a result it is far more efficient to introduce the scalar charge

$$Q_A = \frac{|\phi_{\rm BG} - \phi_A|}{2M_{\rm pl}\Phi_A} \tag{4}$$

where ϕ_A is the value of the scalar at the centre of the source object. For a specific object in a given environment Newton's law is corrected as

$$V_{AB} = -\frac{m_A m_B}{r}(1 + 2Q_A Q_B) \tag{5}$$

when the interaction range of the scalar is much smaller than the distances probed by the experiment, and V_{AB} is the interaction potential between two masses, see

Ref. 854 for a derivation. The two bodies are screened if

$$Q_{A,B} \leq \beta_{\mathrm{BG}} \qquad (6)$$

and when screening is not operating $Q_{A,B} \to \beta_{\mathrm{BG}}$ in V_{AB}. This is what will be used in the following. For instance for atomic interferometry, the atoms will be unscreened whilst the source object will be. The ratio of the charge of the object to the background coupling can be identified with $3\frac{\delta R}{R} = \frac{\delta M}{M}$ where δR is the size of the thin (see Chapter 1 for more detail) shell over which the field varies inside the object and creates the interaction with another object, see the definition of the charge in Chapter 8 where it is directly defined in terms of the mass of the thin shell itself in a way completely equivalent to the way we have introduced it. Deep inside the object the field is constant as the mass of the scalar field is too large to allow for any propagation of the scalar from the inside to the outside of the object. When bodies are not spherical, or nearly spherical, as for the Casimir effect where infinite parallel plates are used, screening operates when

$$m_{\mathrm{plate}}d \gg 1 \qquad (7)$$

where m_{plate} is the mass of the chameleons inside the plates and d their width. For most models this implies that the field effectively vanishes inside the plates and varies between the plates where near vacuum has been realised.

2.4. *Atom interferometry*

Atom interferometry relies on the wave-particle duality of atoms. The wave function of an atom is split into two parts, which describe the centre of mass position traveling along two different paths. The paths start and end at the same point, but are spatially separated otherwise. At the final point the phases of the two parts of the wavefunction interfere, allowing for a difference in the phase accumulated along the two paths to be detected (somewhat analogously to the double slit experiment).

The spatial position of the atoms within the experiment can be controlled using lasers. Once they are cooled sufficiently, the momentum of an atom can be increased by absorption of a laser photon with frequency tuned to a particular atomic excitation, as conservation of momentum requires the atom to inherit the three momentum of the absorbed photon. Similarly the atom can be made to lose momentum by stimulated emission. The probability of whether the atoms absorb (or emit) a photon can also be controlled, so that with three laser pulses it is possible to split the atomic wavefunction into two parts (the analogue of the beam splitter in a classical interferometer) to reverse the direction of motion of the atoms (the analogue of the mirrors) and to recombine the wavefunction (the analogue of the second beam splitter).

Differences in the phase of the wavefunction along the two paths can result either from a difference in the accumulated action along each path, or from differences in the phase inherited from the photon at each interaction. If the atoms are experiencing a constant acceleration along the direction described by the laser pulses, then the difference in phase is particularly simple. As moving the atoms around within the experiment directly correlates with whether the atoms are in the excited or unexcited state it is most convenient to express this in terms of the probability of finding the atom in the excited state at the output of the interferometer,

$$P \propto \cos^2 \left[\frac{akT^2}{\hbar} \right], \tag{8}$$

where k is the photon momentum, T is the time between laser pulses (so that $2T$ is the duration of the experiment), and a is the constant acceleration.

An acceleration a can be caused by the gravitational, or chameleon forces due to a massive source object being placed inside the vacuum chamber. This massive source is typically sufficiently large that it will be screened, but the atomic nuclei are small enough that, as discussed in the previous subsection, they are unscreened over a large range of the parameter space,[851, 855, 856] making them sensitive probes of the chameleon field. Experiments searching for chameleon accelerations with atom interferometry have reached a sensitivity of $10^{-8}g$ ($g \equiv GM_\oplus/R_\oplus$ is the gravitational acceleration at the surface of the Earth).[840, 857, 858]

2.5. *Eöt-Wash*

Torsion balance experiments have a long history of searching for fifth forces, and modifications of gravity. The principle underlying the experiments is to have one or more test masses suspended, and to look for deflections of the test masses towards source masses by measuring the torsion in the suspension of the test masses. Commonly the source and test masses are arranged so that the inverse-square contribution to the total force is canceled, and the experiment is sensitive to deviations from standard gravity.

The current best constraints come from the Eöt-Wash experiment,[838, 859, 860] which uses circular disks for the masses. The disks have holes bored in them, and are arranged one above the other so that if there are no modifications to gravity there is expected to be no net torque between the two plates.

One of the challenges of these experiments is to reduce as far as possible any electromagnetic forces between the plates that could be mistaken for modifications of gravity. One of the ways that this is done in the Eöt-Wash experiment is to place a beryllium-copper membrane between the plates. This still allows the experiment to search for fifth forces that are not screened, but the presence of the plate can act to screen out, for example, chameleon forces between the plates.[87, 761, 762, 861–864] This reduces the sensitivity of the experiment to screened fifth forces.

2.6. *Casimir*

The Casimir force is an effect predicted by quantum electrodynamics, which is absent in classical physics. It is the force that arises between two parallel plates, placed in vacuum, due to the quantum fluctuations of the electromagnetic field in the space between the plates. This force scales as d^{-4}, where d is the distance between the plates, and therefore is most easily detected when the plates are placed close together, current experiments probe sub-mm and sub-micron distance scales.[865, 866]

If fifth forces exist they could also be detected by an experiment searching for Casimir effects. These experiments are particularly sensitive to screening through the thin-shell effect, as close to the surface of a source the field is changing rapidly, giving rise to potentially detectable forces. The chameleon force (per unit area) between two plates scales as[762, 843, 867]

$$\frac{F_{\text{cham}}}{A} \propto d^{-\frac{2n}{n+2}},\tag{9}$$

The experimental challenge for such a search, is to make the two plates perfectly smooth and to keep the plates perfectly parallel. In practice it may be easier to search for the Casimir effect between a plate and a sphere, or between two spheres instead. In this case, the Casimir force scales as d^{-3} and the chameleon force would scale as

$$\frac{F_{\text{cham}}}{A} \propto d^{\frac{2-n}{n+2}}.\tag{10}$$

Current searches for the Casimir force are most constraining for chameleon models with $n = -4$ and $n = -6$ when Λ_c is fixed to the dark energy scale. A new generation of these experiments, specifically tailored to look for the chameleon force with parallel plates and larger separations is currently being developed.[860, 865] Further sensitivity to the chameleon could be obtained by varying the density of the gas between the two plates.[868, 869]

2.7. *Quantum bouncing neutrons*

Neutrons can be used to test for the presence of new interaction with the qBOUNCE experiment.[870–872] They are particularly useful as they are both neutral and with a low polarisabily, hence hardly subject to electromagnetic interactions and therefore well suited for gravitational tests. These experiments use ultra-cold neutrons in the terrestrial gravitational potential above a mirror with a large enough Fermi potential to reflect neutrons totally. As first obtained in Ref. 873 the energy eigenstates of the neutrons are discrete. The basic setup can be found in Ref. 871 where the energy resolution between the level is as low as 3×10^{-15} eV.[874] Recently the transitions between the energy ground state $E_1 = 1.40672$ peV and the excited states $E_3 = 3.32144$ peV as well as $E_4 = 4.08321$ peV have been observed. This can be achieved as follows. First, the neutrons encounter a state selector for the ground state $|1\rangle$ having energy E_1. This combines a polished mirror at the bottom and a rough absorbing scatterer at the top separated by about 20 μm. Neutrons in excited states with a diffuse wave function are scattered out of the system. Then a horizontal mirror undergoes harmonic oscillations with a tunable frequency ω, which drives the system into a coherent superposition of ground and excited states. Finally the neutrons go through a selector which is identical to the first one and act as a ground state selector. When the neutrons are excited to higher levels than the ground state in the second region, nothing is transmitted and a dip in the transmission rate at a given frequency is observed. These dips allow one to measure the energy differences between the ground state and typically the first few levels, e.g. the third and fourth.

The quantum-mechanical description of a neutron above a mirror in the terrestrial gravitational potential is given by the Schrödinger equation,[875]

$$-\frac{\hbar^2}{2m}\frac{\partial^2 \psi_n(z)}{\partial z^2} + mgz\,\psi_n(z) = E_n\psi_n(z)\,. \tag{11}$$

with a characteristic length scale

$$z_0 = \sqrt[3]{\frac{\hbar^2}{2m^2g}} = 5.87\,\mu\text{m}\,, \tag{12}$$

and a typical energy scale $E_0 = ((\hbar^2 mg^2)/2)^{1/3}$ which are given by the mass m of the neutron and the acceleration of the Earth g. Above the mirror the normalized wavefunctions for $z > 0$ read

$$\psi_n^{(0)}(z) = C_n^{(1)}\,\text{Ai}\left(\frac{z - z_n}{z_0}\right)\,, \tag{13}$$

with normalisation $C_n^{(1)} = \dfrac{1}{\sqrt{z_0}\,\text{Ai}'\left(-\frac{z_n}{z_0}\right)}$, and $z_n = \frac{E_n}{mg}$. Here z_n is the n-th zero of the Airy function which characterises the energy levels of the neutrons. Outside this region the wavefunctions vanish as the neutrons do not penetrate inside the mirror. The first few energy levels are given in Table 1 where one can see that the energy levels are not equally spaced like for a harmonic oscillator. When a new interaction of the chameleon type is present, the potential is shifted to

$$V(z) = mgz + m(A(\phi(z)) - 1) \tag{14}$$

where $\phi(z)$ is the profile of the scalar field above the mirror. This can be easily obtained for chameleons or symmetrons for instance. The perturbations to the n-th energy level is obtained as $\delta E_n = m\langle n|(A(\phi(z)) - 1)|n\rangle$ to first order in perturbation theory, where the ket $|n\rangle$ associated to the n-th eigenfunction is defined by $\langle z|n\rangle =$

Table 1: Values of the energy of the lowest six states for the neutrons in the terrestrial gravitational field.

State	Energy [peV]
$|1\rangle$	$E_1 = 1.40672$
$|2\rangle$	$E_2 = 2.45951$
$|3\rangle$	$E_3 = 3.32144$
$|4\rangle$	$E_4 = 4.08321$
$|5\rangle$	$E_5 = 4.77958$
$|6\rangle$	$E_6 = 5.42846$

$\psi_n^{(0)}(z)$. This has to be less than the precision of order 3×10^{-15} eV and leads to interesting constraint on modified gravity models.

2.8. *Precision atomic tests*

As we have discussed above atomic nuclei can be unscreened in a laboratory vacuum. In the atom interferometry experiments discussed above the atoms were test particles probing the chameleon field due to a macroscopic source. But the nuclei can also be considered as the source of a chameleon field that is probed by the orbiting electrons. If the unscreened chameleon force is very strong then this could cause measurable perturbations to atomic energy levels, with an extra contribution to the electron Hamiltonian of the form

$$\delta H = \frac{m_e}{M} \phi_N, \tag{15}$$

where ϕ_N is the chameleon field sourced by the nucleus.

The most precise measurements currently are of the structure of hydrogenic atoms. The shifts to the lowest energy levels due to a chameleon force are[876]

$$\Delta E_{1s} = -\frac{Z m_N m_e}{4\pi a_0 M^2}, \tag{16}$$

$$\Delta E_{2s} = \Delta E_{2p} = -\frac{Z m_N m_e}{16\pi a_0 M^2}, \tag{17}$$

where Z is the atomic number, m_N is the nucleon mass, and a_0 is the Bohr radius. The potential coupling of the chameleons to photons, discussed in the next section, will break the degeneracy between the 2S and 2P levels.

The best measured transition is currently the 1S-2S transition in atomic hydrogen, with a total uncertainty of 10^{-9} eV (at 1σ).[877–879] No signs of a deviation from standard electromagnetism have been found, and so the chameleon coupling must be constrained to be

$$M \gtrsim 10 \text{ TeV.} \tag{18}$$

2.9. *The symmetron*

The constraints from all of the experiments detailed above have also been studied for the symmetron model.[60] The symmetron is similar to the chameleon in that it has canonical kinetic terms, and its screening is through terms that are non-linear in the field. However the difference between the models is that the chameleon can screen because it varies its mass with the environment, and the symmetron because it varies the strength of its coupling to matter. This occurs because the symmetron has a spontaneous symmetry breaking potential, and couples to matter in such a way that regions of high density can restore the symmetry. The resulting effective potential is:

$$V_{\text{eff}}(\phi) = \frac{1}{2}\left(-\mu^2 + \frac{\rho}{M^2}\right)\phi^2 + \frac{\lambda}{4!}\phi^4, \tag{19}$$

where μ is the bare mass of the symmetron, M the energy scale controlling strength of the coupling to matter, and λ the dimensionless constant controlling the self interactions of the field. The form of the coupling to matter also means that the symmetron fifth force experienced by a test particle is $\vec{F} = \phi\vec{\nabla}\phi/M^2$. The consequence of this is that when the symmetry is restored in regions where $\rho > \mu^2 M^2$ the fifth force is switched off. As can be seen from Eq. (19), the mass of the symmetron in the symmetry broken phase is approximately μ. Unlike the chameleon, therefore, the symmetron does not have the ability to adjust its mass in the low density environment of a laboratory vacuum chamber. If the Compton wavelength of the symmetron is larger than the size of the vacuum chamber, $\mu L \ll 1$, the field is not able to vary within the chamber and so no fifth force can be present. Conversely, if the Compton wavelength of the symmetron is smaller than the distances probed in the experiment (for example the distance between test and source masses) then the fifth force will be exponentially suppressed by the Yukawa term e^{-md} where m is the symmetron mass and d the distance between two objects. This means that any experiment is only sensitive to symmetron models whose masses fall between these two limits. This can be seen directly by considering the form of the symmetron field profile around a spherical source of radius R and constant density embedded in a lower density background.

$$\phi = \phi_{\text{out}} - \frac{(\phi_{\text{out}} - \phi_{\text{in}})Re^{m_{\text{out}}(R-r)}}{r} \left(\frac{m_{\text{in}}R - \tanh m_{\text{in}}R}{m_{\text{in}}R + Rm_{\text{out}}\tanh m_{\text{in}}R} \right). \tag{20}$$

Inside the source the field reaches a minimum value $\phi = \phi_{\text{in}}$, and the mass of the field is m_{in}. Similarly in the background surrounding the source the field takes the value $\phi = \phi_{\text{out}}$ and has mass m_{out}. This profile is exponentially suppressed at distances larger than $1/m_{\text{out}}$ away from the source. It is also suppressed when $\phi_{\text{out}} = \phi_{\text{in}}$. This occurs when the mass of the field m_{out} is too large for the field to evolve within the vacuum chamber.

Constraints on the symmetron model have not been computed for all of the experiments described above. But those from atom-interferometry, and from the Eöt-Wash experiment are shown in Fig. 2, with constraints coming from astrophysical observations included for comparison.

3. The Field Profile in a Cylinder

We have mentioned several times already that the chameleon field can "resonate" inside cavities. This can be made completely explicit using a simple model of a cylinder filled with a low density gas surrounded by a dense metallic bore. In this case one can use a "bootstrapping" algorithm whereby the value of the field at the centre of the cylinder is left unknown, then solve the equations for the scalar field and finally impose that the value at the centre is indeed the one postulated initially. This yields a self-consistency condition which turns out to be the "resonance" criterion.[880]

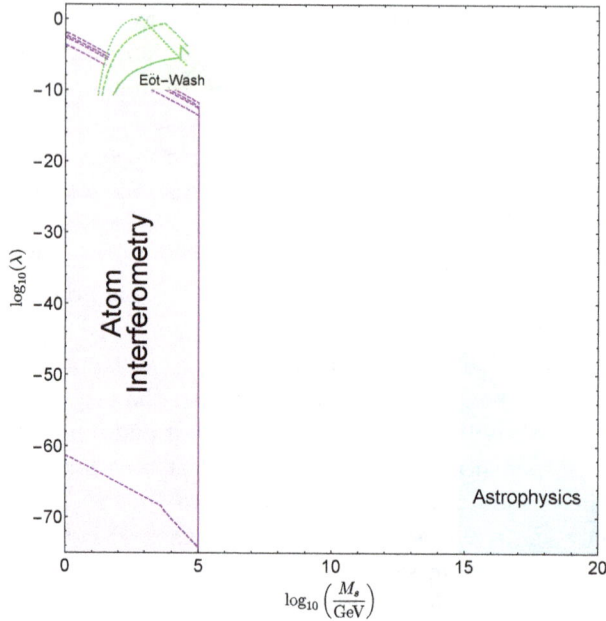

Fig. 2: Experimental constraints on the symmetron parameters M and λ. The Eöt-Wash region corresponds to $\mu = 2.4$ meV; the outlines for values $\mu = \{10^{-4}, 10^{-3}, 10^{-2}\}$ eV are shown by the solid, dashed, and dotted green lines respectively. The atom interferometry lines correspond to the regions excluded for $\mu = \{10^{-4}, 10^{-4.5}, 10^{-5}, 10^{-5}, 2.4 \times 10^{-3}\}$ eV from top to bottom respectively, the latter value corresponding to the dark energy scale. The astrophysical bounds are insensitive to the value of μ for the values considered here.[64]

Outside the vacuum, and far inside the metal, the field settles at ϕ_∞ where the mass is m_∞ and the field minimises the effective potential. Inside the vacuum we assume that the field takes a value ϕ_0 at the centre of the cylinder for $r = 0$. The mass there is defined as the second derivative of the effective potential m_0^2. Hence for $r \geq R$ we have

$$\frac{d^2\phi}{dr^2} + \frac{1}{r}\frac{d\phi}{dr} - m_\infty^2(\phi - \phi_\infty) = 0, \tag{21}$$

whilst inside

$$\frac{d^2\phi}{dr^2} + \frac{1}{r}\frac{d\phi}{dr} - m_0^2(\phi - \phi_0) = V'_{eff}(\phi_0), \tag{22}$$

where $V'_{eff}(\phi_0) = \frac{dV_{eff}}{d\phi}(\phi_0) \neq 0$ if ϕ_0 is not the minimum of the effective potential in vacuum. The solutions can be expressed in terms of Bessel and Neumann functions of zeroth order

$$r < R \quad \phi = CJ_0(im_0 r) + \phi_0 - \frac{V'_{eff}(\phi_0)}{m_0^2},$$

$$r \geq R \quad \phi = A(J_0(im_\infty r) - iN_0(im_\infty r)) + \phi_\infty \tag{23}$$

where A and C are constants obtained by matching the field and its first derivative at $r = R$. When the mass inside the metallic bore is very large, i.e. $m_\infty R \gg 1$, this simplifies to

$$\phi = \frac{\phi_\infty - \phi_0 + \frac{V'_{eff}(\phi_0)}{m_0^2}}{J_0(im_0R)} J_0(im_0r) + \phi_0 - \frac{V'_{eff}(\phi_0)}{m_0^2}, \tag{24}$$

for $r \leq R$. Evaluating this solution at $r = 0$ gives us the resonance condition

$$\phi_\infty - \phi_0 = \frac{V'_{eff}(\phi_0)}{m_0^2}(J_0(im_0R) - 1). \tag{25}$$

Let us consider now two archetypical models. First for inverse power law chameleons and putting the density inside the cylinder to zero, we have $\phi_\infty \ll \phi_0$

$$J_0(im_0R) = n + 2. \tag{26}$$

This is a resonance condition for m_0R which should be of order one, i.e. the mass of the scalar field in the cylinder adapts itself to the radius of the cylinder. A simplified solution to this equation is $m_0R \sim 2\sqrt{n+1}$ which is useful as an order of magnitude estimate. Using

$$m_0^2 = \frac{n(n+1)\Lambda^{n+4}}{\phi_0^{n+2}}, \tag{27}$$

one can easily evaluate the field inside the cavity.

For symmetrons, the solution to the resonance condition is more complex and can be deduced using the resonance condition written as[854]

$$J_0(im_0R) = \frac{1 + \frac{2m_0^2}{m_{\rm vac}^2}}{1 - \frac{m_0^2}{m_{\rm vac}^2}}, \tag{28}$$

where we have defined $m_{\rm vac}^2 = 2\mu^2$ to be the mass of the symmetron in the symmetry breaking phase when to matter density is present, i.e. for $\phi_{\rm vac} = \sqrt{6}\frac{\mu}{\sqrt{\lambda}}$. This condition admits solutions only when

$$m_{\rm vac}R \gtrsim 1, \tag{29}$$

i.e. for values of $m_{\rm vac} \lesssim 1$ the only solution is $\phi = \phi_\infty = 0$ inside the whole apparatus. When a solution exists we have

$$m_0 \sim m_{\rm vac}(1 = \frac{1}{2}\sqrt{\frac{\pi m_{\rm vac}R}{2}}e^{-m_{\rm vac}R}), \tag{30}$$

which implies that

$$\phi_0 \sim \phi_{\rm vac} \tag{31}$$

i.e. in the symmetron case and when the solution inside the chamber exists it is exponentially close to the vacuum solution in the absence of cavity. This is a very convenient criterion which is used when analysing atomic interferometry data.

4. The Dilaton as a Worked Out Example

4.1. *The model*

In this section, we will work out another simple example: the environmentally dependent dilaton.[59] This model has the advantage of being easily tractable and well motivated by string theoretic arguments.[881] The potential for the dilaton is given by

$$V(\phi) = V_0 e^{-\lambda\phi/M_{\rm pl}}, \tag{32}$$

where V_0 is an energy scale related to the dark energy of the Universe and λ a numerical constant. This potential corresponds to the string theory dilaton potential in the strong coupling limit.[881] The coupling function is inspired by the least coupling principle[844] where one assumes that in high density regions, the coupling of the dilaton to matter is driven to zero. It reads

$$A(\phi) = 1 + \frac{A_2}{2M_{\rm pl}^2}\phi^2. \tag{33}$$

where $A_2 \gg 1$ to satisfy the solar system tests, see below. Notice the similarity with the symmetron with

$$A_2 = \frac{M_{\rm pl}^2}{M^2}. \tag{34}$$

In a dense environment with matter density ρ, the effective potential (see Chapter 1)

$$V_{\rm eff}(\phi) = V_0 e^{-\lambda\phi/M_{\rm pl}} + \frac{A_2\rho}{2M_{\rm pl}^2}\phi^2, \tag{35}$$

admits a minimum at

$$\phi_\rho = \frac{\lambda V_0 M_{\rm pl}}{A_2\rho}, \tag{36}$$

where we have assumed that $\lambda\phi \ll M_{\rm pl}$ as can be easily checked. The coupling to matter

$$\beta(\phi) = M_{\rm pl}\frac{\partial \ln A(\phi)}{\partial\phi} = A_2\frac{\phi}{M_{\rm pl}}, \tag{37}$$

becomes

$$\beta_\rho = \frac{\lambda V_0 M_{\rm pl}}{A_2\rho}. \tag{38}$$

The mass at the minimum of the potential is given by

$$m_\rho^2 = \frac{\lambda^2 V_0}{M_{\rm pl}^2} + \frac{A_2}{M_{\rm pl}^2}\rho. \tag{39}$$

The value of the potential at the minimum is given by

$$V_\rho = V_0 + \frac{\lambda^2 V_0^2}{2A_2\rho}, \tag{40}$$

which is always close to V_0 and we choose to tune it to the value of the vacuum energy now

$$V_0 = 3\Omega_{\Lambda 0} M_{pl}^2 H_0^2, \tag{41}$$

where $\Omega_{\Lambda 0} \sim 0.73$. Of course this is a tuning tantamount to the tuning of the vacuum energy which is not guaranteed to be stable under radiative corrections. We only work classically here as no answer to the cosmological constant problem is known yet. Notice that the coupling to matter in the cosmological vacuum is given by

$$\beta_0 = \lambda \frac{\Omega_{\Lambda 0}}{\Omega_{m0}}. \tag{42}$$

which can be arbitrarily small with λ.

The strongest constraint on the dilaton models comes from the Laser Lunar Ranging (LLR) experiment[882] giving

$$\eta \equiv 2 \frac{|a_\oplus - a_{moon}|}{a_\oplus + a_{moon}} \lesssim 10^{-13}, \tag{43}$$

for the Moon and the Earth in the background of the Sun. This is related to the way screened bodies like the Sun, the Earth and the Moon (because of their large densities) couple to the scalar field

$$\eta \sim Q_\odot |Q_\oplus - Q_{moon}|, \tag{44}$$

where the charge $Q_A = \beta$ is the coupling to matter in the environment of an object for unscreened bodies and

$$Q_A = \frac{\phi_G}{2 M_{pl} \Phi_A}, \tag{45}$$

for a screened body of Newtonian potential Φ_A at its surface and embedded in the environment where the scalar field takes a value ϕ_G. For the LLR experiment, ϕ_G is the field value in the galactic medium of density $\rho_G \sim 10^6 \rho_0$ where ρ_0 is the cosmological cold dark matter density. This implies that

$$\eta \sim 10^{-1} \frac{\phi_G^2}{M_{pl,}\Phi_\oplus^2} \tag{46}$$

where $\Phi_\oplus \sim 10^{-9}$, $\Phi_\odot \sim 10^{-6}$, $\Phi_{moon} \sim 10^{-11}$. This leads to the bound

$$\frac{A_2}{\lambda} \gtrsim \Phi_\oplus^{-1} \sim 10^9, \tag{47}$$

and finally for the mass of the dilaton in the cosmological background

$$m_0 \gtrsim \sqrt{\lambda} \Phi_\oplus^{-1/2} H_0, \tag{48}$$

which for $\lambda = 1$ becomes[843]

$$m_0 \gtrsim 34500 H_0. \tag{49}$$

This implies that the effects of the dilaton on the growth of cosmic structure would occur on scales less than 1 Mpc. See also Ref. 814 for promising new tests.

4.2. *Laboratory experiments*

The first step in modeling the behaviour of the dilaton in an experimental context is to solve for the field profile between two infinitely thick plates located at $|z| \geq d$ where $2d$ is the inter-plate distance. Inside the plate which has a density ρ_c, the field converges to ϕ_c deep inside. Between the plates, the field profile would be given by ϕ_b for the density ρ_b if the distance d were infinite. As d is finite, the field reaches a smaller value ϕ_0 for $z = 0$. As $\rho_c \gg \rho_b$ we can approximate $\phi_c \sim 0$ and the field profile between the plates is given by

$$\phi'^2(z) = 2(V_{\text{eff}}(\phi) - V_{\text{eff}}(\phi_0)), \tag{50}$$

where a very good approximation is

$$V_{\text{eff}}(\phi) \sim V_b + \frac{1}{2}m_b^2, (\phi - \phi_b)^2 \tag{51}$$

depending on the density ρ_b. Explicitly this gives

$$\phi(z) = \phi_b \left(1 - \frac{\cosh m_b z}{\cosh m_b d}\right). \tag{52}$$

In particular we have

$$\phi_0 = \phi_b \left(1 - \frac{1}{\cosh m_b d}\right), \tag{53}$$

which converges to ϕ_b when $m_b d \gg 1$.

It turns out that the pressure exerted by one of the plates on the other one, i.e. the scalar equivalent to the Casimir effect, is given by[843]

$$\frac{\Delta F}{A} = V_{\text{eff}}(\phi_b) - V_{\text{eff}}(\phi_0), \tag{54}$$

depending on the potential difference between the energy stored in the field configuration in the absence and in the presence of the plates. In the dilaton case, this becomes

$$\frac{\Delta F}{A} = -\frac{m_b^2 \phi_b^2}{2 \cosh^2 m_b d}, \tag{55}$$

which is attractive. There are two clear regimes. When $m_b d \gg 1$, we have

$$\frac{\Delta F}{A} = -\frac{m_b^2 \phi_b^2}{2} e^{-2m_b d}, \tag{56}$$

corresponding to a Yukawa suppressed interaction (as the distance is $2d$) whilst for $m_b d \ll 1$ we have

$$\frac{\Delta F}{A} = -\frac{m_b^2 \phi_b^2}{2}, \tag{57}$$

corresponding to a pressure given by the amount of energy stored by the scalar field in vacuum.

The most stringent laboratory constraint on dilatons springs from the negative experimental results on the existence of short range scalar interactions by the Eöt-Wash experiment in Seattle.[838] The measured torque between two torsion pendulum is given by[87]

$$T = a_\theta \int_d^\infty dx \left| \frac{\Delta F}{A}(x) \right|, \tag{58}$$

where a_θ is a constant depending on the experiment and the Eöt-Wash constraint is $T \leq a_\theta \Lambda_T^3$, with $\lambda_T = 0.35\Lambda$ and $V_0 = \Lambda^4$, for $d = 55$ μm. As long as $dm_b \ll 1$, the torque is given by

$$T \sim a_\theta \frac{\lambda^2 V_0^2}{2A_2 \rho_b m_b} \sim a_\theta \frac{\lambda^2 V_0^2}{2m_{\rm Pl}^2 m_b^3}, \tag{59}$$

coinciding with (5.6) in Ref. 843. As a result we find that

$$m_b \gtrsim (H_0^2 \Lambda)^{1/3}, \tag{60}$$

and using the background density $\rho_b \sim 10^{-27}$ GeV^4, this gives the weak constraint on the mass m_0 in the cosmological background[843]

$$m_0 \gtrsim 55 \, H_0. \tag{61}$$

The same type of techniques can be applied to the chameleon and symmetron models which we will not detail and refer only to the existing literature. The main difference in the chameleon and symmetron case is that one must take into account the electrostatic shield between the two plates implying that the torque is reduced by the Yukawa suppression

$$T \to e^{-m_s D} T \tag{62}$$

where m_s is the scalar mass in the shield and D its width. For chameleons, this is responsible for the loss of sensitivity of the Eöt-Wash experiment at very large coupling β where the mass becomes large and the torque is essentially zero.

5. Tomographic Parameterisation

The models that we have considered, i.e. chameleons, symmetrons and dilatons can all be described using an implicit definition of the coupling function $A(\phi)$ and $V(\phi)$. This method, which is called tomographic (see Chapter 2), uses the explicit link between the density dependence of the minimum of the effective potential as a function of the matter density in the environment and the shape of the potential and coupling functions. This applies to inverse power law chameleon models, symmetrons and dilatons which are all scalar-tensor theories described by the Lagrangian in the Einstein frame

$$S = \int d^4x \sqrt{-g} \left(\frac{R}{16\pi G_N} - \frac{(\partial\phi)^2}{2} - V(\phi) \right) + S_m(\psi, A^2(\phi)g_{\mu\nu}) \tag{63}$$

where $A(\phi)$ is a function which defines the coupling between matter fields ψ and the scalar ϕ. The coupling to matter itself is

$$\beta(\phi) = M_{\text{pl}} \frac{d \ln A(\phi)}{d\phi}. \tag{64}$$

The scalar field dynamics are determined by an effective potential which takes into account the presence of the conserved matter density ρ in the environment[c]

$$V_{\text{eff}}(\phi) = V(\phi) + (A(\phi) - 1)\rho. \tag{65}$$

where the -1 is introduced for convenience. Chameleon-like theories, e.g. symmetrons and dilatons, are such that the effective potential admits a minimum as a function of the density where the mass function $m(\rho)$ and the coupling $\beta(\rho)$ at the minimum of the effective potential become density dependent.[63,71] It is easier in view of comparing with cosmological tests to characterise the functions $m(\rho)$ and $\beta(\rho)$ using the time evolution of the matter density of the Universe $\rho(a) = \frac{\rho_0}{a^3}$ where $a \leq 1$ is the scale factor of the Universe whose value now is $a_0 = 1$. This allows one to parameterise all chameleon-like theories using simply the a dependence of $\beta(a)$ and $m(a)$. Parametrically we have

$$\frac{\phi(a) - \phi_c}{M_{\text{pl}}} = 9\Omega_{m0} H_0^2 \int_{a_c}^{a} da \frac{\beta(a)}{a^4 m^2(a)}, \tag{66}$$

where the Hubble rate now is $H_0 \sim 10^{-43}$ GeV and the matter fraction is $\Omega_{m0} \sim 0.27$. The mass function is defined as the mass at the minimum of the effective potential $\phi(\rho(a))$

$$m^2(a) = \frac{d^2 V_{\text{eff}}}{d\phi^2}\Big|_{\phi=\phi(\rho(a))}. \tag{67}$$

Similarly the coupling is

$$\beta(a) = M_{\text{pl}} \frac{d \ln A}{d\phi}\Big|_{\phi=\phi(\rho(a))}. \tag{68}$$

and the potential value is given by

$$V(a) - V_c = -27\Omega_{m0}^2 H_0^4 \int_{a_c}^{a} da \frac{\beta^2(a) M_{\text{pl}}^2}{a^7 m^2(a)}. \tag{69}$$

This implicit parameterisation of $V(\phi)$ and $A(\phi)$ is obtained directly from from $m(a)$ and $\beta(a)$, i.e. one can reconstruct the potential by eliminating a between Eq. (66) and Eq. (69).

Familiar models can be easily described using this method. Chameleons with a potential of the type

$$V(\phi) = \Lambda^4 + \frac{\Lambda^{4+n}}{\phi^n} + \ldots, \tag{70}$$

---

[c]The conserved energy density is related to the density defined a $\rho_E = -T_0^0$ in the Einstein frame as $\rho_E = A\rho$.

where $n > 0$, and $\Lambda \sim 10^{-3}$ eV is the cosmological vacuum energy now, and the coupling function is $A(\phi) = \exp(\frac{\beta\phi}{M_{\mathrm{pl}}})$, can be reconstructed using $\beta(a) = \beta$ and

$$m(a) = m_0 a^{-r}, \tag{71}$$

where $r = \frac{3(n+2)}{2(n+1)}$. The mass scale m_0 is determined by

$$m_0^{2(n+1)} = \frac{(n+1)^{n+1}}{3n} \frac{(3\beta\Omega_{m0}H_0^2 M_{\mathrm{pl}})^{n+2}}{\Lambda^{4+n}}. \tag{72}$$

For $f(R)$ models in the large curvature regime, i.e. $R \gtrsim R_0$[72]

$$f(R) = \Lambda_0 + R - \frac{f_{R_0}}{n} \frac{R_0^{n+1}}{R^n}. \tag{73}$$

where Λ_0 is the cosmological constant and R_0 is the present day curvature, we have $\beta(a) = 1/\sqrt{6}$ and the mass function

$$m(a) = m_0 \left(\frac{4\Omega_{\Lambda 0} + \Omega_{m0}a^{-3}}{4\Omega_{\Lambda 0} + \Omega_{m0}} \right)^{(n+2)/2}, \tag{74}$$

where the mass on large cosmological scale is given by $m_0 = H_0 \sqrt{\frac{4\Omega_{\Lambda 0}+\Omega_{m0}}{(n+1)f_{R_0}}}$, and $\Omega_{\Lambda 0} \approx 0.73$ is the dark energy fraction today.[63] When $a \ll 1$, i.e. physical situations where the environment is dense, the mass dependence on a is a power law $m(a) \sim m_0 a^{-r}$ where $r = \frac{3(n+2)}{2}$. Dilatons are described by

$$\beta(a) = \beta_0 a^3, \tag{75}$$

and the mass function

$$m^2(a) = 3A_2 \frac{H_0^2}{a^3}. \tag{76}$$

Finally symmetrons are defined by the potential

$$V(\phi) = V_0 + \frac{\lambda}{4}\phi^4 - \frac{\mu^2}{2}\phi^2, \tag{77}$$

and a coupling function which can be rewritten as

$$A(\phi) = 1 + \frac{\beta_\star}{2\phi_\star M_{\mathrm{pl}}}\phi^2, \tag{78}$$

where the transition from the minimum of the effective potential at the origin to a non-zero value happens for a density $\rho_\star$. Defining

$$m_\star = \sqrt{2}\mu, \quad \phi_\star = \frac{2\beta_\star \rho_\star}{m_\star^2 M_{\mathrm{pl}}}, \quad \lambda = \frac{\mu^2}{\phi_\star^2}, \tag{79}$$

where $\rho_\star = \frac{\rho_0}{a_\star^3}$, the model can be reconstructed using

$$m(a) = m_\star \sqrt{1 - (\frac{a_\star}{a})^3}, \tag{80}$$

and

$$\beta(a) = \beta_\star \sqrt{1 - (\frac{a_\star}{a})^3}, \tag{81}$$

for $a > a_\star$ and $\beta(a) = 0$ for $a < a_\star$. For all these models, the comparison with laboratory tests can be done using the same techniques as the ones outlined in the previous section for the dilaton.

6. Vainshtein Mechanism

The Vainshtein screening is difficult to constrain with laboratory tests, at least for cosmologically interesting models.[82] Theories which contain Vainshtein screening rely on non-linear kinetic terms. This means that the variation in the field is extremely slow, and so the field can only respond to spatial variations in the local density over long distance scales. As with other types of screening the fifth force can be suppressed by the environment so that, for example, fifth forces in the solar system are screened by the galactic density. However, unlike theories with chameleon screening, fifth forces with Vainshtein screening do not respond quickly enough to changes in density that they can be unscreened in a laboratory vacuum. However notice that theories with the Vainshtein screening are shift symmetric and therefore are sensitive to linear gradients. In particular the cosmological Hubble flow has an influence on the time variation of Newton's constant.[883] Hence dynamical tests may help to unravel some of the features of these models. On the other hand, the recent observation of a neutron star merger by the LIGO/VIRGO experiments[2, 23, 884] has enormously restricted the number of viable theories with the Vainshtein property, for instance the models of the Horndeski type and beyond are mostly excluded as self-accelerating dark energy theories.[41–44] In a different part of their parameter space, when not related directly to the acceleration of the expansion of the Universe, these theories are perfectly valid models of modified gravity which are worth investigating.

It is still possible to place bounds on theories with Vainshtein screening from laboratory tests, for example on the Galileon models.[82] However these bounds are weak, and the relationship between the parameters that can be constrained in these experiments, and the fundamental parameters relevant for the cosmology is non-trivial.

Both Eöt-Wash and Casimir experiments use planar geometry. Considering the Galileon with planar symmetry one finds the non-linear terms cancel order by order, so the fifth force is completely unscreened.[763] One finds the ratio of the Galileon to Newtonian force is[d]

$$\frac{F_\phi}{F_G} = 2\beta^2 \, . \tag{82}$$

Similarly one can compute the screening for a cylindrical object and show the screening is reduced. Here the ratio of the Galileon to Newtonian force for the cubic Galileon is

$$\frac{F_\phi}{F_G} = 2\beta^2 \frac{r}{r_v} \, , \tag{83}$$

compared to the spherically symmetric case of

$$\frac{F_\phi}{F_G} = 2\beta^2 \left(\frac{r}{r_v} \right)^{3/2} \, . \tag{84}$$

---

[d]In Chapter 1 $\alpha \equiv \beta$, we have kept β as this is the same coupling as for chameleons.

In both cases r_v is the Vainshtein radius. Thus searching for Galileons in planar symmetric objects would yield a strong fifth force. The Galileon force around a plate of density ρ and thickness Δ when there is no screening is

$$\phi' = \frac{\rho}{2c_2} \begin{cases} \Delta & r > \Delta/2 \\ 2z & \Delta/2 > r > -\Delta/2 \\ -\Delta & -\Delta/2 > r \end{cases} , \tag{85}$$

where we have approximated the plate as a one-dimensional object and imposed continuity at the boundary. Here c_2 is the normalisation of the Galileon field which is canonically normalised when $c_2 = 1$. When one considers laboratory experiments one needs to consider the environment as well as the physical set-up. Both Eöt-Wash and Casimir experiments are performed on Earth, which must be taken into consideration since the radius of the Earth is less than the typical Vainshtein radius. In order to take this into account one writes $\phi(\vec{x}) = \phi_\oplus(r) + \delta\phi(z)$, where $\delta\phi$ is the perturbation due to the plates used in the experiment and the background field due to the Earth is $\phi_\oplus(r)$. Substituting this into the equation of motion one finds that it depends on a function $Z(r)$ which is approximately constant at $r = R_\oplus$,

$$Z_\oplus = \frac{R_\oplus}{\phi'_\oplus} \left[\frac{2\rho_\oplus}{3} + 8c_4 \left(\frac{\phi'_\oplus}{R_\oplus^3} \right)^3 + 32c_5 \left(\frac{\phi'_\oplus}{R_\oplus^3} \right)^4 \right] , \tag{86}$$

where the c_i's are the coefficients of the five Galileon terms in the Lagrangian.[82] The exact size of $Z_\oplus$ depends on the parameters c_4 and c_5. Laboratory experiments can be used to constrain the combination of coefficients in $Z_\oplus$. Of course the effect of the cavity needs to be taken into account as well. However, after detailed calculation it was found that in the background field of the Earth the cavity and plates behave as a linear theory with c_2 replaced by $Z_\oplus$.

The experiments we consider consist of two plates, aligned perpendicular to the z-direction, whose extent in the x,y-directions is much larger that their separation so we can approximate them as infinite. The plates have density ρ. The lower edge of one plate and the upper edge of the other is positioned at $z = d$ and $z = -d$ respectively, and the plates have width Δ, i.e. they have a finite thickness Δ. Therefore ϕ', the strength of the Galileon force due to the configuration, is given by

$$\phi' = \frac{\rho}{Z_\oplus} \begin{cases} z - d & d < z < d + \Delta \\ 0 & -d < z < d \\ z + d - (d + \Delta) < z < -d \end{cases} , \tag{87}$$

where we have imposed continuity of ϕ' at the boundary of the plates, and $\phi'(z) = -\phi'(-z)$. The approximation that the plates are infinite is valid whenever their extent in the x,y-directions is much larger than the distance $2d$ between the plates. Applying this to the Eöt-Wash experiment described earlier we find the torque induced by the Galileon is

$$T = \frac{\rho^2 \Delta^3}{6Z_\oplus} a_\theta , \tag{88}$$

where $a_\theta = dA/d\theta$ is a constant which depends on the experimental setup $a_\theta = 3 \times 10^{-3}$ m^2. The width of the plates is $\Delta = 1$ mm and the plates are made of molybdenum with a density $\rho = 10.28$ gcm^{-3}. The constraint on the Galileon force thus becomes

$$Z_\oplus > 6.05 \times 10^{40} \text{GeV}^2 \,, \tag{89}$$
$$> (20 m_P)^2 \,, \tag{90}$$

which translates into a constraint on the coupling

$$\beta < 0.05 \,. \tag{91}$$

Hence we find that in the context of the Eöt-Wash experiment, the Galileon force between the plates must be much weaker than the gravitational one. This sets a bound on a previously unconstrained combination of the Galileon parameters.

For Casimir experiments the formalism described above is not relevant to the experiments of Decca *et al.*[885, 886] where the force between a plate and a sphere is measured, It can be applied to the force between two parallel plates.[887, 888] However, following the procedure outlined for Eöt-Wash we find the Galileon force could not be detected in these experiments. In Refs. 868, 869 it was proposed that a modified parallel plate Casimir experiment could be used to search for the chameleon by exploiting the change in the chameleon force as the density of the inter-plate medium changes. As the Galileon force also depends on the local energy density this experiment could also provide useful constraints on the Galileon model.

7. Coupling to Photons

Conformally coupled scalar fields do not, classically, interact with photons. This can be seen directly from the conformal invariance of the photon terms of the standard model Lagrangian;

$$\tilde{g}_{\mu\nu} = A^2(\phi) g_{\mu\nu}, \tag{92}$$

where $g_{\mu\nu}$ is the Einstein frame metric. Assuming no coupling between the scalar and photons in the Jordan frame,

$$S_F = -\frac{1}{4} \int d^4x \sqrt{-\tilde{g}} \tilde{g}^{\mu\rho} \tilde{g}^{\nu\lambda} F_{\mu\nu} F_{\rho\lambda} \tag{93}$$

with $F_{\mu\nu} = \partial_\mu A_\nu - \partial_\nu A_\mu$. We see that in the Einstein frame

$$S_F = -\frac{1}{4} \int d^4x \sqrt{-g} g^{\mu\rho} g^{\nu\lambda} F_{\mu\nu} F_{\rho\lambda} \tag{94}$$

which makes explicit the scale invariance of the photon Lagrangian.

However there is no reason to forbid an interaction between the scalar and photons, and there are a number of reasons to expect such a coupling to emerge, which we will describe in the next subsection.

7.1. *Quantum coupling*

Whilst a classical Weyl rescaling does not induce a coupling between the scalar and photons, it can be shown that, after quantisation of the fields such a coupling does indeed emerge.[832] The conformal invariance of the photon Lagrangian is broken by a quantum anomaly which comes from the rescaling of the fermions $\psi \to A^{3/2}(\phi)\psi$ from the Jordan to the Einstein frame and the subsequent change of the measure in the path integral defining the quantum theory.

Additionally, if there are heavy charged fermions beyond the Standard Model then the fermions, which will couple directly to the scalar, will be able to mediate interactions between the scalar and photons through a triangle loop diagram. Integrating these heavy fermions out, leaves a low energy effective theory which possesses a contact interaction between the conformally coupled scalar and two photons,[832, 889] the Einstein frame action picks up a field dependent coupling to photons

$$S_F = -\frac{1}{4} \int d^4 x \sqrt{-g} B(\phi) g^{\mu\rho} g^{\nu\lambda} F_{\mu\nu} F_{\rho\lambda}, \tag{95}$$

such that the dimensionless constant controlling the strength of the coupling

$$\beta_\gamma = M_{\text{pl}} \frac{\partial \ln B}{\partial \phi}, \tag{96}$$

becomes

$$\beta_\gamma = (3N_f + \frac{N_{\tilde{f}}^>}{3}) \frac{\alpha}{4\pi} \beta_m, \tag{97}$$

where we have introduced

$$\beta_m = M_{\text{pl}} \frac{\partial \ln A}{\partial \phi}. \tag{98}$$

Here N_f is the total number of fermions in the model, for instance the fermions of the standard model, and $N_{\tilde{f}}^>$ the number of fermions which have been integrated out, e.g. the fermions with a mass at the grand unification scale. Of course as $N_{\tilde{f}}^>$ is not known, the precise value of β_γ cannot be inferred and in general is taken to be a non-vanishing parameter of the model.

7.2. *Photon-scalar mixing*

The presence of the coupling β_γ implies that the photons and scalars mix and therefore that the mass eigenstates do not coincide with the propagating fields. Physically this can be seen as the effect that a propagating photon has a non-zero probability of becoming a scalar before reverting back to a photon state. Hence this will have two consequences: the effective speed of the photons is affected by the coupling β_γ as the photon wave function picks up a non-vanishing phase shift after a finite distance and the amplitude of the wave function is also altered. This happens when a magnetic field $\vec{B}$ is present where the scalar-Maxwell equations become

$$\Box\vec{A} = \frac{\beta_\gamma}{M_{\rm pl}}\vec{\nabla}\phi \wedge \vec{B}, \tag{99}$$

and

$$\Box\phi - m^2\phi = \frac{\beta_\gamma}{M_{\rm pl}}\vec{B}.\vec{\nabla} \wedge \vec{A}, \tag{100}$$

Here $\vec{A}$ is the vector field associated to the photons in the background magnetic field $\vec{B}$, i.e. the fluctuations of the electromagnetic field around the background. Assuming that the magnetic field is perpendicular to the propagation of the photon, only the parallel polarisation of the photon along the magnetic field mixes with the scalar. Define the mixing angle as

$$\tan 2\theta = \frac{2B\omega}{M_\gamma m^2}, \tag{101}$$

where $\beta_\gamma = \frac{M_{\rm pl}}{M_\gamma}$ and m is the mass of the scalar in the magnetised region where the photon propagates. The energy of the initial photon is ω and the two propagating modes of the system have momenta along the photon direction[880]

$$k_- = \omega + \frac{m^2\theta^2}{2\omega}, \quad k_+ = \omega - \frac{m^2}{2\omega}, \tag{102}$$

where we assume that the mixing angle $\theta \ll 1$ is small. After a distance z, the photon wave function becomes

$$a_\parallel(z) = (1 - a(z))\cos(\omega z + \delta), \tag{103}$$

where

$$a(z) = 2\theta^2 \sin^2 \frac{m^2 z}{4\omega}, \tag{104}$$

and

$$\delta = \theta^2\frac{m^2}{2\omega}z - \theta^2 \sin\frac{m^2 z}{2\omega}. \tag{105}$$

This is the result for the free propagation of photons. Inside a cavity, the photons are reflected N times before leaving and being detected. Introducing the coherence length

$$z_{coh} = \frac{2\omega}{m^2}, \tag{106}$$

and the number of coherent passes P in a cavity of length L

$$PL = 2\pi z_{coh}, \tag{107}$$

the overall change of amplitude and phase shift become[880]

$$a_T = \theta^2, \quad \delta_T = \frac{\pi N\theta^2}{P}. \tag{108}$$

It is possible to measure the rotation of the initial polarisation per pass and the induced ellipticity of the polarisation which does not remain linear

$$\text{rotation/pass} = \frac{\theta^2}{2N}, \quad \text{ellipticity/pass} = \frac{\pi\theta^2}{2P}. \tag{109}$$

Laser experiments constrain these observables and therefore give constraints on the coupling to photons.

7.3. *Scalar reflection*

As the mass of the scalar jumps from a low value inside the cavity to a large value inside the cavity's walls, the scalar wave function is distorted by the presence of the wall. This induces another phase shift compared to the one calculated in the previous section. The phenomenology of scalars inside cavities has been thoroughly investigated for chameleons and we refer to Ref. 880 for details. Here we will elaborate on the dilaton case as the calculations are simpler.

For a dilaton in a cavity with a vacuum density ρ_b and walls of density ρ_c and associated minima of the effective potential $\phi_{b,c}$, the static profile of the scalar is obtained by solving

$$\phi'^2(z) = V_{\text{eff}}(\phi) - V_{\text{eff}}(\phi_b), \tag{110}$$

for $z \geq 0$ where the wall is at $z = 0$ here. For a dilaton we find

$$\phi(z) = \phi_0 + (\phi_b - \phi_0)(1 - e^{-m_b z}), \tag{111}$$

where $\phi_0 = \phi(0) \approx \phi_c$ as $\rho_c \gg \rho_b$. The mass of the scalar evolves away from the wall as

$$m^2(z) = m_b^2 \left(1 - \frac{\lambda^3 V_0}{A_2 \rho_b} \frac{\phi(z) - \phi_b}{M_{\text{pl}}}\right) = m_b^2 \left(1 + \frac{\lambda^3 V_0}{A_2 \rho_b} \frac{\phi_b - \phi_c}{M_{\text{pl}}} e^{-m_b z}\right), \tag{112}$$

i.e., the mass decreases exponentially. Let us consider a scalar wave which is a solution of

$$\frac{d^2 \delta\phi}{dz^2} + (\omega^2 - m^2(z))\delta\phi = 0. \tag{113}$$

There are three regions to consider. Inside the cavity where $m(z) \sim m_b$, the scalar propagates as a wave alongside the photons. In the walls when $w \leq m_c$, the field is attenuated and the scalar is therefore reflected. This has been tested in afterglow experiments where a laser beam is turned off and one expects that the trapped scalars in the cavity will regenerate photons which would then be seen as afterglows. Finally when w is not much larger than m_b, the scalar is reflected at z_ω where $m(z_\omega) = w$ implying that this induces a phase shift compared to the photons, i.e. the scalar is reflected before the photons. This plays a crucial role for chameleons and was taken into account in Ref. 880.

7.4. *Experimental results*

There are three types of experiments which must be taken into account. We will give the bounds for the chameleon model as this is the most studied case. In the symmetron case, the phenomenology has not been worked out and should be closer to the dilaton case presented in the previous section than to the chameleon case.

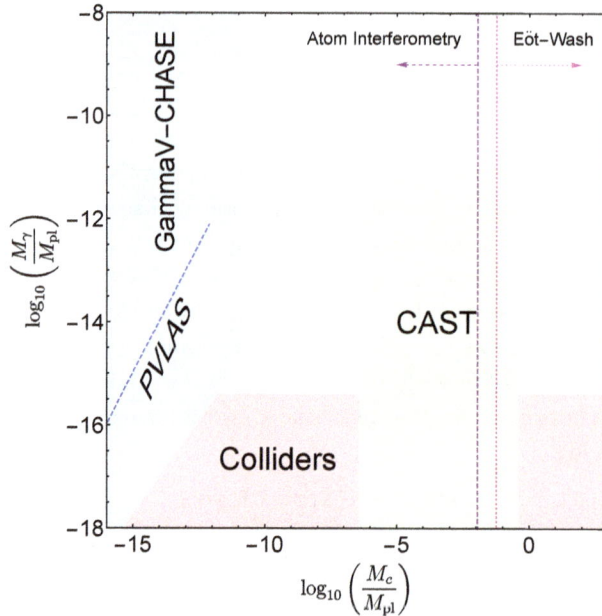

Fig. 3: Experimental constraints on the coupling of the chameleon to photons as a function of the chameleon coupling to matter, for the chameleon potential $V(\phi) = \Lambda^5/\phi$.[64]

The first type of experiments measure the induced ellipticity and rotation angle of a laser beam where a transverse magnetic field is present inside a high quality cavity. From the 2007 PVLAS results[890] with $N = 45000$ passes, the total ellipticity is constrained to be

$$\text{ellipticity} \leq 1.4 \times 10^{-8}, \tag{114}$$

and the rotation

$$\text{rotation} \leq 10^{-8}, \tag{115}$$

with a cavity length of $L = 1$ m, a beam with $\omega = 1.17$ eV and a magnetic field $B = 2.3$ T. When $\beta_\gamma = \frac{M_{pl}}{M_\gamma}$ and $\beta = \frac{M_{pl}}{M_c}$ are such that $M_\gamma = M_c$ this yields $M_\gamma \geq 2 \times 10^6$ GeV. The ALPS experiment at DESY[891] has performed a light-shining-through-walls experiment, where a laser beam with $\omega = 2.33$ eV faces a wall after $L = 4.3$ m in a magnetic field $B = 5$ T. The probability that a photon converts into a scalar after L

$$P_{\gamma \to \phi} = \sin^2 2\theta \sin^2 \lambda \omega L \tag{116}$$

where $\lambda = \frac{m^2}{2\omega^2}(1 + \tan^2 2\theta)$ is constrained to be

$$P_{\gamma \to \phi} \leq 2.08 \times 10^{-25}. \tag{117}$$

This competes with the CHASE experiment[892] where the afterglow phenomenon has been investigated and a bound $M_\gamma \geq 3 \times 10^7$ GeV was found. Similarly inside

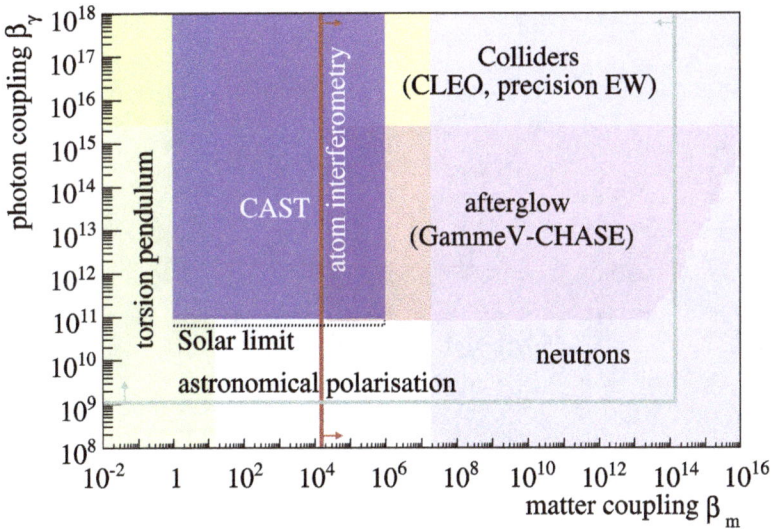

Fig. 4: Experimental constraints on chameleons with $n = 1$ in the $\beta_m - \beta_\gamma$ plane.[895]

the Sun in the tachocline where the solar magnetic field is assumed to emerge, chameleons could be created and then observed with the CAST experiment where X-rays would be back-converted from solar chameleons which would have escaped the Sun.[893, 894] The production and regeneration of X-rays depend on the coupling to photons β_γ. All these constraints are summarised in Fig. 1.

8. Conclusions

In summary, laboratory experiments have a huge amount to tell us about possible modifications of gravity. For theories with screening, these experiments necessarily probe the non-linear parts of the theory, although we have shown how the effects of local screening due to the galactic environment can be 'un-screened' in a laboratory vacuum chamber. As they probe the non-linear regime, there is no model independent way to connect these constraints to the parametrized linear and quasi-linear theories used to obtain cosmological constraints. However, once a model is specified, we have shown the power of combining experimental searches on all scales from the sub-atomic to the cosmological.

Bibliography

1. Planck Collaboration, P. A. R. Ade, N. Aghanim *et al.*, Planck 2015 results. XIII. Cosmological parameters, *Astron. Astrophys.* **594**, A13 (2016). doi: 10.1051/0004-6361/201525830.

2. B. P. Abbott, R. Abbott, T. D. Abbott *et al.*, Gravitational waves and gamma-rays from a binary neutron star merger: GW170817 and GRB 170817A, *Astrophys. J. Lett.* **848**, L13 (2017). doi: 10.3847/2041-8213/aa920c.

3. Planck Collaboration and N. Aghanim, Planck 2018 results. VI. Cosmological parameters (2018). arXiv:1807.06209.

4. Planck Collaboration, P. A. R. Ade, N. Aghanim *et al.*, Planck 2013 results. XVI. Cosmological parameters, *Astron. Astrophys.* **571**, A16 (2014). doi: 10.1051/0004-6361/201321591.

5. A. G. Riess, A. V. Filippenko, P. Challis *et al.*, Observational evidence from supernovae for an accelerating universe and a cosmological constant, *Astron. J.* **116**, 1009–1038 (1998). doi: 10.1086/300499.

6. S. Perlmutter, G. Aldering, G. Goldhaber *et al.*, Measurements of Ω and Λ from 42 high-redshift supernovae, *Astrophys. J.* **517**, 565–586 (1999). doi: 10.1086/307221.

7. D. J. Eisenstein *et al.*, Detection of the baryon acoustic peak in the large-scale correlation function of SDSS luminous red galaxies, *Astrophys. J.* **633**, 560–574 (2005). doi: 10.1086/466512.

8. W. J. Percival, S. Cole, D. J. Eisenstein *et al.*, Measuring the baryon acoustic oscillation scale using the Sloan Digital Sky Survey and 2dF Galaxy Redshift Survey, *Mon. Not. Roy. Astron. Soc.* **381**, 1053–1066 (2007). doi: 10.1111/j.1365-2966.2007.12268.x.

9. Planck Collaboration, P. A. R. Ade, N. Aghanim *et al.*, Planck 2015 results. XIV. Dark energy and modified gravity, *Astron. Astrophys.* **594**, A14 (2016). doi: 10.1051/0004-6361/201525814.

10. H. Hildebrandt *et al.*, KiDS-450: Cosmological parameter constraints from tomographic weak gravitational lensing, *Mon. Not. Roy. Astron. Soc.* **465**, 1454 (2017). doi: 10.1093/mnras/stw2805.

11. T. M. C. Abbott *et al.*, Dark Energy Survey year 1 results: Cosmological constraints from galaxy clustering and weak lensing, *Phys. Rev.* **D98**(4), 043526 (2018). doi: 10.1103/PhysRevD.98.043526.

12. A. G. Riess, S. Casertano, W. Yuan *et al.*, New parallaxes of galactic cepheids from spatially scanning the Hubble Space Telescope: Implications for the Hubble constant (2018). arXiv:1801.01120 [astro-ph.SR].

13. B. P. Abbott, R. Abbott, T. D. Abbott *et al.*, A gravitational-wave standard siren

measurement of the Hubble constant, *Nature.* **551**, 85–88 (2017). doi: 10.1038/nature24471.

14. S. Weinberg, The cosmological constant problem, *Rev. Mod. Phys.* **61**, 1–23 (1989). doi: 10.1103/RevModPhys.61.1.

15. S. R. Green and R. M. Wald, How well is our Universe described by an FLRW model?, *Class. Quant. Grav.* **31**(23), 234003 (2014). doi: 10.1088/0264-9381/31/23/234003.

16. T. Buchert, M. Carfora, G. F. R. Ellis *et al.*, Is there proof that backreaction of inhomogeneities is irrelevant in cosmology?, *Class. Quant. Grav.* **32**(21), 215021 (2015). doi: 10.1088/0264-9381/32/21/215021.

17. K. Tomita, On astrophysical explanations due to cosmological inhomogeneities for the observational acceleration (2009). arXiv:0906.1325 [astro-ph.CO].

18. T. Clifton, P. G. Ferreira, A. Padilla *et al.*, Modified gravity and cosmology, *Phys. Rep.* **513**, 1–189 (2012). doi: 10.1016/j.physrep.2012.01.001.

19. A. Joyce, B. Jain, J. Khoury *et al.*, Beyond the cosmological standard model, *Phys. Rep.* **568**, 1–98 (2015). doi: 10.1016/j.physrep.2014.12.002.

20. K. Koyama, Cosmological tests of modified gravity, *Rep. Prog. Phys.* **79**(4), 046902 (2016). doi: 10.1088/0034-4885/79/4/046902.

21. J. M. Weisberg, D. J. Nice and J. H. Taylor, Timing measurements of the relativistic binary pulsar PSR B1913+16, *Astrophys. J.* **722**, 1030–1034 (2010). doi: 10.1088/0004-637X/722/2/1030.

22. B. P. Abbott, R. Abbott, T. D. Abbott *et al.*, Observation of gravitational waves from a binary black hole merger, *Phys. Rev. Lett.* **116**(6), 061102 (2016). doi: 10.1103/PhysRevLett.116.061102.

23. B. P. Abbott, R. Abbott, T. D. Abbott *et al.*, GW170817: Observation of gravitational waves from a binary neutron star inspiral, *Phys. Rev. Lett.* **119**(16), 161101 (2017). doi: 10.1103/PhysRevLett.119.161101.

24. G.-B. Zhao, M. Raveri, L. Pogosian *et al.*, Dynamical dark energy in light of the latest observations, *Nature Astronomy.* **1**, 627–632 (2017). doi: 10.1038/s41550-017-0216-z.

25. B. Ratra and P. J. E. Peebles, Cosmological consequences of a rolling homogeneous scalar field, *Phys. Rev.* **D37**, 3406–3427 (1988). doi: 10.1103/PhysRevD.37.3406.

26. E. J. Copeland, M. Sami and S. Tsujikawa, Dynamics of dark energy, *Int. J. Mod. Phys. D.* **15**, 1753–1935 (2006). doi: 10.1142/S021827180600942X.

27. A. Joyce, L. Lombriser and F. Schmidt, Dark energy versus modified gravity, *Ann. Rev. Nucl. Part. Sci.* **66**, 95–122 (2016). doi: 10.1146/annurev-nucl-102115-044553.

28. R. Maartens and K. Koyama, Brane-world gravity, *Living Rev. Rel.* **13**, 5 (2010). doi: 10.12942/lrr-2010-5.

29. C. P. Burgess, The cosmological constant problem: Why it's hard to get dark energy from micro-physics (2013). arXiv:1309.4133 [hep-th].

30. A. Padilla, Lectures on the cosmological constant problem (2015). arXiv:1502.05296 [hep-th].

31. G. Dvali, G. Gabadadze and M. Porrati, 4D gravity on a brane in 5D Minkowski space, *Phys. Lett. B.* **485**, 208–214 (2000). doi: 10.1016/S0370-2693(00)00669-9.

32. M. A. Luty, M. Porrati and R. Rattazzi, Strong interactions and stability in the DGP model, *J. High Energy Phys.* **9**, 029 (2003). doi: 10.1088/1126-6708/2003/09/029.

33. D. Gorbunov, K. Koyama and S. Sibiryakov, More on ghosts in the Dvali-Gabadaze-Porrati model, *Phys. Rev.* **D73**(4), 044016 (2006). doi: 10.1103/PhysRevD.73.044016.

34. C. M. Will, The confrontation between general relativity and experiment, *Living Rev. Rel.* **17**, 4 (2014). doi: 10.12942/lrr-2014-4.

35. S. M. Merkowitz, Tests of gravity using lunar laser ranging, *Living Rev. Rel.* **13**, 7 (2010). doi: 10.12942/lrr-2010-7.

36. R. A. Hulse and J. H. Taylor, Discovery of a pulsar in a binary system, *Astrophys. J. Lett.* **195**, L51–L53 (1975). doi: 10.1086/181708.

37. G. D. Moore and A. E. Nelson, Lower bound on the propagation speed of gravity from gravitational Cherenkov radiation, *J. High Energy Phys.* **9**, 023 (2001). doi: 10.1088/1126-6708/2001/09/023.

38. L. Lombriser and A. Taylor, Breaking a dark degeneracy with gravitational waves, *J. Cosmo. Astropart. Phys.* **3**, 031 (2016). doi: 10.1088/1475-7516/2016/03/031.

39. L. Lombriser and N. A. Lima, Challenges to self-acceleration in modified gravity from gravitational waves and large-scale structure, *Phys. Lett. B.* **765**, 382–385 (2017). doi: 10.1016/j.physletb.2016.12.048.

40. D. Bettoni, J. M. Ezquiaga, K. Hinterbichler *et al.*, Speed of gravitational waves and the fate of scalar-tensor gravity, *Phys. Rev.* **D95**(8), 084029 (2017). doi: 10.1103/PhysRevD.95.084029.

41. P. Creminelli and F. Vernizzi, Dark energy after GW170817 and GRB170817A, *Phys. Rev. Lett.* **119**(25), 251302 (2017). doi: 10.1103/PhysRevLett.119.251302.

42. J. Sakstein and B. Jain, Implications of the neutron star merger GW170817 for cosmological scalar-tensor theories, *Phys. Rev. Lett.* **119**(25), 251303 (2017). doi: 10.1103/PhysRevLett.119.251303.

43. J. M. Ezquiaga and M. Zumalacárregui, Dark energy after GW170817: Dead ends and the road ahead, *Phys. Rev. Lett.* **119**(25), 251304 (2017). doi: 10.1103/PhysRevLett.119.251304.

44. T. Baker, E. Bellini, P. G. Ferreira *et al.*, Strong constraints on cosmological gravity from GW170817 and GRB 170817A, *Phys. Rev. Lett.* **119**(25), 251301 (2017). doi: 10.1103/PhysRevLett.119.251301.

45. F. Sbisà, Classical and quantum ghosts, *Euro. J. Phys.* **36**(1), 015009 (2015). doi: 10.1088/0143-0807/36/1/015009.

46. A. Adams, N. Arkani-Hamed, S. Dubovsky *et al.*, Causality, analyticity and an IR obstruction to UV completion, *J. High Energy Phys.* **10**, 014 (2006). doi: 10.1088/1126-6708/2006/10/014.

47. C. de Rham, S. Melville, A. J. Tolley *et al.*, UV complete me: Positivity bounds for particles with spin, *J. High Energy Phys.* **03**, 011 (2018). doi: 10.1007/JHEP03(2018)011.

48. R. Woodard. Avoiding dark energy with 1/R modifications of gravity. In ed. L. Papantonopoulos, *The Invisible Universe: Dark Matter and Dark Energy*, Vol. 720, *Lecture Notes in Physics* (Berlin Springer Verlag) p. 403, (2007).

49. D. Langlois and K. Noui, Degenerate higher derivative theories beyond Horndeski: Evading the Ostrogradski instability, *J. Cosmo. Astropart. Phys.* **2**, 034 (2016). doi: 10.1088/1475-7516/2016/02/034.

50. C. Wetterich, Effective nonlocal Euclidean gravity, *Gen. Rel. Grav.* **30**, 159–172 (1998). doi: 10.1023/A:1018837319976.

51. S. Deser and R. P. Woodard, Nonlocal cosmology, *Phys. Rev. Lett.* **99**(11), 111301 (2007). doi: 10.1103/PhysRevLett.99.111301.

52. M. Maggiore and M. Mancarella, Nonlocal gravity and dark energy, *Phys. Rev.* **D90** (2), 023005 (2014). doi: 10.1103/PhysRevD.90.023005.

53. E. Belgacem, Y. Dirian, S. Foffa *et al.*, Nonlocal gravity. Conceptual aspects and cosmological predictions, *J. Cosmo. Astropart. Phys.* **1803**(03), 002 (2018). doi: 10.1088/1475-7516/2018/03/002.

54. A. Emir Gümrükçüoğlu, M. Saravani and T. P. Sotiriou, Hořava gravity after GW170817, *Phys. Rev.* **D97**(2), 024032 (2018). doi: 10.1103/PhysRevD.97.024032.

55. J. Oost, S. Mukohyama and A. Wang, Constraints on Einstein-aether theory after GW170817, *Phys. Rev.* **D97**(12), 124023 (2018). doi: 10.1103/PhysRevD.97.124023.

56. C. de Rham, Massive Gravity, *Living Rev. Rel.* **17**, 7 (2014). doi: 10.12942/lrr-2014-7.

57. J. Khoury and A. Weltman, Chameleon fields: Awaiting surprises for tests of gravity in space, *Phys. Rev. Lett.* **93**, 171104 (2004). doi: 10.1103/PhysRevLett.93.171104.

58. J. Khoury and A. Weltman, Chameleon cosmology, *Phys. Rev.* **D69**, 044026 (2004). doi: 10.1103/PhysRevD.69.044026.

59. P. Brax, C. van de Bruck, A.-C. Davis *et al.*, Dilaton and modified gravity, *Phys. Rev.* **D82**(6), 063519 (2010). doi: 10.1103/PhysRevD.82.063519.

60. K. Hinterbichler and J. Khoury, Screening long-range forces through local symmetry restoration, *Phys. Rev. Lett.* **104**(23), 231301 (2010). doi: 10.1103/PhysRevLett.104.231301.

61. E. Babichev, C. Deffayet and R. Ziour, k-mouflage gravity, *Int. J. Mod. Phys. D.* **18**, 2147–2154 (2009). doi: 10.1142/S0218271809016107.

62. A. I. Vainshtein, To the problem of nonvanishing gravitation mass, *Phys. Lett. B.* **39**, 393–394 (1972). doi: 10.1016/0370-2693(72)90147-5.

63. P. Brax, A.-C. Davis, B. Li *et al.*, Unified description of screened modified gravity, *Phys. Rev.* **D86**(4), 044015 (2012). doi: 10.1103/PhysRevD.86.044015.

64. C. Burrage and J. Sakstein, Tests of chameleon gravity, *Living Rev. Rel.* **21**(1), 1 (2018). doi: 10.1007/s41114-018-0011-x.

65. F. Rondeau and B. Li, Equivalence of cosmological observables in conformally related scalar tensor theories, *Phys. Rev.* **D96**(12), 124009 (2017). doi: 10.1103/PhysRevD.96.124009.

66. L. Amendola, Coupled quintessence, *Phys. Rev.* **D62**(4), 043511 (2000). doi: 10.1103/PhysRevD.62.043511.

67. C. Llinares and L. Pogosian, Domain walls coupled to matter: The symmetron example, *Phys. Rev.* **D90**(12), 124041 (2014). doi: 10.1103/PhysRevD.90.124041.

68. G.-B. Zhao, B. Li and K. Koyama, Testing gravity using the environmental dependence of dark matter halos, *Phys. Rev. Lett.* **107**(7), 071303 (2011). doi: 10.1103/PhysRevLett.107.071303.

69. A. Cabré, V. Vikram, G.-B. Zhao *et al.*, Astrophysical tests of gravity: A screening map of the nearby universe, *J. Cosmo. Astropart. Phys.* **7**, 034 (2012). doi: 10.1088/1475-7516/2012/07/034.

70. H. Desmond, P. G. Ferreira, G. Lavaux *et al.*, The fifth force in the local cosmic web, *Mon. Not. Roy. Astron. Soc.* **483**(1), L64–L68 (2019). doi: 10.1093/mnrasl/sly221.

71. P. Brax, A.-C. Davis and B. Li, Modified gravity tomography, *Phys. Lett. B.* **715**, 38–43 (2012). doi: 10.1016/j.physletb.2012.08.002.

72. W. Hu and I. Sawicki, Models of f(R) cosmic acceleration that evade solar system tests, *Phys. Rev.* **D76**(6), 064004 (2007). doi: 10.1103/PhysRevD.76.064004.

73. A. Barreira, P. Brax, S. Clesse *et al.*, K-mouflage gravity models that pass solar system and cosmological constraints, *Phys. Rev.* **D91**(12), 123522 (2015). doi: 10. 1103/PhysRevD.91.123522.

74. V. I. Zakharov, Linearized gravitation theory and the graviton mass, *Soviet J. Exp. Theo. Phys. Lett.* **12**, 312 (1970).

75. H. van Dam and M. J. G. Veltman, Massive and massless Yang-Mills and gravitational fields, *Nucl. Phys.* **B22**, 397–411 (1970). doi: 10.1016/0550-3213(70)90416-5.

76. D. G. Boulware and S. Deser, Can gravitation have a finite range?, *Phys. Rev.* **D6**, 3368–3382 (1972). doi: 10.1103/PhysRevD.6.3368.

77. C. de Rham, G. Gabadadze and A. J. Tolley, Resummation of massive gravity, *Phys. Rev. Lett.* **106**(23), 231101 (2011). doi: 10.1103/PhysRevLett.106.231101.

78. B. Li, G.-B. Zhao and K. Koyama, Exploring Vainshtein mechanism on adaptively refined meshes, *J. Cosmo. Astropart. Phys.* **5**, 023 (2013). doi: 10.1088/1475-7516/ 2013/05/023.

79. A. Lue, R. Scoccimarro and G. D. Starkman, Probing Newton's constant on vast scales: Dvali-Gabadadze-Porrati gravity, cosmic acceleration, and large scale structure, *Phys. Rev.* **D69**(12), 124015 (2004). doi: 10.1103/PhysRevD.69.124015.

80. G. Dvali, A. Gruzinov and M. Zaldarriaga, The accelerated universe and the Moon, *Phys. Rev.* **D68**, 024012 (2003). doi: 10.1103/PhysRevD.68.024012.

81. T. Hiramatsu, W. Hu, K. Koyama *et al.*, Equivalence principle violation in Vainshtein screened two-body systems, *Phys. Rev.* **D87**(6), 063525 (2013). doi: 10.1103/ PhysRevD.87.063525.

82. P. Brax, C. Burrage and A.-C. Davis, Laboratory tests of the Galileon, *J. Cosmo. Astropart. Phys.* **9**, 020 (2011). doi: 10.1088/1475-7516/2011/09/020.

83. A. A. Starobinsky, A new type of isotropic cosmological models without singularity, *Phys. Lett. B.* **91**, 99–102 (1980). doi: 10.1016/0370-2693(80)90670-X.

84. S. Nojiri and S. D. Odintsov, Dark energy, inflation and dark matter from modified F(R) gravity, *TSPU Bulletin.* **N8(110)**, 7–19 (2011). arXiv:0807.0685 [hep-th].

85. T. P. Sotiriou and V. Faraoni, f(R) theories of gravity, *Rev. Mod. Phys.* **82**, 451–497 (2010). doi: 10.1103/RevModPhys.82.451.

86. A. De Felice and S. Tsujikawa, f(R) theories, *Liv. Rev. Rel.* **13**, 3 (2010). doi: 10. 12942/lrr-2010-3.

87. P. Brax, C. van de Bruck, A.-C. Davis *et al.*, f(R) gravity and chameleon theories, *Phys. Rev.* **D78**(10), 104021 (2008). doi: 10.1103/PhysRevD.78.104021.

88. A. A. Starobinsky, Disappearing cosmological constant in f(R) gravity, *Soviet J. Exp. Theo. Phys. Lett.* **86**, 157–163 (2007). doi: 10.1134/S0021364007150027.

89. S. Appleby and R. Battye, Do consistent F(R) models mimic general relativity plus Λ?, *Phys. Lett. B.* **654**, 7–12 (2007). doi: 10.1016/j.physletb.2007.08.037.

90. L. Pogosian and A. Silvestri, Pattern of growth in viable f(R) cosmologies, *Phys. Rev.* **D77**(2), 023503 (2008). doi: 10.1103/PhysRevD.77.023503.

91. J. Wang, L. Hui and J. Khoury, No-go theorems for generalized chameleon field theories, *Phys. Rev. Lett.* **109**(24), 241301 (2012). doi: 10.1103/PhysRevLett.109. 241301.

92. J. J. Ceron-Hurtado, J.-h. He and B. Li, Can background cosmology hold the key for modified gravity tests?, *Phys. Rev.* **D94**(6), 064052 (2016). doi: 10.1103/PhysRevD. 94.064052.

93. Y.-S. Song, W. Hu and I. Sawicki, Large scale structure of f(R) gravity, *Phys. Rev.* **D75**(4), 044004 (2007). doi: 10.1103/PhysRevD.75.044004.

94. C. Deffayet, G. Dvali and G. Gabadadze, Accelerated universe from gravity leaking to extra dimensions, *Phys. Rev.* **D65**(4), 044023 (2002). doi: 10.1103/PhysRevD.65. 044023.

95. A. Lue and G. Starkman, Gravitational leakage into extra dimensions: Probing dark energy using local gravity, *Phys. Rev.* **D67**(6), 064002 (2003). doi: 10.1103/ PhysRevD.67.064002.

96. K. Koyama and F. P. Silva, Nonlinear interactions in a cosmological background in the Dvali-Gabadadze-Porrati braneworld, *Phys. Rev.* **D75**(8), 084040 (2007). doi: 10.1103/PhysRevD.75.084040.

97. A. Nicolis and R. Rattazzi, Classical and quantum consistency of the DGP model, *J. High Energy Phys.* **6**, 059 (2004). doi: 10.1088/1126-6708/2004/06/059.

98. K. Koyama, Ghosts in the self-accelerating brane universe, *Phys. Rev.* **D72**(12), 123511 (2005). doi: 10.1103/PhysRevD.72.123511.

99. C. Charmousis, R. Gregory, N. Kaloper *et al.*, DGP specteroscopy, *J. High Energy Phys.* **10**, 066 (2006). doi: 10.1088/1126-6708/2006/10/066.

100. C. Deffayet, G. Dvali, G. Gabadadze *et al.*, Nonperturbative continuity in graviton mass versus perturbative discontinuity, *Phys. Rev.* **D65**(4), 044026 (2002). doi: 10. 1103/PhysRevD.65.044026.

101. F. Schmidt, Self-consistent cosmological simulations of DGP braneworld gravity, *Phys. Rev.* **D80**(4), 043001 (2009). doi: 10.1103/PhysRevD.80.043001.

102. H. Ishihara, Causality of the brane universe, *Phys. Rev. Lett.* **86**, 381–384 (2001). doi: 10.1103/PhysRevLett.86.381.

103. G. W. Horndeski, Second-order scalar-tensor field equations in a four-dimensional space, *Int. J. Theor. Phys.* **10**, 363–384 (1974). doi: 10.1007/BF01807638.

104. A. Nicolis, R. Rattazzi and E. Trincherini, Galileon as a local modification of gravity, *Phys. Rev.* **D79**(6), 064036 (2009). doi: 10.1103/PhysRevD.79.064036.

105. C. Deffayet, G. Esposito-Farèse and A. Vikman, Covariant Galileon, *Phys. Rev.* **D79** (8), 084003 (2009). doi: 10.1103/PhysRevD.79.084003.

106. C. Deffayet, X. Gao, D. A. Steer *et al.*, From k-essence to generalized Galileons, *Phys. Rev.* **D84**(6), 064039 (2011). doi: 10.1103/PhysRevD.84.064039.

107. C. Deffayet, S. Deser and G. Esposito-Farèse, Generalized Galileons: All scalar models whose curved background extensions maintain second-order field equations and stress tensors, *Phys. Rev.* **D80**(6), 064015 (2009). doi: 10.1103/PhysRevD.80.064015.

108. T. Kobayashi, M. Yamaguchi and J. Yokoyama, Inflation driven by the Galileon field, *Phys. Rev. Lett.* **105**(23), 231302 (2010). doi: 10.1103/PhysRevLett.105.231302.

109. C. Charmousis, E. J. Copeland, A. Padilla *et al.*, General second-order scalar-tensor theory and self-tuning, *Phys. Rev. Lett.* **108**(5), 051101 (2012). doi: 10.1103/ PhysRevLett.108.051101.

110. C. Deffayet and D. A. Steer, A formal introduction to Horndeski and Galileon theories and their generalizations, *Class. Quant. Grav.* **30**(21), 214006 (2013). doi: 10.1088/ 0264-9381/30/21/214006.

111. E. Bellini and I. Sawicki, Maximal freedom at minimum cost: Linear large-scale structure in general modifications of gravity, *J. Cosmo. Astropart. Phys.* **7**, 050 (2014). doi: 10.1088/1475-7516/2014/07/050.

112. C. Deffayet, O. Pujolàs, I. Sawicki *et al.*, Imperfect dark energy from kinetic gravity braiding, *J. Cosmo. Astropart. Phys.* **10**, 026 (2010). doi: 10.1088/1475-7516/2010/10/026.

113. R. Kimura and K. Yamamoto, Large scale structures in the kinetic gravity braiding model that can be unbraided, *J. Cosmo. Astropart. Phys.* **4**, 025 (2011). doi: 10.1088/1475-7516/2011/04/025.

114. R. Kimura, T. Kobayashi and K. Yamamoto, Observational constraints on kinetic gravity braiding from the integrated Sachs-Wolfe effect, *Phys. Rev.* **D85**(12), 123503 (2012). doi: 10.1103/PhysRevD.85.123503.

115. Z. Zhai, M. Blanton, A. Slosar *et al.*, An evaluation of cosmological models from the expansion and growth of structure measurements, *Astrophys. J.* **850**, 183 (2017). doi: 10.3847/1538-4357/aa9888.

116. H. Motohashi, K. Noui, T. Suyama *et al.*, Healthy degenerate theories with higher derivatives, *J. Cosmo. Astropart. Phys.* **7**, 033 (2016). doi: 10.1088/1475-7516/2016/07/033.

117. R. Klein and D. Roest, Exorcising the Ostrogradsky ghost in coupled systems, *J. High Energy Phys.* **7**, 130 (2016). doi: 10.1007/JHEP07(2016)130.

118. M. Crisostomi, R. Klein and D. Roest, Higher derivative field theories: Degeneracy conditions and classes, *J. High Energy Phys.* **6**, 124 (2017). doi: 10.1007/JHEP06(2017)124.

119. H. Motohashi, T. Suyama and M. Yamaguchi, Ghost-free theory with third-order time derivatives, *J. Phys. Soc. Jap.* **87**(6), 063401 (2018). doi: 10.7566/JPSJ.87.063401.

120. M. Zumalacárregui and J. García-Bellido, Transforming gravity: From derivative couplings to matter to second-order scalar-tensor theories beyond the Horndeski Lagrangian, *Phys. Rev.* **D89**(6), 064046 (2014). doi: 10.1103/PhysRevD.89.064046.

121. J. Gleyzes, D. Langlois, F. Piazza *et al.*, New class of consistent scalar-tensor theories, *Phys. Rev. Lett.* **114**(21), 211101 (2015). doi: 10.1103/PhysRevLett.114.211101.

122. J. Gleyzes, D. Langlois, F. Piazza *et al.*, Exploring gravitational theories beyond Horndeski, *J. Cosmo. Astropart. Phys.* **2**, 018 (2015). doi: 10.1088/1475-7516/2015/02/018.

123. D. Langlois and K. Noui, Hamiltonian analysis of higher derivative scalar-tensor theories, *J. Cosmo. Astropart. Phys.* **7**, 016 (2016). doi: 10.1088/1475-7516/2016/07/016.

124. M. Crisostomi, K. Koyama and G. Tasinato, Extended scalar-tensor theories of gravity, *J. Cosmo. Astropart. Phys.* **4**, 044 (2016). doi: 10.1088/1475-7516/2016/04/044.

125. J. Ben Achour, D. Langlois and K. Noui, Degenerate higher order scalar-tensor theories beyond Horndeski and disformal transformations, *Phys. Rev.* **D93**(12), 124005 (2016). doi: 10.1103/PhysRevD.93.124005.

126. J. Ben Achour, M. Crisostomi, K. Koyama *et al.*, Degenerate higher order scalar-tensor theories beyond Horndeski up to cubic order, *J. High Eenergy Phys.* **12**, 100 (2016). doi: 10.1007/JHEP12(2016)100.

127. C. de Rham and A. Matas, Ostrogradsky in theories with multiple fields, *J. Cosmo. Astropart. Phys.* **6**, 041 (2016). doi: 10.1088/1475-7516/2016/06/041.

128. T. Kobayashi, Y. Watanabe and D. Yamauchi, Breaking of Vainshtein screening

in scalar-tensor theories beyond Horndeski, *Phys. Rev.* **D91**(6), 064013 (2015). doi: 10.1103/PhysRevD.91.064013.

129. M. Crisostomi and K. Koyama, Vainshtein mechanism after GW170817, *Phys. Rev.* **D97**(2), 021301 (2018). doi: 10.1103/PhysRevD.97.021301.

130. D. Langlois, R. Saito, D. Yamauchi *et al.*, Scalar-tensor theories and modified gravity in the wake of GW170817, *Phys. Rev.* **D97**(6), 061501 (2018). doi: 10.1103/PhysRevD.97.061501.

131. K. Koyama and J. Sakstein, Astrophysical probes of the Vainshtein mechanism: Stars and galaxies, *Phys. Rev.* **D91**(12), 124066 (2015). doi: 10.1103/PhysRevD.91.124066.

132. J. Beltrán Jiménez, F. Piazza and H. Velten, Evading the Vainshtein mechanism with anomalous gravitational wave speed: Constraints on modified gravity from binary pulsars, *Phys. Rev. Lett.* **116**(6), 061101 (2016). doi: 10.1103/PhysRevLett.116.061101.

133. A. Dima and F. Vernizzi, Vainshtein screening in scalar-tensor theories before and after GW170817: Constraints on theories beyond Horndeski, *Phys. Rev.* **D97**(10), 101302 (2018). doi: 10.1103/PhysRevD.97.101302.

134. R. Saito, D. Yamauchi, S. Mizuno *et al.*, Modified gravity inside astrophysical bodies, *J. Cosmo. Astropart. Phys.* **6**, 008 (2015). doi: 10.1088/1475-7516/2015/06/008.

135. J. Sakstein, Hydrogen burning in low mass stars constrains scalar-tensor theories of gravity, *Phys. Rev. Lett.* **115**(20), 201101 (2015). doi: 10.1103/PhysRevLett.115.201101.

136. J. Sakstein, Testing gravity using dwarf stars, *Phys. Rev.* **D92**(12), 124045 (2015). doi: 10.1103/PhysRevD.92.124045.

137. M. Crisostomi and K. Koyama, Self-accelerating universe in scalar-tensor theories after GW170817, *Phys. Rev.* **D97**(8), 084004 (2018). doi: 10.1103/PhysRevD.97.084004.

138. M. Crisostomi, K. Koyama, D. Langlois *et al.*, Cosmological evolution in DHOST theories, *J. Cosmo. Astropart. Phys.* **1901**(01), 030 (2019). doi: 10.1088/1475-7516/2019/01/030.

139. N. Frusciante, R. Kase, K. Koyama *et al.*, Tracker and scaling solutions in DHOST theories, *Phys. Lett. B.* **790**, 167–175 (2019). doi: 10.1016/j.physletb.2019.01.009.

140. C. de Rham and S. Melville, Gravitational rainbows: LIGO and dark energy at its cutoff, *Phys. Rev. Lett.* **121**(22), 221101 (2018). doi: 10.1103/PhysRevLett.121.221101.

141. P. Creminelli, M. Lewandowski, G. Tambalo *et al.*, Gravitational wave decay into dark energy, *J. Cosmo. Astropart. Phys.* **2018**(12), 025 (2018). doi: 10.1088/1475-7516/2018/12/025.

142. P. Bull, Y. Akrami, J. Adamek *et al.*, Beyond ΛCDM: Problems, solutions, and the road ahead, *Physics of the Dark Universe.* **12**, 56–99 (2016). doi: 10.1016/j.dark.2016.02.001.

143. M. Ishak, Testing general relativity in cosmology, *Living Rev. Rel.* **22**(1), 1 (2019). doi: 10.1007/s41114-018-0017-4.

144. C. M. Will, Theoretical frameworks for testing relativistic gravity. II. Parametrized post-Newtonian hydrodynamics, and the Nordtvedt effect, *Astrophys. J.* **163**, 611 (1971). doi: 10.1086/150804.

145. C. M. Will, *Theory and Experiment in Gravitational Physics* (Cambridge University Press, 1993).

146. A. Avilez-Lopez, A. Padilla, P. M. Saffin *et al.*, The parametrized post-Newtonian-Vainshteinian formalism, *J. Cosmo. Astropart. Phys.* **6**, 044 (2015). doi: 10.1088/1475-7516/2015/06/044.

147. R. McManus, L. Lombriser and J. Peñarrubia, Parameterised post-Newtonian expansion in screened regions, *J. Cosmo. Astropart. Phys.* **12**, 031 (2017). doi: 10.1088/1475-7516/2017/12/031.

148. J.-P. Uzan, The acceleration of the universe and the physics behind it, (2006). arXiv:astro-ph/0605313.

149. R. Caldwell, A. Cooray and A. Melchiorri, Constraints on a new post-general relativity cosmological parameter, *Phys. Rev.* **D76**(2), 023507 (2007). doi: 10.1103/PhysRevD.76.023507.

150. P. Zhang, M. Liguori, R. Bean *et al.*, Probing gravity at cosmological scales by measurements which test the relationship between gravitational lensing and matter overdensity, *Phys. Rev. Lett.* **99**(14), 141302 (2007). doi: 10.1103/PhysRevLett.99.141302.

151. L. Amendola, M. Kunz and D. Sapone, Measuring the dark side (with weak lensing), *J. Cosmo. Astropart. Phys.* **4**, 013 (2008). doi: 10.1088/1475-7516/2008/04/013.

152. W. Hu and I. Sawicki, Parametrized post-Friedmann framework for modified gravity, *Phys. Rev.* **D76**(10), 104043 (2007). doi: 10.1103/PhysRevD.76.104043.

153. M. Tegmark, Measuring the metric: A Parametrized post-Friedmanian approach to the cosmic dark energy problem, *Phys. Rev.* **D66**, 103507 (2002). doi: 10.1103/PhysRevD.66.103507.

154. T. Baker, P. G. Ferreira and C. Skordis, The parameterized post-Friedmann framework for theories of modified gravity: Concepts, formalism, and examples, *Phys. Rev.* **D87**(2), 024015 (2013). doi: 10.1103/PhysRevD.87.024015.

155. P. J. E. Peebles, *The Large-Scale Structure of the Universe* (Princeton University Press, 1980).

156. E. V. Linder, Cosmic growth history and expansion history, *Phys. Rev.* **D72**(4), 043529 (2005). doi: 10.1103/PhysRevD.72.043529.

157. P. Creminelli, G. D'Amico, J. Noreña *et al.*, The effective theory of quintessence: The $w < -1$ side unveiled, *J. Cosmo. Astropart. Phys.* **2**, 018 (2009). doi: 10.1088/1475-7516/2009/02/018.

158. M. Park, K. M. Zurek and S. Watson, Unified approach to cosmic acceleration, *Phys. Rev.* **D81**(12), 124008 (2010). doi: 10.1103/PhysRevD.81.124008.

159. G. Gubitosi, F. Piazza and F. Vernizzi, The effective field theory of dark energy, *J. Cosmo. Astropart. Phys.* **2**, 032 (2013). doi: 10.1088/1475-7516/2013/02/032.

160. J. Bloomfield, É. É. Flanagan, M. Park *et al.*, Dark energy or modified gravity? An effective field theory approach, *J. Cosmo. Astropart. Phys.* **8**, 010 (2013). doi: 10.1088/1475-7516/2013/08/010.

161. J. Gleyzes, D. Langlois, F. Piazza *et al.*, Essential building blocks of dark energy, *J. Cosmo. Astropart. Phys.* **8**, 025 (2013). doi: 10.1088/1475-7516/2013/08/025.

162. S. Tsujikawa, The effective field theory of inflation/dark energy and the Horndeski theory, *Lect. Notes Phys.* **892**, 97–136 (2015). doi: 10.1007/978-3-319-10070-8_4.

163. J. Gleyzes, D. Langlois and F. Vernizzi, A unifying description of dark energy, *Int. J. Mod. Phys.* **D23**(13), 1443010 (2015). doi: 10.1142/S021827181443010X.

164. M. Lagos, E. Bellini, J. Noller *et al.*, A general theory of linear cosmological perturbations: Stability conditions, the quasistatic limit and dynamics, *J. Cosmo. Astropart. Phys.* **1803**(03), 021 (2018). doi: 10.1088/1475-7516/2018/03/021.

165. K. Koyama, A. Taruya and T. Hiramatsu, Nonlinear evolution of the matter power spectrum in modified theories of gravity, *Phys. Rev.* **D79**(2), 123512 (2009). doi: 10.1103/PhysRevD.79.123512.

166. E. Bellini, R. Jimenez and L. Verde, Signatures of Horndeski gravity on the dark matter bispectrum, *J. Cosmo. Astropart. Phys.* **5**, 057 (2015). doi: 10.1088/1475-7516/2015/05/057.

167. A. Taruya, Constructing perturbation theory kernels for large-scale structure in generalized cosmologies, *Phys. Rev.* **D94**(2), 023504 (2016). doi: 10.1103/PhysRevD.94.023504.

168. B. Bose and K. Koyama, A perturbative approach to the redshift space power spectrum: Beyond the standard model, *J. Cosmo. Astropart. Phys.* **2016**(08), 032 (2016). doi: 10.1088/1475-7516/2016/08/032.

169. D. Yamauchi, S. Yokoyama and H. Tashiro, Constraining modified theories of gravity with the galaxy bispectrum, *Phys. Rev.* **D96**(12), 123516 (2017). doi: 10.1103/PhysRevD.96.123516.

170. N. Frusciante and G. Papadomanolakis, Tackling non-linearities with the effective field theory of dark energy and modified gravity, *J. Cosmo. Astropart. Phys.* **12**, 014 (2017). doi: 10.1088/1475-7516/2017/12/014.

171. B. Bose, K. Koyama, M. Lewandowski *et al.*, Towards precision constraints on gravity with the effective field theory of large-scale structure, *J. Cosmo. Astropart. Phys.* **2018**(04), 063 (2018). doi: 10.1088/1475-7516/2018/04/063.

172. S. Hirano, T. Kobayashi, H. Tashiro *et al.*, Matter bispectrum beyond Horndeski theories, *Phys. Rev.* **D97**(10), 103517 (2018). doi: 10.1103/PhysRevD.97.103517.

173. G.-B. Zhao, B. Li and K. Koyama, N-body simulations for f(R) gravity using a self-adaptive particle-mesh code, *Phys. Rev.* **D83**, 044007 (2011). doi: 10.1103/PhysRevD.83.044007.

174. L. Lombriser, A parametrisation of modified gravity on nonlinear cosmological scales, *J. Cosmo. Astropart. Phys.* **2016**(11), 039 (2016). doi: 10.1088/1475-7516/2016/11/039.

175. B. Hu, X.-W. Liu and R.-G. Cai, CHAM: A fast algorithm of modelling non-linear matter power spectrum in the sCreened HAlo Model, *Mon. Not. Roy. Astron. Soc.* **476**(1), L65–L68 (2018). doi: 10.1093/mnrasl/sly032.

176. V. A. A. Sanghai and T. Clifton, Parameterized post-Newtonian cosmology, *Class. Quant. Grav.* **34**(6), 065003 (2017). doi: 10.1088/1361-6382/aa5d75.

177. I. Milillo, D. Bertacca, M. Bruni *et al.*, Missing link: A nonlinear post-Friedmann framework for small and large scales, *Phys. Rev.* **D92**(2), 023519 (2015). doi: 10.1103/PhysRevD.92.023519.

178. B. P. Abbott *et al.*, GW170814: A three-detector observation of gravitational waves from a binary black hole coalescence, *Phys. Rev. Lett.* **119**(14), 141101 (2017). doi: 10.1103/PhysRevLett.119.141101.

179. N. Yunes and F. Pretorius, Fundamental theoretical bias in gravitational wave astro-

physics and the parametrized post-Einsteinian framework, *Phys. Rev.* **D80**(12), 122003 (2009). doi: 10.1103/PhysRevD.80.122003.

180. T. G. F. Li, W. Del Pozzo, S. Vitale *et al.*, Towards a generic test of the strong field dynamics of general relativity using compact binary coalescence, *Phys. Rev.* **D85**(8), 082003 (2012). doi: 10.1103/PhysRevD.85.082003.

181. S. Mirshekari and C. M. Will, Compact binary systems in scalar-tensor gravity: Equations of motion to 2.5 post-Newtonian order, *Phys. Rev.* **D87**(8), 084070 (2013). doi: 10.1103/PhysRevD.87.084070.

182. R. N. Lang, Compact binary systems in scalar-tensor gravity. II. Tensor gravitational waves to second post-Newtonian order, *Phys. Rev.* **D89**(8), 084014 (2014). doi: 10. 1103/PhysRevD.89.084014.

183. B. P. Abbott, R. Abbott, T. D. Abbott *et al.*, Tests of general relativity with GW150914, *Phys. Rev. Lett.* **116**(22), 221101 (2016). doi: 10.1103/PhysRevLett.116. 221101.

184. B. P. Abbott *et al.*, Binary black hole mergers in the first Advanced LIGO observing run, *Phys. Rev.* **X6**(4), 041015 (2016). doi: 10.1103/PhysRevX.6.041015.

185. I. D. Saltas, I. Sawicki, L. Amendola *et al.*, Anisotropic stress as a signature of nonstandard propagation of gravitational waves, *Phys. Rev. Lett.* **113**(19), 191101 (2014). doi: 10.1103/PhysRevLett.113.191101.

186. A. Nishizawa, Generalized framework for testing gravity with gravitational-wave propagation. I. Formulation, *Phys. Rev.* **D97**(10), 104037 (2018). doi: 10.1103/ PhysRevD.97.104037.

187. M. Chevallier and D. Polarski, Accelerating universes with scaling dark matter, *Int. J.Mod. Phys. D.* **10**, 213–223 (2001). doi: 10.1142/S0218271801000822.

188. E. V. Linder, Exploring the expansion history of the universe, *Phys. Rev. Lett.* **90** (9), 091301 (2003). doi: 10.1103/PhysRevLett.90.091301.

189. C. Wetterich, Cosmology and the fate of dilatation symmetry, *Nucl. Phys. B.* **302**, 668–696 (1988). doi: 10.1016/0550-3213(88)90193-9.

190. R. G. Crittenden, L. Pogosian and G.-B. Zhao, Investigating dark energy experiments with principal components, *J. Cosmo. Astropart. Phys.* **12**, 025 (2009). doi: 10.1088/ 1475-7516/2009/12/025.

191. L. Lombriser, K. Koyama and B. Li, Halo modelling in chameleon theories, *J. Cosmo. Astropart. Phys.* **1403**, 021 (2014). doi: 10.1088/1475-7516/2014/03/021.

192. R. A. Battye, B. Bolliet and F. Pace, Do cosmological data rule out $f(\mathcal{R})$ with $w \neq -1$?, *Phys. Rev.* **D97**(10), 104070 (2018). doi: 10.1103/PhysRevD.97.104070.

193. C. Armendáriz-Picón, T. Damour and V. Mukhanov, k-inflation, *Phys. Lett. B.* **458**, 209–218 (1999). doi: 10.1016/S0370-2693(99)00603-6.

194. E. V. Linder, G. Sengör and S. Watson, Is the effective field theory of dark energy effective?, *J. Cosmo. Astropart. Phys.* **5**, 053 (2016). doi: 10.1088/1475-7516/2016/ 05/053.

195. J. Gleyzes, D. Langlois, M. Mancarella *et al.*, Effective theory of interacting dark energy, *J. Cosmo. Astropart. Phys.* **8**, 054 (2015). doi: 10.1088/1475-7516/2015/08/ 054.

196. L. Lombriser and A. Taylor, Classifying linearly shielded modified gravity models in effective field theory, *Phys. Rev. Lett.* **114**(3), 031101 (2015). doi: 10.1103/ PhysRevLett.114.031101.

197. A. Nishizawa and T. Nakamura, Measuring speed of gravitational waves by observations of photons and neutrinos from compact binary mergers and supernovae, *Phys. Rev.* **D90**(4), 044048 (2014). doi: 10.1103/PhysRevD.90.044048.

198. M. Ishak, A. Upadhye and D. N. Spergel, Probing cosmic acceleration beyond the equation of state: Distinguishing between dark energy and modified gravity models, *Phys. Rev.* **D74**(4), 043513 (2006). doi: 10.1103/PhysRevD.74.043513.

199. M. J. Mortonson, W. Hu and D. Huterer, Falsifying paradigms for cosmic acceleration, *Phys. Rev.* **D79**(2), 023004 (2009). doi: 10.1103/PhysRevD.79.023004.

200. E. J. Ruiz and D. Huterer, Testing the dark energy consistency with geometry and growth, *Phys. Rev.* **D91**(6), 063009 (2015). doi: 10.1103/PhysRevD.91.063009.

201. L. Lombriser, Consistency check of ΛCDM phenomenology, *Phys. Rev.* **D83**(6), 063519 (2011). doi: 10.1103/PhysRevD.83.063519.

202. L. Lombriser, J. Yoo and K. Koyama, Relativistic effects in galaxy clustering in a parametrized post-Friedmann universe, *Phys. Rev.* **D87**(10), 104019 (2013). doi: 10.1103/PhysRevD.87.104019.

203. L. Lombriser and A. Taylor, Semi-dynamical perturbations of unified dark energy, *J. Cosmo. Astropart. Phys.* **11**, 040 (2015). doi: 10.1088/1475-7516/2015/11/040.

204. G.-B. Zhao, L. Pogosian, A. Silvestri *et al.*, Searching for modified growth patterns with tomographic surveys, *Phys. Rev.* **D79**(8), 083513 (2009). doi: 10.1103/PhysRevD.79.083513.

205. S. F. Daniel, E. V. Linder, T. L. Smith *et al.*, Testing general relativity with current cosmological data, *Phys. Rev.* **D81**(12), 123508 (2010). doi: 10.1103/PhysRevD.81.123508.

206. J. N. Dossett, M. Ishak and J. Moldenhauer, Testing general relativity at cosmological scales: Implementation and parameter correlations, *Phys. Rev.* **D84**(12), 123001 (2011). doi: 10.1103/PhysRevD.84.123001.

207. C. Bonvin and P. Fleury, Testing the equivalence principle on cosmological scales, *J. Cosmo. Astropart. Phys.* **5**, 061 (2018). doi: 10.1088/1475-7516/2018/05/061.

208. E. Bertschinger, On the growth of perturbations as a test of dark energy and gravity, *Astrophys. J.* **648**, 797–806 (2006). doi: 10.1086/506021.

209. R. A. Battye and J. A. Pearson, Effective action approach to cosmological perturbations in dark energy and modified gravity, *J. Cosmo. Astropart. Phys.* **7**, 019 (2012). doi: 10.1088/1475-7516/2012/07/019.

210. B. Hu, M. Raveri, N. Frusciante *et al.*, Effective field theory of cosmic acceleration: An implementation in CAMB, *Phys. Rev.* **D89**(10), 103530 (2014). doi: 10.1103/PhysRevD.89.103530.

211. M. Zumalacárregui, E. Bellini, I. Sawicki *et al.*, hi_class: Horndeski in the cosmic linear anisotropy solving system, *J. Cosmo. Astropart. Phys.* **8**, 019 (2017). doi: 10.1088/1475-7516/2017/08/019.

212. J. Bloomfield, A simplified approach to general scalar-tensor theories, *J. Cosmo. Astropart. Phys.* **12**, 044 (2013). doi: 10.1088/1475-7516/2013/12/044.

213. P. Hořava, Quantum gravity at a Lifshitz point, *Phys. Rev.* **D79**(8), 084008 (2009). doi: 10.1103/PhysRevD.79.084008.

214. C. M. Caves, Gravitational radiation and the ultimate speed in Rosen's bimetric theory of gravity, *Anna. Phys.* **125**, 35–52 (1980). doi: 10.1016/0003-4916(80)90117-7.

215. R. A. Battye, F. Pace and D. Trinh, Gravitational wave constraints on dark sector models, *Phys. Rev.* **D98**(2), 023504 (2018). doi: 10.1103/PhysRevD.98.023504.

216. P. Brax, C. Burrage and A.-C. Davis, The speed of Galileon gravity, *J. Cosmo. Astropart. Phys.* **3**, 004 (2016). doi: 10.1088/1475-7516/2016/03/004.

217. A. Barreira, B. Li, C. M. Baugh *et al.*, The observational status of Galileon gravity after Planck, *J. Cosmo. Astropart. Phys.* **8**, 059 (2014). doi: 10.1088/1475-7516/2014/08/059.

218. L. Lombriser, W. Hu, W. Fang *et al.*, Cosmological constraints on DGP braneworld gravity with brane tension, *Phys. Rev.* **D80**(6), 063536 (2009). doi: 10.1103/PhysRevD.80.063536.

219. R. McManus, L. Lombriser and J. Peñarrubia, Finding Horndeski theories with Einstein gravity limits, *J. Cosmo. Astropart. Phys.* **11**, 006 (2016). doi: 10.1088/1475-7516/2016/11/006.

220. N. Cornish, D. Blas and G. Nardini, Bounding the speed of gravity with gravitational wave observations, *Phys. Rev. Lett.* **119**(16), 161102 (2017). doi: 10.1103/PhysRevLett.119.161102.

221. D. Blas, M. M. Ivanov, I. Sawicki *et al.*, On constraining the speed of gravitational waves following GW150914, *Sov. J. Exp. Theor. Phys. Lett.* **103**, 624–626 (2016). doi: 10.1134/S0021364016100040.

222. M. Raveri, C. Baccigalupi, A. Silvestri *et al.*, Measuring the speed of cosmological gravitational waves, *Phys. Rev.* **D91**(6), 061501 (2015). doi: 10.1103/PhysRevD.91.061501.

223. L. Amendola, G. Ballesteros and V. Pettorino, Effects of modified gravity on B-mode polarization, *Phys. Rev.* **D90**(4), 043009 (2014). doi: 10.1103/PhysRevD.90.043009.

224. B. F. Schutz, Determining the Hubble constant from gravitational wave observations, *Nature.* **323**, 310 (1986). doi: 10.1038/323310a0.

225. D. E. Holz and S. A. Hughes, Using gravitational-wave standard sirens, *Astrophys. J.* **629**, 15–22 (2005). doi: 10.1086/431341.

226. E. Belgacem, Y. Dirian, S. Foffa *et al.*, Gravitational-wave luminosity distance in modified gravity theories, *Phys. Rev.* **D97**(10), 104066 (2018). doi: 10.1103/PhysRevD.97.104066.

227. L. Amendola, I. Sawicki, M. Kunz *et al.*, Direct detection of gravitational waves can measure the time variation of the Planck mass, *J. Cosmo. Astropart. Phys.* **2018** (08), 030 (2018). doi: 10.1088/1475-7516/2018/08/030.

228. C. L. MacLeod and C. J. Hogan, Precision of Hubble constant derived using black hole binary absolute distances and statistical redshift information, *Phys. Rev.* **D77** (4), 043512 (2008). doi: 10.1103/PhysRevD.77.043512.

229. N. Tamanini, C. Caprini, E. Barausse *et al.*, Science with the space-based interferometer eLISA. III: Probing the expansion of the universe using gravitational wave standard sirens, *J. Cosmo. Astropart. Phys.* **4**, 002 (2016). doi: 10.1088/1475-7516/2016/04/002.

230. A. Silvestri, L. Pogosian and R. V. Buniy, Practical approach to cosmological perturbations in modified gravity, *Phys. Rev.* **D87**(10), 104015 (2013). doi: 10.1103/PhysRevD.87.104015.

231. E. Bertschinger and P. Zukin, Distinguishing modified gravity from dark energy, *Phys. Rev.* **D78**(2), 024015 (2008). doi: 10.1103/PhysRevD.78.024015.

232. F. Simpson *et al.*, CFHTLenS: Testing the laws of gravity with tomographic weak lensing and redshift space distortions, *Mon. Not. Roy. Astron. Soc.* **429**, 2249 (2013). doi: 10.1093/mnras/sts493.

233. N. A. Lima, V. Smer-Barreto and L. Lombriser, Constraints on decaying early modified gravity from cosmological observations, *Phys. Rev.* **D94**(8), 083507 (2016). doi: 10.1103/PhysRevD.94.083507.

234. J. Kennedy, L. Lombriser and A. Taylor, Reconstructing Horndeski theories from phenomenological modified gravity and dark energy models on cosmological scales, *Phys. Rev.* **D98**(4), 044051 (2018). doi: 10.1103/PhysRevD.98.044051.

235. R. Laureijs, J. Amiaux, S. Arduini *et al.*, Euclid definition study report (2011). arXiv:1110.3193 [astro-ph.CO].

236. Ž. Ivezić, S. M. Kahn, J. A. Tyson *et al.*, LSST: From science drivers to reference design and anticipated data products (2008). arXiv:0805.2366.

237. H. A. Buchdahl, Non-linear Lagrangians and cosmological theory, *Mon. Not. Roy. Astron. Soc.* **150**, 1 (1970). doi: 10.1093/mnras/150.1.1.

238. P. Brax and P. Valageas, Impact on the power spectrum of screening in modified gravity scenarios, *Phys. Rev.* **D88**(2), 023527 (2013). doi: 10.1103/PhysRevD.88.023527.

239. L. Lombriser, Constraining chameleon models with cosmology, *Annalen der Physik.* **526**, 259–282 (2014). doi: 10.1002/andp.201400058.

240. F. Schmidt, M. V. Lima, H. Oyaizu *et al.*, Non-linear evolution of f(R) cosmologies III: Halo statistics, *Phys. Rev.* **D79**, 083518 (2009). doi: 10.1103/PhysRevD.79.083518.

241. F. Pace, J.-C. Waizmann and M. Bartelmann, Spherical collapse model in dark-energy cosmologies, *Mon. Not. Roy. Astron. Soc.* **406**, 1865–1874 (2010). doi: 10.1111/j.1365-2966.2010.16841.x.

242. A. Borisov, B. Jain and P. Zhang, Spherical collapse in f(R) gravity, *Phys. Rev.* **D85**, 063518 (2012). doi: 10.1103/PhysRevD.85.063518.

243. M. Kopp, S. A. Appleby, I. Achitouv *et al.*, Spherical collapse and halo mass function in $f(R)$ theories, *Phys. Rev.* **D88**(8), 084015 (2013). doi: 10.1103/PhysRevD.88.084015.

244. B. Li and G. Efstathiou, An extended excursion set approach to structure formation in chameleon models, *Mon. Not. Roy. Astron. Soc.* **421**, 1431–1442 (2012). doi: 10.1111/j.1365-2966.2011.20404.x.

245. B. Li and T. Lam, Excursion set theory for modified gravity: Eulerian versus Lagrangian environments, *Mon. Not. Roy. Astron. Soc.* **425**, 730–739 (2012). doi: 10.1111/j.1365-2966.2012.21592.x.

246. L. Lombriser, B. Li, K. Koyama *et al.*, Modeling halo mass functions in chameleon f(R) gravity, *Phys. Rev.* **D87**(12), 123511 (2013). doi: 10.1103/PhysRevD.87.123511.

247. L. Lombriser, F. Schmidt, T. Baldauf *et al.*, Cluster density profiles as a test of modified gravity, *Phys. Rev.* **D85**, 102001 (2012). doi: 10.1103/PhysRevD.85.102001.

248. L. Lombriser, K. Koyama, G.-B. Zhao *et al.*, Chameleon f(R) gravity in the virialized cluster, *Phys. Rev.* **D85**, 124054 (2012). doi: 10.1103/PhysRevD.85.124054.

249. M. Gronke, D. F. Mota and H. A. Winther, Universal predictions of screened modified gravity on cluster scales, *Astron. Astrophys.* **583**, A123 (2015). doi: 10.1051/0004-6361/201526611.

250. M. A. Mitchell, J.-h. He, C. Arnold *et al.*, A general framework to test gravity using

galaxy clusters – I. Modelling the dynamical mass of haloes in $f(R)$ gravity, *Mon. Not. Roy. Astron. Soc.* **477**(1), 1133–1152 (2018). doi: 10.1093/mnras/sty636.

251. F. Schmidt, W. Hu and M. Lima, Spherical collapse and the halo model in braneworld gravity, *Phys. Rev.* **D81**, 063005 (2010). doi: 10.1103/PhysRevD.81.063005.

252. L. Lombriser, A. Slosar, U. Seljak *et al.*, Constraints on f(R) gravity from probing the large-scale structure, *Phys. Rev.* **D85**(12), 124038 (2012). doi: 10.1103/PhysRevD.85.124038.

253. A. Barreira, B. Li, W. A. Hellwing *et al.*, Halo model and halo properties in Galileon gravity cosmologies, *J. Cosmo. Astropart. Phys.* **1404**, 029 (2014). doi: 10.1088/1475-7516/2014/04/029.

254. M. Cataneo, D. Rapetti, L. Lombriser *et al.*, Cluster abundance in chameleon $f(R)$ gravity I: Toward an accurate halo mass function prediction, *J. Cosmo. Astropart. Phys.* **2016**(12), 024 (2016). doi: 10.1088/1475-7516/2016/12/024.

255. S. Hagstotz, M. Costanzi, M. Baldi *et al.*, Joint halo-mass function for modified gravity and massive neutrinos – I. Simulations and cosmological forecasts, *Mon. Not. Roy. Astron. Soc.* **486**(3), 3927–3941 (2019). doi: 10.1093/mnras/stz1051.

256. J. Clampitt, Y.-C. Cai and B. Li, Voids in modified gravity: Excursion set predictions, *Mon. Not. Roy. Astron. Soc.* **431**, 749–766 (2013). doi: 10.1093/mnras/stt219.

257. T. Y. Lam, J. Clampitt, Y.-C. Cai *et al.*, Voids in modified gravity reloaded: Eulerian void assignment, *Mon. Not. Roy. Astron. Soc.* **450**(3), 3319–3330 (2015). doi: 10.1093/mnras/stv797.

258. J. A. Peacock and R. E. Smith, Halo occupation numbers and galaxy bias, *Mon. Not. Roy. Astron. Soc.* **318**, 1144 (2000). doi: 10.1046/j.1365-8711.2000.03779.x.

259. U. Seljak, Analytic model for galaxy and dark matter clustering, *Mon. Not. Roy. Astron. Soc.* **318**, 203 (2000). doi: 10.1046/j.1365-8711.2000.03715.x.

260. A. Cooray and R. K. Sheth, Halo models of large scale structure, *Phys. Rept.* **372**, 1–129 (2002). doi: 10.1016/S0370-1573(02)00276-4.

261. I. Achitouv, M. Baldi, E. Puchwein *et al.*, Imprint of f (R) gravity on nonlinear structure formation, *Phys. Rev.* **D93**(10), 103522 (2016). doi: 10.1103/PhysRevD.93.103522.

262. Y. Li and W. Hu, Chameleon halo modeling in f(R) gravity, *Phys. Rev.* **D84**(8), 084033 (2011). doi: 10.1103/PhysRevD.84.084033.

263. A. Mead, C. Heymans, L. Lombriser *et al.*, Accurate halo-model matter power spectra with dark energy, massive neutrinos and modified gravitational forces, *Mon. Not. Roy. Astron. Soc.* **459**(2), 1468–1488 (2016). doi: 10.1093/mnras/stw681.

264. C. Fedeli, The clustering of baryonic matter. I: A halo-model approach, *J. Cosmo. Astropart. Phys.* **4**, 028 (2014). doi: 10.1088/1475-7516/2014/04/028.

265. A. Terukina, L. Lombriser, K. Yamamoto *et al.*, Testing chameleon gravity with the Coma cluster, *J. Cosmo. Astropart. Phys.* **4**, 013 (2014). doi: 10.1088/1475-7516/2014/04/013.

266. J. Sakstein, H. Wilcox, D. Bacon *et al.*, Testing gravity using galaxy clusters: New constraints on beyond Horndeski theories, *J. Cosmo. Astropart. Phys.* **7**, 019 (2016). doi: 10.1088/1475-7516/2016/07/019.

267. L. Lombriser, F. Simpson and A. Mead, Unscreening modified gravity in the matter power spectrum, *Phys. Rev. Lett.* **114**(25), 251101 (2015). doi: 10.1103/PhysRevLett.114.251101.

268. H. A. Winther and P. G. Ferreira, Fast route to nonlinear clustering statistics in modified gravity theories, *Phys. Rev.* **D91**(12), 123507 (2015). doi: 10.1103/PhysRevD. 91.123507.

269. A. J. Mead, J. A. Peacock, L. Lombriser *et al.*, Rapid simulation rescaling from standard to modified gravity models, *Mon. Not. Roy. Astron. Soc.* **452**(4), 4203–4221 (2015). doi: 10.1093/mnras/stv1484.

270. B. Bertotti, L. Iess and P. Tortora, A test of general relativity using radio links with the Cassini spacecraft, *Nature.* **425**, 374–376 (2003). doi: 10.1038/nature01997.

271. X. Zhang, W. Zhao, H. Huang *et al.*, Post-Newtonian parameters and cosmological constant of screened modified gravity, *Phys. Rev.* **D93**(12), 124003 (2016). doi: 10. 1103/PhysRevD.93.124003.

272. G. Gabadadze, K. Hinterbichler and D. Pirtskhalava, Classical duals of derivatively self-coupled theories, *Phys. Rev.* **D85**(12), 125007 (2012). doi: 10.1103/PhysRevD. 85.125007.

273. A. Padilla and P. M. Saffin, Classical duals, Legendre transforms and the Vainshtein mechanism, *J. High Energy Phys.* **7**, 122 (2012). doi: 10.1007/JHEP07(2012)122.

274. R. Epstein and R. V. Wagoner, Post-Newtonian generation of gravitational waves, *Astrophys. J.* **197**, 717–723 (1975). doi: 10.1086/153561.

275. C. M. Will and A. G. Wiseman, Gravitational radiation from compact binary systems: Gravitational waveforms and energy loss to second post-Newtonian order, *Phys. Rev.* **D54**, 4813–4848 (1996). doi: 10.1103/PhysRevD.54.4813.

276. M. E. Pati and C. M. Will, Post-Newtonian gravitational radiation and equations of motion via direct integration of the relaxed Einstein equations. II. Two-body equations of motion to second post-Newtonian order, and radiation reaction to 3.5 post-Newtonian order, *Phys. Rev.* **D65**(10), 104008 (2002). doi: 10.1103/PhysRevD. 65.104008.

277. D. B. Thomas, M. Bruni and D. Wands, Relativistic weak lensing from a fully nonlinear cosmological density field, *J. Cosmo. Astropart. Phys.* **9**, 021 (2015). doi: 10. 1088/1475-7516/2015/09/021.

278. R. Kase, L. Á. Gergely and S. Tsujikawa, Effective field theory of modified gravity on the spherically symmetric background: Leading order dynamics and the odd-type perturbations, *Phys. Rev.* **D90**(12), 124019 (2014). doi: 10.1103/PhysRevD.90. 124019.

279. P. Brax, A.-C. Davis, B. Li *et al.*, Systematic simulations of modified gravity: Symmetron and dilaton models, *J. Cosmo. Astropart. Phys.* **10**, 002 (2012). doi: 10.1088/1475-7516/2012/10/002.

280. J. Kennedy, L. Lombriser and A. Taylor, Reconstructing Horndeski models from the effective field theory of dark energy, *Phys. Rev.* **D96**(8), 084051 (2017). doi: 10.1103/PhysRevD.96.084051.

281. J. Adamek, D. Daverio, R. Durrer *et al.*, gevolution: A cosmological N-body code based on General Relativity, *J. Cosmo. Astropart. Phys.* **7**, 053 (2016). doi: 10.1088/1475-7516/2016/07/053.

282. J. Adamek, R. Durrer and M. Kunz, Relativistic N-body simulations with massive neutrinos, *J. Cosmo. Astropart. Phys.* **11**, 004 (2017). doi: 10.1088/1475-7516/2017/11/004.

283. B. Li, *Simulating Large-Scale Structure for Models of Cosmic Acceleration* (IOP Publishing, 2018). doi: 10.1088/978-0-7503-1587-6.

284. E. Puchwein, M. Baldi and V. Springel, Modified-Gravity-GADGET: A new code for cosmological hydrodynamical simulations of modified gravity models, *Mon. Not. Roy. Astron. Soc.* **436**, 348–360 (2013). doi: 10.1093/mnras/stt1575.

285. C. Llinares, D. F. Mota and H. A. Winther, ISIS: A new N-body cosmological code with scalar fields based on RAMSES. Code presentation and application to the shapes of clusters, *Astron. Astrophys.* **562**, A78 (2014). doi: 10.1051/0004-6361/201322412.

286. H. A. Winther *et al.*, Modified gravity N-body code comparison project, *Mon. Not. Roy. Astron. Soc.* **454**(4), 4208–4234 (2015). doi: 10.1093/mnras/stv2253.

287. A. Klypin, A. V. Macciò, R. Mainini *et al.*, Halo properties in models with dynamical dark energy, *Astrophys. J.* **599**, 31–37 (2003). doi: 10.1086/379237.

288. M. Baldi, V. Pettorino, G. Robbers *et al.*, Hydrodynamical N-body simulations of coupled dark energy cosmologies, *Mon. Not. Roy. Astron. Soc.* **403**, 1684–1702 (2010). doi: 10.1111/j.1365-2966.2009.15987.x.

289. A. V. Macciò, C. Quercellini, R. Mainini *et al.*, Coupled dark energy: Parameter constraints from N-body simulations, *Phys. Rev.* **D69**(12), 123516 (2004). doi: 10.1103/PhysRevD.69.123516.

290. A. Barreira, B. Li, W. A. Hellwing *et al.*, Nonlinear structure formation in nonlocal gravity, *J. Cosmo. Astropart. Phys.* **9**, 031 (2014). doi: 10.1088/1475-7516/2014/09/031.

291. E. Carlesi, A. Knebe, G. Yepes *et al.*, N-body simulations with a cosmic vector for dark energy, *Mon. Not. Roy. Astron. Soc.* **424**, 699–715 (2012). doi: 10.1111/j.1365-2966.2012.21258.x.

292. H. Oyaizu, Nonlinear evolution of f(R) cosmologies. I. Methodology, *Phys. Rev.* **D78**(12), 123523 (2008). doi: 10.1103/PhysRevD.78.123523.

293. B. Li, G.-B. Zhao, R. Teyssier *et al.*, ECOSMOG: An Efficient COde for Simulating MOdified Gravity, *J. Cosmo. Astropart. Phys.* **1**, 051 (2012). doi: 10.1088/1475-7516/2012/01/051.

294. B. Li and H. Zhao, Structure formation by a fifth force: N-body versus linear simulations, *Phys. Rev.* **D80**(4), 044027 (2009). doi: 10.1103/PhysRevD.80.044027.

295. A.-C. Davis, B. Li, D. F. Mota *et al.*, Structure formation in the symmetron model, *Astrophys. J.* **748**, 61 (2012). doi: 10.1088/0004-637X/748/1/61.

296. P. Brax, C. van de Bruck, A.-C. Davis *et al.*, Nonlinear structure formation with the environmentally dependent dilaton, *Phys. Rev.* **D83**, 104026 (2011). doi: 10.1103/PhysRevD.83.104026.

297. J. Khoury and M. Wyman, N-body simulations of DGP and degravitation theories, *Phys. Rev.* **D80**(6), 064023 (2009). doi: 10.1103/PhysRevD.80.064023.

298. I. Laszlo and R. Bean, Nonlinear growth in modified gravity theories of dark energy, *Phys. Rev.* **D77**(2), 024048 (2008). doi: 10.1103/PhysRevD.77.024048.

299. K. C. Chan and R. Scoccimarro, Large-scale structure in brane-induced gravity. II. Numerical simulations, *Phys. Rev.* **D80**(10), 104005 (2009). doi: 10.1103/PhysRevD.80.104005.

300. A. Barreira, B. Li, W. A. Hellwing *et al.*, Nonlinear structure formation in the Cubic Galileon gravity model, *J. Cosmo. Astropart. Phys.* **1310**, 027 (2013). doi: 10.1088/1475-7516/2013/10/027.

301. B. Li, A. Barreira, C. M. Baugh *et al.*, Simulating the quartic Galileon gravity model on adaptively refined meshes, *J. Cosmo. Astropart. Phys.* **11**, 012 (2013). doi: 10.1088/1475-7516/2013/11/012.

302. C. Llinares and D. F. Mota, Releasing scalar fields: Cosmological simulations of scalar-tensor theories for gravity beyond the static approximation, *Phys. Rev. Lett.* **110**(16), 161101 (2013). doi: 10.1103/PhysRevLett.110.161101.

303. C. Llinares and D. F. Mota, Cosmological simulations of screened modified gravity out of the static approximation: Effects on matter distribution, *Phys. Rev.* **D89**(8), 084023 (2014). doi: 10.1103/PhysRevD.89.084023.

304. S. Bose, W. A. Hellwing and B. Li, Testing the quasi-static approximation in f(R) gravity simulations, *J. Cosmo. Astropart. Phys.* **2**, 034 (2015). doi: 10.1088/1475-7516/2015/02/034.

305. R. Hagala, C. Llinares and D. F. Mota, Cosmological simulations with disformally coupled symmetron fields, *Astron. Astrophys.* **585**, A37 (2016). doi: 10.1051/0004-6361/201526439.

306. C. Llinares, The shrinking domain framework I: A new, faster, more efficient approach to cosmological simulations (2017). arXiv:1709.04703.

307. R. W. Hockney and J. W. Eastwood, *Computer Simulation Using Particles* (CRC Press, 1988).

308. A. Nusser, Modified Newtonian dynamics of large-scale structure, *Mon. Not. Roy. Astron. Soc.* **331**, 909–916 (2002). doi: 10.1046/j.1365-8711.2002.05235.x.

309. A. Knebe and B. K. Gibson, Galactic haloes in MONDian cosmological simulations, *Mon. Not. Roy. Astron. Soc.* **347**, 1055–1064 (2004). doi: 10.1111/j.1365-2966.2004.07182.x.

310. R. Brada and M. Milgrom, Stability of disk galaxies in the modified dynamics, *Astrophys. J.* **519**, 590–598 (1999). doi: 10.1086/307402.

311. O. Tiret and F. Combes, Evolution of spiral galaxies in modified gravity, *Astron. Astrophys.* **464**, 517–528 (2007). doi: 10.1051/0004-6361:20066446.

312. C. Llinares, A. Knebe and H. Zhao, Cosmological structure formation under MOND: A new numerical solver for Poisson's equation, *Mon. Not. Roy. Astron. Soc.* **391**, 1778–1790 (2008). doi: 10.1111/j.1365-2966.2008.13961.x.

313. C. Llinares. *On the linear and non-linear evolution of dust density perturbations with MOND*. PhD thesis, Rijksuniversiteit Groningen, (2011).

314. P. Londrillo and C. Nipoti, N-MODY: A code for collisionless N-body simulations in modified Newtonian dynamics., *Memo. Soc. Astron. Italiana Suppl.* **13**, 89 (2009). arXiv:0803.4456.

315. P. Londrillo and C. Nipoti. N-MODY: A code for collisionless N-body simulations in modified Newtonian dynamics. Astrophysics Source Code Library, (2011).

316. G. W. Angus, K. J. van der Heyden, B. Famaey *et al.*, A QUMOND galactic N-body code – I. Poisson solver and rotation curve fitting, *Mon. Not. Roy. Astron. Soc.* **421**, 2598–2609 (2012). doi: 10.1111/j.1365-2966.2012.20532.x.

317. G. N. Candlish, R. Smith and M. Fellhauer, RAyMOND: An N-body and hydrodynamics code for MOND, *Mon. Not. Roy. Astron. Soc.* **446**, 1060–1070 (2015). doi: 10.1093/mnras/stu2158.

318. F. Lüghausen, B. Famaey and P. Kroupa, Phantom of RAMSES (POR): A new

Milgromian dynamics N-body code, *Can. J. Phys.* **93**, 232–241 (2015). doi: 10.1139/cjp-2014-0168.

319. S. M. Carroll, *Spacetime and Geometry. An Introduction to General Relativity* (Pearson, 2004).

320. H. Martel and P. R. Shapiro, A convenient set of comoving cosmological variables and their application, *Mon. Not. Roy. Astron. Soc.* **297**, 467–485 (1998). doi: 10.1046/j.1365-8711.1998.01497.x.

321. N. E. Chisari and M. Zaldarriaga, Connection between Newtonian simulations and general relativity, *Phys. Rev.* **D83**(12), 123505 (2011). doi: 10.1103/PhysRevD.83.123505.

322. S. R. Green and R. M. Wald, Newtonian and relativistic cosmologies, *Phys. Rev.* **D85**(6), 063512 (2012). doi: 10.1103/PhysRevD.85.063512.

323. C. Fidler, C. Rampf, T. Tram *et al.*, General relativistic corrections to N -body simulations and the Zel'dovich approximation, *Phys. Rev.* **D92**(12), 123517 (2015). doi: 10.1103/PhysRevD.92.123517.

324. C. Fidler, T. Tram, C. Rampf *et al.*, General relativistic weak-field limit and Newtonian N-body simulations, *J. Cosmo. Astropart. Phys.* **12**, 022 (2017). doi: 10.1088/1475-7516/2017/12/022.

325. E. Bertschinger, Simulations of structure formation in the universe, *Ann. Rev. Astron. Astrophys.* **36**, 599–654 (1998). doi: 10.1146/annurev.astro.36.1.599.

326. J. S. Bagla, Cosmological N-Body simulation: Techniques, scope and status, *Cur. Sci.* **88**, 1088–1100 (2005). astro-ph/0411043.

327. K. Dolag, S. Borgani, S. Schindler *et al.*, Simulation techniques for cosmological simulations, *Space Sci. Rev..* **134**, 229–268 (2008). doi: 10.1007/s11214-008-9316-5.

328. S. Borgani and A. Kravtsov, Cosmological simulations of galaxy clusters, *Adv. Sci. Lett.* **4**, 204–227 (2011). doi: 10.1166/asl.2011.1209.

329. M. Zumalacárregui, T. S. Koivisto and D. F. Mota, DBI Galileons in the Einstein frame: Local gravity and cosmology, *Phys. Rev.* **D87**(8), 083010 (2013). doi: 10.1103/PhysRevD.87.083010.

330. R. P. Woodard, Nonlocal models of cosmic acceleration, *Found. Phys.* **44**, 213–233 (2014). doi: 10.1007/s10701-014-9780-6.

331. Y. Dirian, S. Foffa, N. Khosravi *et al.*, Cosmological perturbations and structure formation in nonlocal infrared modifications of general relativity, *J. Cosmo. Astropart. Phys.* **6**, 033 (2014). doi: 10.1088/1475-7516/2014/06/033.

332. S. Bose, B. Li, A. Barreira *et al.*, Speeding up N-body simulations of modified gravity: Chameleon screening models, *J. Cosmo. Astropart. Phys.* **2017**(02), 050 (2017). doi: 10.1088/1475-7516/2017/02/050.

333. F. Schmidt, Cosmological simulations of normal-branch braneworld gravity, *Phys. Rev.* **D80**(12), 123003 (2009). doi: 10.1103/PhysRevD.80.123003.

334. M. Milgrom, A modification of the Newtonian dynamics as a possible alternative to the hidden mass hypothesis, *Astrophys. J.* **270**, 365–370 (1983). doi: 10.1086/161130.

335. J. Bekenstein and M. Milgrom, Does the missing mass problem signal the breakdown of Newtonian gravity?, *Astrophys. J.* **286**, 7–14 (1984). doi: 10.1086/162570.

336. M. Milgrom, Quasi-linear formulation of MOND, *Mon. Not. Roy. Astron. Soc.* **403**, 886–895 (2010). doi: 10.1111/j.1365-2966.2009.16184.x.

337. M. Milgrom, Bimetric MOND gravity, *Phys. Rev.* (12), 123536 (2009). doi: 10.1103/PhysRevD.80.123536.

338. A. Brandt, Multi-level adaptive solutions to boundary-value problems, *Math. Comp.* **31**, 333–390 (1977).

339. P. Wesseling, *An Introduction to Multigrid Methods* (John Wiley and Sons Inc, —c1992, 2nd ed., 1992).

340. U. Trottenberg, C. Oosterlee and A. Scholler, *Multigrid* (Academic Press, 2000).

341. A. Brandt and O. Livne, *Multigrid Techniques: 1984 Guide with Applications to Fluid Dynamics, Revised Edition*, Classics in Applied Mathematics (Society for Industrial and Applied Mathematics, 2011). ISBN 9781611970746.

342. A. Barreira, B. Li, C. M. Baugh *et al.*, Spherical collapse in Galileon gravity: Fifth force solutions, halo mass function and halo bias, *J. Cosmo. Astropart. Phys.* **1311**, 056 (2013). doi: 10.1088/1475-7516/2013/11/056.

343. H. A. Winther and P. G. Ferreira, Vainshtein mechanism beyond the quasistatic approximation, *Phys. Rev.* **D92**(6), 064005 (2015). doi: 10.1103/PhysRevD.92.064005.

344. A. Barreira, S. Bose and B. Li, Speeding up N-body simulations of modified gravity: Vainshtein screening models, *J. Cosmo. Astropart. Phys.* **2015**(12), 059 (2015). doi: 10.1088/1475-7516/2015/12/059.

345. II. Weinberger, *A First Course in Partial Differential Equations with Complex Variables and Transform Methods*, Blaisdell book in pure and applied mathematics (Dover Publications, 1995). ISBN 9780486686400.

346. R. Agarwal and D. O'Regan, *Ordinary and Partial Differential Equations*, Universitext (Springer-Verlag New York, 2009). ISBN 978-0-387-79146-3.

347. J. Noller, F. von Braun-Bates and P. G. Ferreira, Relativistic scalar fields and the quasistatic approximation in theories of modified gravity, *Phys. Rev.* **D89**(2), 023521 (2014). doi: 10.1103/PhysRevD.89.023521.

348. L. M. Widrow, Dynamics of thick domain walls, *Phys. Rev.* **D40**, 1002–1010 (1989). doi: 10.1103/PhysRevD.40.1002.

349. W. H. Press, B. S. Ryden and D. N. Spergel, Dynamical evolution of domain walls in an expanding universe, *Astrophys. J.* **347**, 590–604 (1989). doi: 10.1086/168151.

350. R. Hagala, C. Llinares and D. F. Mota, Cosmic tsunamis in modified gravity: Disruption of screening mechanisms from scalar waves, *Phys. Rev. Lett.* **118**(10), 101301 (2017). doi: 10.1103/PhysRevLett.118.101301.

351. E. F. Toro, *Riemann Solvers and Numerical Methods for Fluid Dynamics – A Practical Introduction, 2nd edition* (Springer, Berlin, 1999).

352. W. H. Press, S. A. Teukolsky, W. T. Vetterling *et al.*, *Numerical Recipes: The Art of Scientific Computing*, 3rd edition (Cambridge University Press, New York, NY, USA, 2007). ISBN 0521880688, 9780521880688.

353. M. Vogelsberger, S. Genel, D. Sijacki *et al.*, A model for cosmological simulations of galaxy formation physics, *Mon. Not. Roy. Astron. Soc.* **436**, 3031–3067 (2013). doi: 10.1093/mnras/stt1789.

354. M. Vogelsberger, S. Genel, V. Springel *et al.*, Introducing the Illustris Project: Simulating the coevolution of dark and visible matter in the Universe, *Mon. Not. Roy. Astron. Soc.* **444**, 1518–1547 (2014). doi: 10.1093/mnras/stu1536.

355. J. Schaye, R. A. Crain, R. G. Bower *et al.*, The EAGLE project: Simulating the

evolution and assembly of galaxies and their environments, *Mon. Not. Roy. Astron. Soc.* **446**, 521–554 (2015). doi: 10.1093/mnras/stu2058.

356. A. Pillepich, V. Springel, D. Nelson *et al.*, Simulating galaxy formation with the IllustrisTNG model, *Mon. Not. Roy. Astron. Soc.* **473**, 4077–4106 (2018). doi: 10.1093/mnras/stx2656.

357. L. Casarini, A. V. Macciò, S. A. Bonometto *et al.*, High-accuracy power spectra including baryonic physics in dynamical Dark Energy models, *Mon. Not. Roy. Astron. Soc.* **412**, 911–920 (2011). doi: 10.1111/j.1365-2966.2010.17948.x.

358. A. Hammami, C. Llinares, D. F. Mota *et al.*, Hydrodynamic effects in the symmetron and f(R)-gravity models, *Mon. Not. Roy. Astron. Soc.* **449**, 3635–3644 (2015). doi: 10.1093/mnras/stv529.

359. J.-h. He and B. Li, Accurate method of modeling cluster scaling relations in modified gravity, *Phys. Rev.* **D93**(12), 123512 (2016). doi: 10.1103/PhysRevD.93.123512.

360. C. Penzo, A. V. Macciò, L. Casarini *et al.*, Dark MaGICC: The effect of dark energy on disc galaxy formation. Cosmology does matter, *Mon. Not. Roy. Astron. Soc.* **442**, 176–186 (2014). doi: 10.1093/mnras/stu857.

361. C. C. Moran, R. Teyssier and B. Li, Chameleon f(R) gravity on the virgo cluster scale, *Mon. Not. Roy. Astron. Soc.* **448**(1), 307–327 (2015). doi: 10.1093/mnras/stu2757. URL http://dx.doi.org/10.1093/mnras/stu2757.

362. A. V. Macciò, R. Mainini, C. Penzo *et al.*, Strongly coupled dark energy cosmologies: Preserving ΛCDM success and easing low-scale problems – II. Cosmological simulations, *Mon. Not. Roy. Astron. Soc.* **453**, 1371–1378 (2015). doi: 10.1093/mnras/stv1680.

363. C. Penzo, A. V. Macciò, M. Baldi *et al.*, Effects of coupled dark energy on the Milky Way and its satellites, *Mon. Not. Roy. Astron. Soc.* **461**, 2490–2501 (2016). doi: 10.1093/mnras/stw1502.

364. F. Fontanot, E. Puchwein, V. Springel *et al.*, Semi-analytic galaxy formation in f(R)-gravity cosmologies, *Mon. Not. Roy. Astron. Soc.* **436**, 2672–2679 (2013). doi: 10.1093/mnras/stt1763.

365. F. Fontanot, M. Baldi, V. Springel *et al.*, Semi-analytic galaxy formation in coupled dark energy cosmologies, *Mon. Not. Roy. Astron. Soc.* **452**, 978–985 (2015). doi: 10.1093/mnras/stv1345.

366. B. Li and J. D. Barrow, N-body simulations for coupled scalar-field cosmology, *Phys. Rev.* **D83**(2), 024007 (2011). doi: 10.1103/PhysRevD.83.024007.

367. D. Shi, B. Li, J. Han *et al.*, Exploring the liminality: Properties of haloes and subhaloes in borderline $f(R)$ gravity, *Mon. Not. Roy. Astron. Soc.* **452**(3), 3179–3191 (2015). doi: 10.1093/mnras/stv1549.

368. C. Arnold, P. Fosalba, V. Springel *et al.*, The modified gravity light-cone simulation project – I. Statistics of matter and halo distributions, *Mon. Not. Roy. Astron. Soc.* **483**(1), 790–805 (2019). doi: 10.1093/mnras/sty3044.

369. M. A. Mitchell, C. Arnold, J.-h. He *et al.*, A general framework to test gravity using galaxy clusters II: A universal model for the halo concentration in $f(R)$ gravity, *Mon. Not. Roy. Astron. Soc.* **487**, 1410 (2019). doi: 10.1093/mnras/stz1389.

370. S. Tassev, M. Zaldarriaga and D. Eisenstein, Solving large scale structure in ten easy steps with COLA, *J. Cosmo. Astropart. Phys.* **1306**, 036 (2013). doi: 10.1088/1475-7516/2013/06/036.

371. G. Valogiannis and R. Bean, Efficient simulations of large scale structure in modified gravity cosmologies with comoving Lagrangian acceleration, *Phys. Rev.* **D95**(10), 103515 (2017). doi: 10.1103/PhysRevD.95.103515.

372. H. A. Winther, K. Koyama, M. Manera *et al.*, COLA with scale-dependent growth: Applications to screened modified gravity models, *J. Cosmo. Astropart. Phys.* **1708** (08), 006 (2017). doi: 10.1088/1475-7516/2017/08/006.

373. B. S. Wright, H. A. Winther and K. Koyama, COLA with massive neutrinos, *J. Cosmo. Astropart. Phys.* **1710**(10), 054 (2017). doi: 10.1088/1475-7516/2017/10/054.

374. R. E. Angulo and S. D. M. White, One simulation to fit them all – changing the background parameters of a cosmological N-body simulation, *Mon. Not. Roy. Astron. Soc.* **405**, 143–154 (2010). doi: 10.1111/j.1365-2966.2010.16459.x.

375. R. E. Angulo and S. Hilbert, Cosmological constraints from the CFHTLenS shear measurements using a new, accurate, and flexible way of predicting non-linear mass clustering, *Mon. Not. Roy. Astron. Soc.* **448**(1), 364–375 (2015). doi: 10.1093/mnras/stv050.

376. A. Mead and J. Peacock, Remapping dark matter halo catalogues between cosmological simulations, *Mon. Not. Roy. Astron. Soc.* **440**(2), 1233–1247 (2014). doi: 10.1093/mnras/stu345.

377. A. Mead and J. Peacock, Remapping simulated halo catalogues in redshift space, *Mon. Not. Roy. Astron. Soc.* **445**(4), 3453–3465 (2014). doi: 10.1093/mnras/stu1964.

378. A. A. Berlind, D. H. Weinberg, A. J. Benson *et al.*, The Halo occupation distribution and the physics of galaxy formation, *Astrophys. J.* **593**, 1–25 (2003). doi: 10.1086/376517.

379. Z. Zheng, A. A. Berlind, D. H. Weinberg *et al.*, Theoretical models of the halo occupation distribution: Separating central and satellite galaxies, *Astrophys. J.* **633**, 791–809 (2005). doi: 10.1086/466510.

380. W. H. Press and P. Schechter, Formation of galaxies and clusters of galaxies by self-similar gravitational condensation, *Astrophys. J.* **187**, 425–438 (1974). doi: 10.1086/152650.

381. J. R. Bond, S. Cole, G. Efstathiou *et al.*, Excursion set mass functions for hierarchical Gaussian fluctuations, *Astrophys. J.* **379**, 440–460 (1991). doi: 10.1086/170520.

382. R. K. Sheth and G. Tormen, Large scale bias and the peak background split, *Mon. Not. Roy. Astron. Soc.* **308**, 119 (1999). doi: 10.1046/j.1365-8711.1999.02692.x.

383. A. Jenkins, C. S. Frenk, S. D. M. White *et al.*, The mass function of dark matter halos, *Mon. Not. Roy. Astron. Soc.* **321**, 372 (2001). doi: 10.1046/j.1365-8711.2001.04029.x.

384. M. S. Warren, K. Abazajian, D. E. Holz *et al.*, Precision determination of the mass function of dark matter halos, *Astrophys. J.* **646**, 881–885 (2006). doi: 10.1086/504962.

385. D. Reed, R. Bower, C. Frenk *et al.*, The halo mass function from the dark ages through the present day, *Mon. Not. Roy. Astron. Soc.* **374**, 2–15 (2007). doi: 10.1111/j.1365-2966.2006.11204.x.

386. J. L. Tinker, A. V. Kravtsov, A. Klypin *et al.*, Toward a halo mass function for precision cosmology: The limits of universality, *Astrophys. J.* **688**, 709–728 (2008). doi: 10.1086/591439.

387. H. J. Mo and S. D. M. White, An analytic model for the spatial clustering of dark

matter halos, *Mon. Not. Roy. Astron. Soc.* **282**, 347 (1996). doi: 10.1093/mnras/282. 2.347.

388. J. N. Fry and E. Gaztanaga, Biasing and hierarchical statistics in large scale structure, *Astrophys. J.* **413**, 447–452 (1993). doi: 10.1086/173015.

389. R. K. Sheth, H. J. Mo and G. Tormen, Ellipsoidal collapse and an improved model for the number and spatial distribution of dark matter haloes, *Mon. Not. Roy. Astron. Soc.* **323**, 1 (2001). doi: 10.1046/j.1365-8711.2001.04006.x.

390. R. K. Sheth and G. Tormen, An excursion set model of hierarchical clustering: Ellipsoidal collapse and the moving barrier, *Mon. Not. Roy. Astron. Soc.* **329**, 61 (2002). doi: 10.1046/j.1365-8711.2002.04950.x.

391. J. Zhang and L. Hui, On random walks with a general moving barrier, *Astrophys. J.* **641**, 641–646 (2006). doi: 10.1086/499802.

392. B. Li, G. Zhao and K. Kazuya, Haloes and voids in f(R) gravity, *Mon. Not. Roy. Astron. Soc.* **421**, 3481–3487 (2012). doi: 10.1111/j.1365-2966.2012.20573.x.

393. T. Lam and B. Li, Excursion set theory for modified gravity: Correlated steps, mass functions and halo bias, *Mon. Not. Roy. Astron. Soc.* **426**, 3260–3270 (2012). doi: 10.1111/j.1365-2966.2012.21746.x.

394. F. Bernardeau, The nonlinear evolution of rare events, *Astrophys. J.* **427**, 51 (1994). doi: 10.1086/174121.

395. R. K. Sheth, An excursion set model for the distribution of dark matter and dark matter halos, *Mon. Not. Roy. Astron. Soc.* **300**, 1057–1070 (1998). doi: 10.1046/j. 1365-8711.1998.01976.x.

396. T. Y. Lam and R. K. Sheth, Perturbation theory and excursion set estimates of the probability distribution function of dark matter, and a method for reconstructing the initial distribution function, *Mon. Not. Roy. Astron. Soc.* **386**, 407 (2008). doi: 10.1111/j.1365-2966.2008.13038.x.

397. K. Jones-Smith and F. Ferrer, Detecting chameleon dark energy via electrostatic analogy, *Phys. Rev. Lett.* **108**, 221101 (2012). doi: 10.1103/PhysRevLett.108.221101.

398. C. Burrage, E. J. Copeland and J. Stevenson, Ellipticity weakens chameleon screening, *Phys. Rev.* **D91**, 065030 (2015). doi: 10.1103/PhysRevD.91.065030.

399. R. Pourhasan, N. Afshordi, R. B. Mann *et al.*, Chameleon gravity, electrostatics, and kinematics in the outer galaxy, *J. Cosmo. Astropart. Phys.* **1112**, 005 (2011). doi: 10.1088/1475-7516/2011/12/005.

400. F. Bernardeau, S. Colombi, E. Gaztanaga *et al.*, Large scale structure of the universe and cosmological perturbation theory, *Phys. Rept.* **367**, 1–248 (2002). doi: 10.1016/ S0370-1573(02)00135-7.

401. D. J. Eisenstein, H.-j. Seo, E. Sirko *et al.*, Improving cosmological distance measurements by reconstruction of the baryon acoustic peak, *Astrophys. J.* **664**, 675–679 (2007). doi: 10.1086/518712.

402. H.-M. Zhu, Y. Yu, U.-L. Pen *et al.*, Nonlinear reconstruction, *Phys. Rev.* **D96**(12), 123502 (2017). doi: 10.1103/PhysRevD.96.123502.

403. M. Schmittfull, T. Baldauf and M. Zaldarriaga, Iterative initial condition reconstruction, *Phys. Rev.* **D96**(2), 023505 (2017). doi: 10.1103/PhysRevD.96.023505.

404. Y. Shi, M. Cautun and B. Li, New method for initial density reconstruction, *Phys. Rev.* **D97**(2), 023505 (2018). doi: 10.1103/PhysRevD.97.023505.

405. J. Birkin, B. Li, M. Cautun *et al.*, Reconstructing the baryon acoustic oscillations

using biased tracers, *Mon. Not. Roy. Astron. Soc.* **483**(4), 5267–5280 (2019). doi: 10.1093/mnras/sty3365.

406. B. Li, W. A. Hellwing, K. Koyama *et al.*, The non-linear matter and velocity power spectra in f(R) gravity, *Mon. Not. Roy. Astron. Soc.* **428**, 743–755 (2013). doi: 10.1093/mnras/sts072.

407. B. Bose. *Cosmological Tests of Gravity.* PhD thesis, University of Portsmouth (2018).

408. N. Bartolo, E. Bellini, D. Bertacca *et al.*, Matter bispectrum in cubic Galileon cosmologies, *J. Cosmo. Astropart. Phys.* **1303**, 034 (2013). doi: 10.1088/1475-7516/2013/03/034.

409. F. Bernardeau, M. Crocce and R. Scoccimarro, Multi-point propagators in cosmological gravitational instability, *Phys. Rev.* **D78**, 103521 (2008). doi: 10.1103/PhysRevD.78.103521.

410. A. Taruya, F. Bernardeau, T. Nishimichi *et al.*, RegPT: Direct and fast calculation of regularized cosmological power spectrum at two-loop order, *Phys. Rev.* **D86**, 103528 (2012). doi: 10.1103/PhysRevD.86.103528.

411. A. Taruya, T. Nishimichi and F. Bernardeau, Precision modeling of redshift-space distortions from a multipoint propagator expansion, *Phys. Rev.* **D87**(8), 083509 (2013). doi: 10.1103/PhysRevD.87.083509.

412. A. Taruya, T. Nishimichi, F. Bernardeau *et al.*, Regularized cosmological power spectrum and correlation function in modified gravity models, *Phys. Rev.* **D90**(12), 123515 (2014). doi: 10.1103/PhysRevD.90.123515.

413. B. Bose, K. Koyama, W. A. Hellwing *et al.*, Theoretical accuracy in cosmological growth estimation, *Phys. Rev.* **D96**(2), 023519 (2017). doi: 10.1103/PhysRevD.96.023519.

414. B. Bose and K. Koyama, A perturbative approach to the redshift space correlation function: Beyond the standard model, *J. Cosmo. Astropart. Phys.* **2017**(08), 029 (2017). doi: 10.1088/1475-7516/2017/08/029.

415. B. T., A class of solutions in Newtonian cosmology and the pancake theory, *Astron. Astrophys.* **223**, 9–24 (1989).

416. P. Catelan, Lagrangian dynamics in nonflat universes and nonlinear gravitational evolution, *Mon. Not. Roy. Astron. Soc.* **276**, 115 (1995). doi: 10.1093/mnras/276.1.115.

417. F. R. Bouchet, S. Colombi, E. Hivon *et al.*, Perturbative Lagrangian approach to gravitational instability, *Astron. Astrophys.* **296**, 575 (1995). arXiv:astro-ph/9406013 [astro-ph].

418. A. N. Taylor and A. J. S. Hamilton, Nonlinear cosmological power spectra in real and redshift space, *Mon. Not. Roy. Astron. Soc.* **282**, 767 (1996). doi: 10.1093/mnras/282.3.767.

419. T. Matsubara, Resumming cosmological perturbations via the lagrangian picture: One-loop results in real space and in redshift space, *Phys. Rev.* **D77**, 063530 (2008). doi: 10.1103/PhysRevD.77.063530.

420. A. Aviles and J. L. Cervantes-Cota, Lagrangian perturbation theory for modified gravity, *Phys. Rev.* **D96**(12), 123526 (2017). doi: 10.1103/PhysRevD.96.123526.

421. J. F. Navarro, C. S. Frenk and S. D. M. White, A Universal density profile from hierarchical clustering, *Astrophys. J.* **490**, 493–508 (1997). doi: 10.1086/304888.

422. J. S. Bullock, T. S. Kolatt, Y. Sigad *et al.*, Profiles of dark haloes. Evolution, scatter,

and environment, *Mon. Not. Roy. Astron. Soc.* **321**, 559–575 (2001). doi: 10.1046/j. 1365-8711.2001.04068.x.

423. A. F. Neto, L. Gao, P. Bett *et al.*, The statistics of lambda CDM halo concentrations, *Mon. Not. Roy. Astron. Soc.* **381**, 1450–1462 (2007). doi: 10.1111/j.1365-2966.2007. 12381.x.

424. A. V. Macciò, A. A. Dutton, F. C. van den Bosch *et al.*, Concentration, spin and shape of dark matter haloes: Scatter and the dependence on mass and environment, *Mon. Not. Roy. Astron. Soc.* **378**, 55–71 (2007). doi: 10.1111/j.1365-2966.2007.11720.x.

425. A. R. Duffy, J. Schaye, S. T. Kay *et al.*, Dark matter halo concentrations in the Wilkinson Microwave Anisotropy Probe year 5 cosmology, *Mon. Not. Roy. Astron. Soc.* **390**, L64 (2008). doi: 10.1111/j.1745-3933.2008.00537.x. [Erratum: *ibid.* **415**, L85 (2011)].

426. A. D. Ludlow, J. F. Navarro, R. E. Angulo *et al.*, The mass-concentration-redshift relation of cold dark matter haloes, *Mon. Not. Roy. Astron. Soc.* **441**(1), 378–388 (2014). doi: 10.1093/mnras/stu483.

427. P. Valageas and T. Nishimichi, Combining perturbation theories with halo models, *Astron. Astrophys.* **527**(12), 87 (2015). doi: 10.1051/0004-6361/201015685.

428. U. Seljak and Z. Vlah, Halo Zel'dovich model and perturbation theory: Dark matter power spectrum and correlation function, *Phys. Rev.* **D91**(12), 123516 (2015). doi: 10.1103/PhysRevD.91.123516.

429. A. Mead, J. Peacock, C. Heymans *et al.*, An accurate halo model for fitting non-linear cosmological power spectra and baryonic feedback models, *Mon. Not. Roy. Astron. Soc.* **454**(2), 1958–1975 (2015). doi: 10.1093/mnras/stv2036.

430. R. E. Smith, J. A. Peacock, A. Jenkins *et al.*, Stable clustering, the halo model and nonlinear cosmological power spectra, *Mon. Not. Roy. Astron. Soc.* **341**, 1311 (2003). doi: 10.1046/j.1365-8711.2003.06503.x.

431. R. Takahashi, M. Sato, T. Nishimichi *et al.*, Revising the halofit model for the nonlinear matter power spectrum, *Astrophys. J.* **761**, 152 (2012). doi: 10.1088/0004-637X/ 761/2/152.

432. S. Bird, M. Viel and M. G. Haehnelt, Massive neutrinos and the non-linear matter power spectrum, *Mon. Not. Roy. Astron. Soc.* **420**(2), 2551–2561 (2012). doi: 10. 1111/j.1365-2966.2011.20222.x.

433. A. Lewis, A. Challinor and A. Lasenby, Efficient computation of CMB anisotropies in closed FRW models, *Astrophys. J.* **538**, 473–476 (2000). doi: 10.1086/309179.

434. G.-B. Zhao, Modeling the nonlinear clustering in modified gravity models. I. A fitting formula for the matter power spectrum of $f(R)$ gravity, *Astrophys. J. Suppl.* **211**, 23 (2014). doi: 10.1088/0067-0049/211/2/23.

435. M. Shirasaki, T. Hamana and N. Yoshida, Probing cosmology with weak lensing selected clusters II: Dark energy and f(R) gravity models, *Publ. Astron. Soc. Jap.* **68** (1), 4 (2016). doi: 10.1093/pasj/psv105.

436. B. Li and M. Shirasaki, Galaxy-galaxy weak gravitational lensing in $f(R)$ gravity, *Mon. Not. Roy. Astron. Soc.* **474**, 3599 (2018). doi: 10.1093/mnras/stx3006.

437. A. Taruya, T. Nishimichi and S. Saito, Baryon acoustic oscillations in 2D: Modeling redshift-space power spectrum from perturbation theory, *Phys. Rev.* **D82**, 063522 (2010). doi: 10.1103/PhysRevD.82.063522.

438. R. Scoccimarro, Redshift-space distortions, pairwise velocities and nonlinearities, *Phys. Rev.* **D70**, 083007 (2004). doi: 10.1103/PhysRevD.70.083007.

439. N. Kaiser, Clustering in real space and in redshift space, *Mon. Not. Roy. Astron. Soc.* **227**, 1–21 (1987). doi: 10.1093/mnras/227.1.1.

440. K. B. Fisher, On the validity of the streaming model for the redshift space correlation function in the linear regime, *Astrophys. J.* **448**, 494–499 (1995). doi: 10.1086/175980.

441. B. A. Reid and M. White, Towards an accurate model of the redshift-space clustering of haloes in the quasi-linear regime, *Mon. Not. Roy. Astron. Soc.* **417**, 1913–1927 (2011). doi: 10.1111/j.1365-2966.2011.19379.x.

442. R. van de Weygaert and E. Platen, *Int. J. Mod. Phys. Conf. Ser.* **1**, 41–66 (2011). doi: 10.1142/S2010194511000092.

443. R. K. Sheth and R. van de Weygaert, A hierarchy of voids: Much ado about nothing, *Mon. Not. Roy. Astron. Soc.* **350**, 517 (2004). doi: 10.1111/j.1365-2966.2004.07661.x.

444. A. Paranjape, T. Y. Lam and R. K. Sheth, A hierarchy of voids: More ado about nothing, *Mon. Not. Roy. Astron. Soc.* **420**, 1648–1655 (2011). doi: 10.1111/j.1365-2966.2011.20154.x.

445. J. M. Colberg *et al.*, The Aspen–Amsterdam Void Finder Comparison Project, *Mon. Not. Roy. Astron. Soc.* **387**, 933 (2008). doi: 10.1111/j.1365-2966.2008.13307.x.

446. M. Cautun, E. Paillas, Y.-C. Cai *et al.*, The Santiago–Harvard–Edinburgh–Durham void comparison – I. SHEDding light on chameleon gravity tests, *Mon. Not. Roy. Astron. Soc.* **476**(3), 3195–3217 (2018). doi: 10.1093/mnras/sty463.

447. M. Kilbinger, Cosmology with cosmic shear observations: A review, *Rep. Prog. Phys.* **78**(8), 086901 (2015). doi: 10.1088/0034-4885/78/8/086901.

448. LSST Science Collaboration, P. A. Abell, J. Allison *et al.*, LSST Science Book, Version 2.0 (2009). arXiv:0912.0201 [astro-ph.IM].

449. D. Huterer and D. L. Shafer, Dark energy two decades after: Observables, probes, consistency tests, *Rep. Prog. Phys.* **81**(1), 016901 (2018). doi: 10.1088/1361-6633/aa997e.

450. J. Harnois-Déraps, D. Munshi, P. Valageas *et al.*, Testing modified gravity with cosmic shear, *Mon. Not. Roy. Astron. Soc.* **454**, 2722–2735 (2015). doi: 10.1093/mnras/stv2120.

451. E. Semboloni, H. Hoekstra, J. Schaye *et al.*, Quantifying the effect of baryon physics on weak lensing tomography, *Mon. Not. Roy. Astron. Soc.* **417**, 2020–2035 (2011). doi: 10.1111/j.1365-2966.2011.19385.x.

452. R. Mandelbaum, Weak lensing for precision cosmology, *Ann. Rev. Astron. Astrophys.* **56**, 393–433 (2018). doi: 10.1146/annurev-astro-081817-051928.

453. J. A. Peacock, S. Cole, P. Norberg *et al.*, A measurement of the cosmological mass density from clustering in the 2dF Galaxy Redshift Survey, *Nature.* **410**, 169–173 (2001). astro-ph/0103143.

454. E.-M. Mueller, W. Percival, E. Linder *et al.*, The clustering of galaxies in the completed SDSS-III Baryon Oscillation Spectroscopic Survey: Constraining modified gravity, *Mon. Not. Roy. Astron. Soc.* **475**, 2122–2131 (2018). doi: 10.1093/mnras/stx3232.

455. Y.-S. Song, G.-B. Zhao, D. Bacon *et al.*, Complementarity of weak lensing and peculiar velocity measurements in testing general relativity, *Phys. Rev.* **D84**(8), 083523 (2011). doi: 10.1103/PhysRevD.84.083523.

456. G.-B. Zhao *et al.*, The clustering of the SDSS-IV extended Baryon Oscillation Spectroscopic Survey DR14 quasar sample: A tomographic measurement of cosmic structure growth and expansion rate based on optimal redshift weights, *Mon. Not. Roy. Astron. Soc.* **482**(3), 3497–3513 (2019). doi: 10.1093/mnras/sty2845.

457. G.-B. Zhao, T. Giannantonio, L. Pogosian *et al.*, Probing modifications of general relativity using current cosmological observations, *Phys. Rev.* **D81**(10), 103510 (2010). doi: 10.1103/PhysRevD.81.103510.

458. C. D. Leonard, T. Baker and P. G. Ferreira, Exploring degeneracies in modified gravity with weak lensing, *Phys. Rev.* **D91**(8), 083504 (2015). doi: 10.1103/PhysRevD. 91.083504.

459. C. Heymans, L. Van Waerbeke, L. Miller *et al.*, CFHTLenS: the Canada-France-Hawaii Telescope Lensing Survey, *Mon. Not. Roy. Astron. Soc.* **427**, 146–166 (2012). doi: 10.1111/j.1365-2966.2012.21952.x.

460. L. Anderson, E. Aubourg, S. Bailey *et al.*, The clustering of galaxies in the SDSS-III Baryon Oscillation Spectroscopic Survey: Baryon acoustic oscillations in the Data Release 9 spectroscopic galaxy sample, *Mon. Not. Roy. Astron. Soc.* **427**, 3435–3467 (2012). doi: 10.1111/j.1365-2966.2012.22066.x.

461. D. Larson, J. Dunkley, G. Hinshaw *et al.*, Seven-year Wilkinson Microwave Anisotropy Probe (WMAP) Observations: Power spectra and WMAP-derived parameters, *Astrophys. J. Suppl.* **192**, 16 (2011). doi: 10.1088/0067-0049/192/2/16.

462. S. Joudaki, C. Blake, C. Heymans *et al.*, CFHTLenS revisited: Assessing concordance with Planck including astrophysical systematics, *Mon. Not. Roy. Astron. Soc.* **465**, 2033–2052 (2017). doi: 10.1093/mnras/stw2665.

463. M. A. Troxel *et al.*, Dark Energy Survey Year 1 results: Cosmological constraints from cosmic shear, *Phys. Rev.* **D98**(4), 043528 (2018). doi: 10.1103/PhysRevD.98.043528.

464. R. Reyes, R. Mandelbaum, U. Seljak *et al.*, Confirmation of general relativity on large scales from weak lensing and galaxy velocities, *Nature.* **464**, 256–258 (2010). doi: 10.1038/nature08857.

465. A. Amon, C. Blake, C. Heymans *et al.*, KiDS+2dFLenS+GAMA: Testing the cosmological model with the E_G statistic, *Mon. Not. Roy. Astron. Soc.* **479**, 3422–3437 (2018). doi: 10.1093/mnras/sty1624.

466. C. D. Leonard, P. G. Ferreira and C. Heymans, Testing gravity with E_G: Mapping theory onto observations, *J. Cosmo. Astropart. Phys.* **12**, 051 (2015). doi: 10.1088/ 1475-7516/2015/12/051.

467. S. Singh, S. Alam, R. Mandelbaum *et al.*, Probing gravity with a joint analysis of galaxy and CMB lensing and SDSS spectroscopy, *Mon. Not. Roy. Astron. Soc.* **482** (1), 785–806 (2019). doi: 10.1093/mnras/sty2681.

468. G.-B. Zhao, H. Li, E. V. Linder *et al.*, Testing Einstein gravity with cosmic growth and expansion, *Phys. Rev.* **D85**(12), 123546 (2012). doi: 10.1103/PhysRevD.85. 123546.

469. A. Hojjati, G.-B. Zhao, L. Pogosian *et al.*, Cosmological tests of general relativity: A principal component analysis, *Phys. Rev.* **85**(4), 043508 (2012). doi: 10.1103/PhysRevD.85.043508.

470. D. Kirk, I. Laszlo, S. Bridle *et al.*, Optimizing cosmic shear surveys to measure modifications to gravity on cosmic scales, *Mon. Not. Roy. Astron. Soc.* **430**, 197–208 (2013). doi: 10.1093/mnras/sts571.

471. A. Hojjati, L. Pogosian and G.-B. Zhao, Testing gravity with CAMB and CosmoMC, *J. Cosmo. Astropart. Phys.* **8**, 005 (2011). doi: 10.1088/1475-7516/2011/08/005.

472. L. Pogosian and A. Silvestri, What can cosmology tell us about gravity? Constraining Horndeski gravity with Σ and μ, *Phys. Rev.* **D94**(10), 104014 (2016). doi: 10.1103/PhysRevD.94.104014.

473. S. Peirone, K. Koyama, L. Pogosian *et al.*, Large-scale structure phenomenology of viable Horndeski theories, *Phys. Rev.* **D97**(4), 043519 (2018). doi: 10.1103/PhysRevD.97.043519.

474. E. Bellini, A. J. Cuesta, R. Jimenez *et al.*, Constraints on deviations from ΛCDM within Horndeski gravity, *J. Cosmo. Astropart. Phys.* **2**, 053 (2016). doi: 10.1088/1475-7516/2016/02/053.

475. A. Spurio Mancini, R. Reischke, V. Pettorino *et al.*, Testing (modified) gravity with 3D and tomographic cosmic shear, *Mon. Not. Roy. Astron. Soc.* **480**, 3725 (2018). doi: 10.1093/mnras/sty2092.

476. W. A. Hellwing, K. Koyama, B. Bose *et al.*, Revealing modified gravity signals in matter and halo hierarchical clustering, *Phys. Rev.* **D96**, 023515 (2017). doi: 10.1103/PhysRevD.96.023515. URL https://link.aps.org/doi/10.1103/PhysRevD.96.023515.

477. A. Peel, V. Pettorino, C. Giocoli *et al.*, Breaking degeneracies in modified gravity with higher (than 2nd) order weak-lensing statistics, *Astron. Astrophys.* **610**, A38 (2018). doi: 10.1051/0004-6361/201833481.

478. J. Armijo, Y.-C. Cai, N. Padilla *et al.*, Testing modified gravity using a marked correlation function, *Mon. Not. Roy. Astron. Soc.* **478**, 3627–3632 (2018). doi: 10.1093/mnras/sty1335.

479. C. Hernández-Aguayo, C. M. Baugh and B. Li, Marked clustering statistics in f(R) gravity cosmologies, *Mon. Not. Roy. Astron. Soc.* **479**, 4824–4835 (2018). doi: 10.1093/mnras/sty1822.

480. E. Semboloni, C. Heymans, L. van Waerbeke *et al.*, Sources of contamination to weak lensing three-point statistics: Constraints from N-body simulations, *Mon. Not. Roy. Astron. Soc.* **388**, 991–1000 (2008). doi: 10.1111/j.1365-2966.2008.13478.x.

481. X. Liu, B. Li, G.-B. Zhao *et al.*, Constraining f(R) gravity theory using weak lensing peak statistics from the Canada-France-Hawii-Telescope Lensing Survey, *Phys. Rev. Lett.* **117**(5), 051101 (2016). doi: 10.1103/PhysRevLett.117.051101.

482. A. V. Kravtsov and S. Borgani, Formation of galaxy clusters, *Ann. Rev. Astron. Astrophys.* **50**, 353–409 (2012). doi: 10.1146/annurev-astro-081811-125502.

483. E. Rozo, R. H. Wechsler, E. S. Rykoff *et al.*, Cosmological constraints from the Sloan Digital Sky Survey maxBCG Cluster Catalog, *Astrophys. J.* **708**, 645–660 (2010). doi: 10.1088/0004-637X/708/1/645.

484. B. P. Koester *et al.*, A MaxBCG Catalog of 13,823 Galaxy clusters from the Sloan Digital Sky Survey, *Astrophys. J.* **660**, 239–255 (2007). doi: 10.1086/509599.

485. E. S. Rykoff, E. Rozo, M. T. Busha *et al.*, redMaPPer. I. Algorithm and SDSS DR8 Catalog, *Astrophys. J.* **785**, 104 (2014). doi: 10.1088/0004-637X/785/2/104.

486. E. S. Rykoff *et al.*, The redMaPPer galaxy cluster catalog from DES Science Verification Data, *Astrophys. J. Suppl.* **224**(1), 1 (2016). doi: 10.3847/0067-0049/224/1/1.

487. S. W. Allen, A. E. Evrard and A. B. Mantz, Cosmological parameters from observations of galaxy clusters, *Ann. Rev. Astron. Astrophys.* **49**, 409–470 (2011). doi: 10.1146/annurev-astro-081710-102514.

488. H. Ebeling, A. C. Edge, H. Bohringer *et al.*, The ROSAT Brightest Cluster Sample – I. The compilation of the sample and the cluster log N-log S distribution, *Mon. Not. Roy. Astron. Soc.* **301**, 881–914 (1998). doi: 10.1046/j.1365-8711.1998.01949.x.

489. H. Böhringer, P. Schuecker, L. Guzzo *et al.*, The ROSAT-ESO Flux Limited X-ray (REFLEX) Galaxy cluster survey. V. The cluster catalogue, *Astron. Astrophys.* **425**, 367–383 (2004). doi: 10.1051/0004-6361:20034484.

490. H. Ebeling, A. C. Edge, A. Mantz *et al.*, The X-ray brightest clusters of galaxies from the Massive Cluster Survey, *Mon. Not. Roy. Astron. Soc.* **407**, 83–93 (2010). doi: 10.1111/j.1365-2966.2010.16920.x.

491. R. A. Burenin, A. Vikhlinin, A. Hornstrup *et al.*, The 400 square degree ROSAT PSPC galaxy cluster survey: Catalog and statistical calibration, *Astrophys. J. Suppl.* **172**, 561–582 (2007). doi: 10.1086/519457.

492. L. E. Bleem, B. Stalder, T. de Haan *et al.*, Galaxy clusters discovered via the Sunyaev-Zel'dovich effect in the 2500-square-degree SPT-SZ survey, *Astrophys. J. Suppl.* **216**, 27 (2015). doi: 10.1088/0067-0049/216/2/27.

493. Planck Collaboration, P. A. R. Ade, N. Aghanim *et al.*, Planck 2015 results. XXIV. Cosmology from Sunyaev-Zeldovich cluster counts, *Astron. Astrophys.* **594**, A24 (2016). doi: 10.1051/0004-6361/201525833.

494. M. Hilton *et al.*, The Atacama Cosmology Telescope: The two-season ACTPol Sunyaev-Zel'dovich effect selected cluster catalog, *Astrophys. J. Suppl.* **235**(1), 20 (2018). doi: 10.3847/1538-4365/aaa6cb.

495. A. Mantz, S. W. Allen, H. Ebeling *et al.*, New constraints on dark energy from the observed growth of the most X-ray luminous galaxy clusters, *Mon. Not. Roy. Astron. Soc.* **387**, 1179–1192 (2008). doi: 10.1111/j.1365-2966.2008.13311.x.

496. A. Vikhlinin, A. V. Kravtsov, R. A. Burenin *et al.*, Chandra Cluster Cosmology Project III: Cosmological parameter constraints, *Astrophys. J.* **692**, 1060–1074 (2009). doi: 10.1088/0004-637X/692/2/1060.

497. D. Rapetti, S. W. Allen, A. Mantz *et al.*, Constraints on modified gravity from the observed X-ray luminosity function of galaxy clusters, *Mon. Not. Roy. Astron. Soc.* **400**, 699–704 (2009). doi: 10.1111/j.1365-2966.2009.15510.x.

498. L.-M. Wang and P. J. Steinhardt, Cluster abundance constraints on quintessence models, *Astrophys. J.* **508**, 483–490 (1998). doi: 10.1086/306436.

499. F. Schmidt, A. Vikhlinin and W. Hu, Cluster constraints on $f(R)$ gravity, *Phys. Rev.* **D80**, 083505 (2009). doi: 10.1103/PhysRevD.80.083505. URL https://link.aps.org/doi/10.1103/PhysRevD.80.083505.

500. D. Polarski and R. Gannouji, On the growth of linear perturbations, *Phys. Lett. B.* **660**, 439–443 (2008). doi: 10.1016/j.physletb.2008.01.032.

501. A. B. Mantz, A. von der Linden, S. W. Allen *et al.*, Weighing the giants – IV. Cosmology and neutrino mass, *Mon. Not. Roy. Astron. Soc.* **446**, 2205–2225 (2015). doi: 10.1093/mnras/stu2096.

502. A. Leauthaud, S. Saito, S. Hilbert *et al.*, Lensing is low: Cosmology, galaxy formation or new physics?, *Mon. Not. Roy. Astron. Soc.* **467**, 3024–3047 (2017). doi: 10.1093/mnras/stx258.

503. T. H. Reiprich and H. Böhringer, The mass function of an X-Ray flux-limited sample of galaxy clusters, *Astrophys. J.* **567**, 716–740 (2002). doi: 10.1086/338753.

504. A. Mantz, S. W. Allen, D. Rapetti *et al.*, The observed growth of massive galaxy

clusters – I. Statistical methods and cosmological constraints, *Mon. Not. Roy. Astron. Soc.* **406**, 1759–1772 (2010). doi: 10.1111/j.1365-2966.2010.16992.x.

505. A. Mantz, S. W. Allen, H. Ebeling *et al.*, The observed growth of massive galaxy clusters – II. X-ray scaling relations, *Mon. Not. Roy. Astron. Soc.* **406**, 1773–1795 (2010). doi: 10.1111/j.1365-2966.2010.16993.x.

506. D. Rapetti, S. W. Allen, A. Mantz *et al.*, The observed growth of massive galaxy clusters – III. Testing general relativity on cosmological scales, *Mon. Not. Roy. Astron. Soc.* **406**, 1796–1804 (2010). doi: 10.1111/j.1365-2966.2010.16799.x.

507. A. Mantz, S. W. Allen and D. Rapetti, The observed growth of massive galaxy clusters – IV. Robust constraints on neutrino properties, *Mon. Not. Roy. Astron. Soc.* **406**, 1805–1814 (2010). doi: 10.1111/j.1365-2966.2010.16794.x.

508. S. W. Allen, D. A. Rapetti, R. W. Schmidt *et al.*, Improved constraints on dark energy from Chandra X-ray observations of the largest relaxed galaxy clusters, *Mon. Not. Roy. Astron. Soc.* **383**, 879–896 (2008). doi: 10.1111/j.1365-2966.2007.12610.x.

509. D. J. Spiegelhalter, N. G. Best, B. P. Carlin *et al.*, Bayesian measures of model complexity and fit, *J. Roy. Stat. Soc. Ser. B.* **64**(4), 583–639 (2002).

510. D. Rapetti, C. Blake, S. W. Allen *et al.*, A combined measurement of cosmic growth and expansion from clusters of galaxies, the CMB and galaxy clustering, *Mon. Not. Roy. Astron. Soc.* **432**, 973–985 (2013). doi: 10.1093/mnras/stt514.

511. L. Samushia *et al.*, The clustering of galaxies in the SDSS-III Baryon Oscillation Spectroscopic Survey: Measuring growth rate and geometry with anisotropic clustering, *Mon. Not. Roy. Astron. Soc.* **439**(4), 3504–3519 (2014). doi: 10.1093/mnras/stu197.

512. S. de la Torre *et al.*, The VIMOS Public Extragalactic Redshift Survey (VIPERS). Gravity test from the combination of redshift-space distortions and galaxy-galaxy lensing at $0.5 < z < 1.2$, *Astron. Astrophys.* **608**, A44 (2017). doi: 10.1051/0004-6361/201630276.

513. A. von der Linden, M. T. Allen, D. E. Applegate *et al.*, Weighing the Giants – I. Weak-lensing masses for 51 massive galaxy clusters: Project overview, data analysis methods and cluster images, *Mon. Not. Roy. Astron. Soc.* **439**, 2–27 (2014). doi: 10.1093/mnras/stt1945.

514. D. E. Applegate, A. von der Linden, P. L. Kelly *et al.*, Weighing the Giants – III. Methods and measurements of accurate galaxy cluster weak-lensing masses, *Mon. Not. Roy. Astron. Soc.* **439**, 48–72 (2014). doi: 10.1093/mnras/stt2129.

515. P. L. Kelly, A. von der Linden, D. E. Applegate *et al.*, Weighing the Giants – II. Improved calibration of photometry from stellar colours and accurate photometric redshifts, *Mon. Not. Roy. Astron. Soc.* **439**, 28–47 (2014). doi: 10.1093/mnras/stt1946.

516. A. B. Mantz, S. W. Allen, R. G. Morris *et al.*, Weighing the giants – V. Galaxy cluster scaling relations, *Mon. Not. Roy. Astron. Soc.* **463**, 3582–3603 (2016). doi: 10.1093/mnras/stw2250.

517. J. P. Dietrich *et al.*, Sunyaev–Zel'dovich effect and X-ray scaling relations from weak lensing mass calibration of 32 South Pole Telescope selected galaxy clusters, *Mon. Not. Roy. Astron. Soc.* **483**(3), 2871–2906 (2019). doi: 10.1093/mnras/sty3088.

518. T. Schrabback *et al.*, Cluster mass calibration at high redshift: HST weak lensing analysis of 13 distant galaxy clusters from the South Pole Telescope Sunyaev–

Zel'dovich Survey, *Mon. Not. Roy. Astron. Soc.* **474**(2), 2635–2678 (2018). doi: 10.1093/mnras/stx2666.

519. H. Miyatake *et al.*, Weak-lensing mass calibration of ACTPol Sunyaev-Zel'dovich clusters with the Hyper Suprime-Cam Survey, *Astrophys. J.* **875**(1), 63 (2019). doi: 10.3847/1538-4357/ab0af0.

520. A. Nagarajan *et al.*, Weak-lensing mass calibration of the Sunyaev-Zel'dovich effect using APEX-SZ galaxy clusters *Mon. Not. Roy. Astron. Soc.* **488**(2), 1728 (2019). doi: 10.1093/mnras/sty1904.

521. T. McClintock *et al.*, Dark Energy Survey Year 1 Results: Weak Lensing Mass Calibration of redMaPPer Galaxy Clusters, *Mon. Not. Roy. Astron. Soc.* **482**(1), 1352–1378 (2019). doi: 10.1093/mnras/sty2711.

522. M. Pierre *et al.*, The XXL Survey – I. Scientific motivations – XMM-Newton observing plan – Follow-up observations and simulation programme, *Astron. Astrophys.* **592**, A1 (2016). doi: 10.1051/0004-6361/201526766.

523. F. Pacaud *et al.*, The XXL Survey. II. The bright cluster sample: catalogue and luminosity function, *Astron. Astrophys.* **592**, A2 (2016). doi: 10.1051/0004-6361/201526891.

524. T. de Haan *et al.*, Cosmological constraints from galaxy clusters in the 2500 square-degree SPT-SZ Survey, *Astrophys. J.* **832**(1), 95 (2016). doi: 10.3847/0004-637X/832/1/95.

525. Planck Collaboration, P. A. R. Ade, N. Aghanim, C. Armitage-Caplan *et al.*, Planck 2013 results. XX. Cosmology from Sunyaev-Zeldovich cluster counts, *Astron. Astrophys.* **571**, A20 (2014). doi: 10.1051/0004-6361/201321521.

526. A. von der Linden, A. Mantz, S. W. Allen *et al.*, Robust weak-lensing mass calibration of Planck galaxy clusters, *Mon. Not. Roy. Astron. Soc.* **443**, 1973–1978 (2014). doi: 10.1093/mnras/stu1423.

527. H. Hoekstra, R. Herbonnet, A. Muzzin *et al.*, The Canadian Cluster Comparison Project: Detailed study of systematics and updated weak lensing masses, *Mon. Not. Roy. Astron. Soc.* **449**, 685–714 (2015). doi: 10.1093/mnras/stv275.

528. M. Sereno, G. Covone, L. Izzo *et al.*, PSZ2LenS. Weak lensing analysis of the Planck clusters in the CFHTLenS and in the RCSLenS, *Mon. Not. Roy. Astron. Soc.* **472**, 1946–1971 (2017). doi: 10.1093/mnras/stx2085.

529. Y. Akrami *et al.*, Planck 2018 results. I. Overview and the cosmological legacy of Planck (2018). arXiv:1807.06205.

530. S. Bocquet, A. Saro, J. J. Mohr *et al.*, Mass calibration and cosmological analysis of the SPT-SZ galaxy cluster sample using velocity dispersion σ_v and X-ray Y_X measurements, *Astrophys. J.* **799**, 214 (2015). doi: 10.1088/0004-637X/799/2/214.

531. C. L. Reichardt, B. Stalder, L. E. Bleem *et al.*, Galaxy clusters discovered via the Sunyaev-Zel'dovich effect in the first 720 square degrees of the South Pole Telescope survey, *Astrophys. J.* **763**, 127 (2013). doi: 10.1088/0004-637X/763/2/127.

532. D. S. Y. Mak, E. Pierpaoli, F. Schmidt *et al.*, Constraints on modified gravity from Sunyaev-Zeldovich cluster surveys, *Phys. Rev.* **D85**(12), 123513 (2012). doi: 10.1103/PhysRevD.85.123513.

533. M. Cataneo, D. Rapetti, F. Schmidt *et al.*, New constraints on f(R) gravity from clusters of galaxies, *Phys. Rev.* **D92**(4), 044009 (2015). doi: 10.1103/PhysRevD.92.044009.

534. S. Ferraro, F. Schmidt and W. Hu, Cluster abundance in f(R) gravity models, *Phys. Rev.* **D83**(6), 063503 (2011). doi: 10.1103/PhysRevD.83.063503.

535. C. L. Bennett, D. Larson, J. L. Weiland *et al.*, Nine-year Wilkinson Microwave Anisotropy Probe (WMAP) observations: Final maps and results, *Astrophys. J. Suppl.* **208**, 20 (2013). doi: 10.1088/0067-0049/208/2/20.

536. G. Hinshaw, D. Larson, E. Komatsu *et al.*, Nine-year Wilkinson Microwave Anisotropy Probe (WMAP) observations: Cosmological parameter results, *Astrophys. J. Suppl.* **208**, 19 (2013). doi: 10.1088/0067-0049/208/2/19.

537. Planck Collaboration, P. A. R. Ade, N. Aghanim *et al.*, Planck 2013 results. XV. CMB power spectra and likelihood, *Astron. Astrophys.* **571**, A15 (2014). doi: 10.1051/0004-6361/201321573.

538. S. Peirone, M. Raveri, M. Viel *et al.*, Constraining f(R) gravity with Sunyaev-Zel'dovich clusters detected by the Planck satellite, *Phys. Rev.* **D95**(2), 023521 (2017). doi: 10.1103/PhysRevD.95.023521.

539. A. B. Mantz, S. W. Allen, R. G. Morris *et al.*, Cosmology and astrophysics from relaxed galaxy clusters – II. Cosmological constraints, *Mon. Not. Roy. Astron. Soc.* **440**, 2077–2098 (2014). doi: 10.1093/mnras/stu368.

540. R. Keisler, C. L. Reichardt, K. A. Aird *et al.*, A Measurement of the damping tail of the cosmic microwave background power spectrum with the South Pole Telescope, *Astrophys. J.* **743**, 28 (2011). doi: 10.1088/0004-637X/743/1/28.

541. C. L. Reichardt, L. Shaw, O. Zahn *et al.*, A measurement of secondary cosmic microwave background anisotropies with two years of South Pole Telescope observations, *Astrophys. J.* **755**, 70 (2012). doi: 10.1088/0004-637X/755/1/70.

542. K. T. Story, C. L. Reichardt, Z. Hou *et al.*, A measurement of the cosmic microwave background damping tail from the 2500-square-degree SPT-SZ survey, *Astrophys. J.* **779**, 86 (2013). doi: 10.1088/0004-637X/779/1/86.

543. S. Das *et al.*, The Atacama Cosmology Telescope: Temperature and gravitational lensing power spectrum measurements from three seasons of data, *J. Cosmo. Astropart. Phys.* **1404**, 014 (2014). doi: 10.1088/1475-7516/2014/04/014.

544. A. S. Chudaykin, D. S. Gorbunov, A. A. Starobinsky *et al.*, Cosmology based on f(R) gravity with Script O(1) eV sterile neutrino, *J. Cosmo. Astropart. Phys.* **5**, 004 (2015). doi: 10.1088/1475-7516/2015/05/004.

545. F. Schmidt, Dynamical masses in modified gravity, *Phys. Rev.* **D81**, 103002 (2010). doi: 10.1103/PhysRevD.81.103002.

546. M. Gronke, A. Hammami, D. F. Mota *et al.*, Estimates of cluster masses in screened modified gravity, *Astron. Astrophys.* **595**, A78 (2016). doi: 10.1051/0004-6361/201628644.

547. J. F. Navarro, C. S. Frenk and S. D. M. White, The structure of cold dark matter halos, *Astrophys. J.* **462**, 563–575 (1996). doi: 10.1086/177173.

548. L. Amendola, M. Kunz, M. Motta *et al.*, Observables and unobservables in dark energy cosmologies, *Phys. Rev.* **D87**(2), 023501 (2013). doi: 10.1103/PhysRevD.87.023501.

549. A. de Felice, T. Kobayashi and S. Tsujikawa, Effective gravitational couplings for cosmological perturbations in the most general scalar-tensor theories with second-order field equations, *Phys. Lett. B.* **706**, 123–133 (2011). doi: 10.1016/j.physletb.2011.11.028.

550. A. de Felice, R. Kase and S. Tsujikawa, Matter perturbations in Galileon cosmology, *Phys. Rev.* **D83**(4), 043515 (2011). doi: 10.1103/PhysRevD.83.043515.

551. J. Clampitt, B. Jain and J. Khoury, Halo scale predictions of symmetron modified gravity, *J. Cosmo. Astropart. Phys.* **1**, 030 (2012). doi: 10.1088/1475-7516/2012/01/030.

552. A. E. Evrard, Formation and evolution of X-ray clusters – A hydrodynamic simulation of the intracluster medium, *Astrophys. J.* **363**, 349–366 (1990). doi: 10.1086/169350.

553. Y. Suto, S. Sasaki and N. Makino, Gas density and X-ray surface brightness profiles of clusters of galaxies from dark matter halo potentials: Beyond the isothermal β-model, *Astrophys. J.* **509**, 544–550 (1998). doi: 10.1086/306520.

554. J. E. Carlstrom, G. P. Holder and E. D. Reese, Cosmology with the Sunyaev-Zel'dovich effect, *Ann. Rev. Astron. Astrophys.* **40**, 643–680 (2002). doi: 10.1146/annurev.astro.40.060401.093803.

555. N. Battaglia, J. R. Bond, C. Pfrommer *et al.*, On the cluster physics of Sunyaev-Zel'dovich and X-ray surveys. I. The influence of feedback, non-thermal pressure, and cluster shapes on Y-M scaling relations, *Astrophys. J.* **758**, 74 (2012). doi: 10.1088/0004-637X/758/2/74.

556. M. Bartelmann and P. Schneider, Weak gravitational lensing, *Phys. Rep.* **340**, 291–472 (2001). doi: 10.1016/S0370-1573(00)00082-X.

557. H. Hoekstra, M. Bartelmann, H. Dahle *et al.*, Masses of galaxy clusters from gravitational lensing, *Space Sci. Rev.* **177**, 75–118 (2013). doi: 10.1007/s11214-013-9978-5.

558. D. G. York *et al.*, The Sloan Digital Sky Survey: Technical summary, *Astron. J.* **120**, 1579–1587 (2000). doi: 10.1086/301513.

559. T. Narikawa and K. Yamamoto, Testing gravity with halo density profiles observed through gravitational lensing, *J. Cosmo. Astropart. Phys.* **5**, 016 (2012). doi: 10.1088/1475-7516/2012/05/016.

560. K. Umetsu, T. Broadhurst, A. Zitrin *et al.*, Cluster mass profiles from a bayesian analysis of weak-lensing distortion and magnification measurements: Applications to Subaru data, *Astrophys. J.* **729**, 127 (2011). doi: 10.1088/0004-637X/729/2/127.

561. K. Umetsu, T. Broadhurst, A. Zitrin *et al.*, A precise cluster mass profile averaged from the highest-quality lensing data, *Astrophys. J.* **738**, 41 (2011). doi: 10.1088/0004-637X/738/1/41.

562. M. Postman, D. Coe, N. Benítez *et al.*, The cluster lensing and supernova survey with hubble: An overview, *Astrophys. J. Suppl.* **199**, 25 (2012). doi: 10.1088/0067-0049/199/2/25.

563. K. Umetsu, E. Medezinski, M. Nonino *et al.*, CLASH: Weak-lensing shear-and-magnification analysis of 20 galaxy clusters, *Astrophys. J.* **795**, 163 (2014). doi: 10.1088/0004-637X/795/2/163.

564. J. Merten, M. Meneghetti, M. Postman *et al.*, CLASH: The concentration-mass relation of galaxy clusters, *Astrophys. J.* **806**, 4 (2015). doi: 10.1088/0004-637X/806/1/4.

565. A. Terukina and K. Yamamoto, Gas density profile in dark matter halo in chameleon cosmology, *Phys. Rev.* **D86**(10), 103503 (2012). doi: 10.1103/PhysRevD.86.103503.

566. T. Sato, T. Sasaki, K. Matsushita *et al.*, Suzaku observations of the Hydra A cluster out to the virial radius, *Pub. Astron. Soc. Japan.* **64**, 95 (2012). doi: 10.1093/pasj/64.5.95.

567. S. L. Snowden, R. F. Mushotzky, K. D. Kuntz *et al.*, A catalog of galaxy clusters observed by XMM-Newton, *Astron. Astrophys.* **478**, 615–658 (2008). doi: 10.1051/0004-6361:20077930.

568. D. R. Wik, C. L. Sarazin, A. Finoguenov *et al.*, A Suzaku search for nonthermal emission at hard X-ray energies in the Coma cluster, *Astrophys. J.* **696**, 1700–1711 (2009). doi: 10.1088/0004-637X/696/2/1700.

569. E. Churazov, A. Vikhlinin, I. Zhuravleva *et al.*, X-ray surface brightness and gas density fluctuations in the Coma cluster, *Mon. Not. Roy. Astron. Soc.* **421**, 1123–1135 (2012). doi: 10.1111/j.1365-2966.2011.20372.x.

570. Planck Collaboration, P. A. R. Ade, N. Aghanim *et al.*, Planck intermediate results. X. Physics of the hot gas in the Coma cluster, *Astron. Astrophys.* **554**, A140 (2013). doi: 10.1051/0004-6361/201220247.

571. R. Gavazzi, C. Adami, F. Durret *et al.*, A weak lensing study of the Coma cluster, *Astron. Astrophys.* **498**, L33–L36 (2009). doi: 10.1051/0004-6361/200911841.

572. N. Okabe, Y. Okura and T. Futamase, Weak-lensing mass measurements of substructures in Coma cluster with Subaru/Suprime-cam, *Astrophys. J.* **713**, 291–303 (2010). doi: 10.1088/0004-637X/713/1/291.

573. H. Wilcox *et al.*, The XMM Cluster Survey: Testing chameleon gravity using the profiles of clusters, *Mon. Not. Roy. Astron. Soc.* **452**(2), 1171–1183 (2015). doi: 10.1093/mnras/stv1366.

574. C. Arnold, E. Puchwein and V. Springel, Scaling relations and mass bias in hydrodynamical f(R) gravity simulations of galaxy clusters, *Mon. Not. Roy. Astron. Soc.* **440**, 833–842 (2014). doi: 10.1093/mnras/stu332.

575. A. Terukina, K. Yamamoto, N. Okabe *et al.*, Testing a generalized cubic Galileon gravity model with the Coma Cluster, *J. Cosmo. Astropart. Phys.* **10**, 064 (2015). doi: 10.1088/1475-7516/2015/10/064.

576. K. Matsushita, T. Sato, E. Sakuma *et al.*, Distribution of Si, Fe, and Ni in the intracluster medium of the Coma cluster, *Pub. Astron. Soc. Japan.* **65**, 10 (2013). doi: 10.1093/pasj/65.1.10.

577. N. Okabe, T. Futamase, M. Kajisawa *et al.*, Subaru weak-lensing survey of dark matter subhalos in the coma cluster: Subhalo mass function and statistical properties, *Astrophys. J.* **784**, 90 (2014). doi: 10.1088/0004-637X/784/2/90.

578. H. Wilcox, R. C. Nichol, G.-B. Zhao *et al.*, Simulation tests of galaxy cluster constraints on chameleon gravity, *Mon. Not. Roy. Astron. Soc.* **462**(1), 715–725 (2016). doi: 10.1093/mnras/stw1617.

579. M. Fitchett and R. Webster, Substructure in the Coma cluster, *Astrophys. J.* **317**, 653–667 (1987). doi: 10.1086/165310.

580. U. G. Briel, J. P. Henry and H. Boehringer, Observation of the Coma cluster of galaxies with ROSAT during the all-sky survey, *Astron. Astrophys.* **259**, L31–L34 (1992).

581. M. Colless and A. M. Dunn, Structure and dynamics of the Coma cluster, *Astrophys. J.* **458**, 435 (1996). doi: 10.1086/176827.

582. A. K. Romer, P. T. P. Viana, A. R. Liddle *et al.*, A serendipitous galaxy cluster survey with XMM: Expected catalog properties and scientific applications, *Astrophys. J.* **547**, 594–608 (2001). doi: 10.1086/318382.

583. E. J. Lloyd-Davies, A. K. Romer, N. Mehrtens *et al.*, The XMM Cluster Survey:

X-ray analysis methodology, *Mon. Not. Roy. Astron. Soc.* **418**, 14–53 (2011). doi: 10.1111/j.1365-2966.2011.19117.x.

584. N. Mehrtens, A. K. Romer, M. Hilton *et al.*, The XMM Cluster Survey: Optical analysis methodology and the first data release, *Mon. Not. Roy. Astron. Soc.* **423**, 1024–1052 (2012). doi: 10.1111/j.1365-2966.2012.20931.x.

585. T. Erben, H. Hildebrandt, L. Miller *et al.*, CFHTLenS: The Canada-France-Hawaii Telescope Lensing Survey – imaging data and catalogue products, *Mon. Not. Roy. Astron. Soc.* **433**, 2545–2563 (2013). doi: 10.1093/mnras/stt928.

586. V. Salzano, D. F. Mota, S. Capozziello *et al.*, Breaking the Vainshtein screening in clusters of galaxies, *Phys. Rev.* **D95**(4), 044038 (2017). doi: 10.1103/PhysRevD.95. 044038.

587. M. Donahue *et al.*, CLASH-X: A comparison of lensing and X-ray techniques for measuring the mass profiles of galaxy clusters, *Astrophys. J.* **794**(2), 136 (2014). doi: 10.1088/0004-637X/794/2/136.

588. L. Pizzuti, B. Sartoris, S. Borgani *et al.*, CLASH-VLT: Testing the nature of gravity with galaxy cluster mass profiles, *J. Cosmo. Astropart. Phys.* **4**, 023 (2016). doi: 10.1088/1475-7516/2016/04/023.

589. L. Pizzuti, B. Sartoris, L. Amendola *et al.*, CLASH-VLT: Constraints on f(R) gravity models with galaxy clusters using lensing and kinematic analyses, *J. Cosmo. Astropart. Phys.* **7**, 023 (2017). doi: 10.1088/1475-7516/2017/07/023.

590. P. Rosati, I. Balestra, C. Grillo *et al.*, CLASH-VLT: A VIMOS large programme to map the dark matter mass distribution in galaxy clusters and probe distant lensed galaxies, *The Messenger.* **158**, 48–53 (2014).

591. A. Cappi, Gravitational redshift in galaxy clusters., *Astron. Astrophys.* **301**, 6 (1995).

592. T. Broadhurst and E. Scannapieco, Detecting the gravitational redshift of cluster gas, *Astrophys. J. Lett.* **533**, L93–L97 (2000). doi: 10.1086/312630.

593. Y.-R. Kim and R. A. C. Croft, Gravitational redshifts in simulated galaxy clusters, *Astrophys. J.* **607**, 164–174 (2004). doi: 10.1086/383218.

594. H. Zhao, J. A. Peacock and B. Li, Testing gravity theories via transverse Doppler and gravitational redshifts in galaxy clusters, *Phys. Rev.* **D88**(4), 043013 (2013). doi: 10.1103/PhysRevD.88.043013.

595. N. Kaiser, Measuring gravitational redshifts in galaxy clusters, *Mon. Not. Roy. Astron. Soc.* **435**, 1278–1286 (2013). doi: 10.1093/mnras/stt1370.

596. D. Sakuma, A. Terukina, K. Yamamoto *et al.*, Gravitational redshifts of clusters and voids, *Phys. Rev.* **D97**(6), 063512 (2018). doi: 10.1103/PhysRevD.97.063512.

597. R. Wojtak, S. H. Hansen and J. Hjorth, Gravitational redshift of galaxies in clusters as predicted by general relativity, *Nature.* **477**, 567–569 (2011). doi: 10.1038/nature10445.

598. K. N. Abazajian, J. K. Adelman-McCarthy, M. A. Agüeros *et al.*, The seventh data release of the Sloan Digital Sky Survey, *Astrophys. J. Suppl.* **182**, 543–558 (2009). doi: 10.1088/0067-0049/182/2/543.

599. J. Hao, T. A. McKay, B. P. Koester *et al.*, A GMBCG Galaxy cluster catalog of 55,424 rich clusters from SDSS DR7, *Astrophys. J. Suppl.* **191**, 254–274 (2010). doi: 10.1088/0067-0049/191/2/254.

600. Y.-C. Cai, N. Kaiser, S. Cole *et al.*, Gravitational redshift and asymmetric redshift-

space distortions for stacked clusters, *Mon. Not. Roy. Astron. Soc.* **468**, 1981–1993 (2017). doi: 10.1093/mnras/stx469.

601. I. Sadeh, L. L. Feng and O. Lahav, Gravitational redshift of galaxies in clusters from the Sloan Digital Sky Survey and the Baryon Oscillation Spectroscopic Survey, *Phys. Rev. Lett.* **114**(7), 071103 (2015). doi: 10.1103/PhysRevLett.114.071103.

602. M. B. Gronke, C. Llinares and D. F. Mota, Gravitational redshift profiles in the f(R) and symmetron models, *Astron. Astrophys.* **562**, A9 (2014). doi: 10.1051/0004-6361/201322403.

603. M. C. Weisskopf, H. D. Tananbaum, L. P. Van Speybroeck *et al.*, Chandra X-ray Observatory (CXO): Overview, *Proc. SPIE.* **4012**, 2–16 (2000). doi: 10.1117/12.391545.

604. F. Jansen, D. Lumb, B. Altieri *et al.*, XMM-Newton observatory. I. The space-craft and operations, *Astron. Astrophys.* **365**, L1–L6 (2001). doi: 10.1051/0004-6361:20000036.

605. S. Ettori, G. W. Pratt, J. de Plaa *et al.*, The hot and energetic universe: The astrophysics of galaxy groups and clusters (2013). arXiv:1306.2322 [astro-ph.HE].

606. B. Li, J.-h. He and L. Gao, Cluster gas fraction as a test of gravity, *Mon. Not. Roy. Astron. Soc.* **456**, 146–155 (2016). doi: 10.1093/mnras/stv2650.

607. A. Hammami and D. F. Mota, Probing modified gravity via the mass-temperature relation of galaxy clusters, *Astron. Astrophys.* **598**, A132 (2017). doi: 10.1051/0004-6361/201629003.

608. M. Wyman and J. Khoury, Enhanced peculiar velocities in brane-induced gravity, *Phys. Rev.* **D82**(4), 044032 (2010). doi: 10.1103/PhysRevD.82.044032.

609. E. Jennings, C. M. Baugh, B. Li *et al.*, Redshift-space distortions in f(R) gravity, *Mon. Not. Roy. Astron. Soc.* **425**, 2128–2143 (2012). doi: 10.1111/j.1365-2966.2012.21567.x.

610. M. Wyman, E. Jennings and M. Lima, Simulations of Galileon modified gravity: Clustering statistics in real and redshift space, *Phys. Rev.* **D88**(8), 084029 (2013). doi: 10.1103/PhysRevD.88.084029.

611. T. Y. Lam, T. Nishimichi, F. Schmidt *et al.*, Testing gravity with the stacked phase space around galaxy clusters, *Phys. Rev. Lett.* **109**(5), 051301 (2012). doi: 10.1103/PhysRevLett.109.051301.

612. T. Y. Lam, F. Schmidt, T. Nishimichi *et al.*, Modeling the phase-space distribution around massive halos, *Phys. Rev.* **D88**(2), 023012 (2013). doi: 10.1103/PhysRevD.88.023012.

613. Y. Zu and D. H. Weinberg, The redshift-space cluster-galaxy cross-correlation function – I. Modelling galaxy infall on to Millennium simulation clusters and SDSS groups, *Mon. Not. Roy. Astron. Soc.* **431**, 3319–3337 (2013). doi: 10.1093/mnras/stt411.

614. Y. Zu, D. H. Weinberg, E. Jennings *et al.*, Galaxy infall kinematics as a test of modified gravity, *Mon. Not. Roy. Astron. Soc.* **445**, 1885–1897 (2014). doi: 10.1093/mnras/stu1739.

615. LSST Dark Energy Science Collaboration, Large Synoptic Survey Telescope: Dark Energy Science Collaboration (2012). arXiv:1211.0310 [astro-ph.CO].

616. D. Spergel, N. Gehrels, J. Breckinridge *et al.*, Wide-Field InfraRed Survey Telescope-

Astrophysics Focused Telescope Assets WFIRST-AFTA Final Report (2013). arXiv:1305.5422 [astro-ph.IM].

617. M. Levi, C. Bebek, T. Beers *et al.*, The DESI Experiment, a whitepaper for Snowmass 2013 (2013). arXiv:1308.0847 [astro-ph.CO].

618. N. Tamura *et al.*, Prime Focus Spectrograph (PFS) for the Subaru Telescope: Overview, recent progress, and future perspectives, *Proc. SPIE Int. Soc. Opt. Eng.* **9908**, 99081M (2016). doi: 10.1117/12.2232103.

619. A. P. Hearin, Assembly bias & redshift-space distortions: Impact on cluster dynamics tests of general relativity, *Mon. Not. Roy. Astron. Soc.* **451**, L45–L49 (2015). doi: 10.1093/mnrasl/slv064.

620. A. Diaferio and M. J. Geller, Infall regions of galaxy clusters, *Astrophys. J.* **481**, 633–643 (1997). doi: 10.1086/304075.

621. A. Stark, C. J. Miller, N. Kern *et al.*, Probing theories of gravity with phase space-inferred potentials of galaxy clusters, *Phys. Rev.* **D93**(8), 084036 (2016). doi: 10.1103/PhysRevD.93.084036.

622. V. R. Eke, S. Cole and C. S. Frenk, Cluster evolution as a diagnostic for Omega, *Mon. Not. Roy. Astron. Soc.* **282**. (1996). doi: 10.1093/mnras/282.1.263.

623. V. Pavlidou and T. N. Tomaras, Where the world stands still: Turnaround as a strong test of ΛCDM cosmology, *J. Cosmo. Astropart. Phys.* **9**, 020 (2014). doi: 10.1088/1475-7516/2014/09/020.

624. J. Lee and B. Li, The effect of modified gravity on the odds of the bound violations of the turn-around radii, *Astrophys. J.* **842**, 2 (2017). doi: 10.3847/1538-4357/aa706f.

625. M. Falco, S. H. Hansen, R. Wojtak *et al.*, A new method to measure the mass of galaxy clusters, *Mon. Not. Roy. Astron. Soc.* **442**, 1887–1896 (2014). doi: 10.1093/mnras/stu971.

626. J. Lee, S. Kim and S.-C. Rey, A bound violation on the galaxy group scale: The turn-around radius of NGC 5353/4, *Astrophys. J.* **815**, 43 (2015). doi: 10.1088/0004-637X/815/1/43.

627. B. Falck, K. Koyama, G.-b. Zhao *et al.*, The Vainshtein mechanism in the cosmic web, *J. Cosmo. Astropart. Phys.* **7**, 058 (2014). doi: 10.1088/1475-7516/2014/07/058.

628. B. Falck, K. Koyama and G.-B. Zhao, Cosmic web and environmental dependence of screening: Vainshtein vs. chameleon, *J. Cosmo. Astropart. Phys.* **7**, 049 (2015). doi: 10.1088/1475-7516/2015/07/049.

629. A. Merloni, P. Predehl, W. Becker *et al.*, eROSITA Science Book: Mapping the Structure of the Energetic Universe (2012). arXiv:1209.3114 [astro-ph.HE].

630. K. N. Abazajian, P. Adshead, Z. Ahmed *et al.*, CMB-S4 Science Book, First Edition (2016). arXiv:1610.02743.

631. C. Llinares and D. F. Mota, Shape of clusters of galaxies as a probe of screening mechanisms in modified gravity, *Phys. Rev. Lett.* **110**(15), 151104 (2013). doi: 10.1103/PhysRevLett.110.151104.

632. B. L'Huillier, H. A. Winther, D. F. Mota *et al.*, Dark matter haloes in modified gravity and dark energy: Interaction rate, small- and large-scale alignment, *Mon. Not. Roy. Astron. Soc.* **468**, 3174–3183 (2017). doi: 10.1093/mnras/stx700.

633. M. Oguri, M. Takada, N. Okabe *et al.*, Direct measurement of dark matter halo ellipticity from two-dimensional lensing shear maps of 25 massive clusters, *Mon. Not. Roy. Astron. Soc.* **405**, 2215–2230 (2010). doi: 10.1111/j.1365-2966.2010.16622.x.

634. K. Umetsu, M. Sereno, S.-I. Tam *et al.*, The projected dark and baryonic ellipsoidal structure of 20 CLASH galaxy clusters, *Astrophys. J.* **860**, 104 (2018). doi: 10.3847/1538-4357/aac3d9.

635. E. T. Lau, D. Nagai, A. V. Kravtsov *et al.*, Constraining cluster physics with the shape of X-ray clusters: Comparison of local X-ray clusters versus ΛCDM clusters, *Astrophys. J.* **755**, 116 (2012). doi: 10.1088/0004-637X/755/2/116.

636. S. Zaroubi, G. Squires, G. de Gasperis *et al.*, Deprojection of galaxy cluster X-ray, Sunyaev-Zeldovich temperature decrement, and weak-lensing mass maps, *Astrophys. J.* **561**, 600–620 (2001). doi: 10.1086/323359.

637. E. van Uitert, H. Hoekstra, B. Joachimi *et al.*, Halo ellipticity of GAMA galaxy groups from KiDS weak lensing, *Mon. Not. Roy. Astron. Soc.* **467**, 4131–4149 (2017). doi: 10.1093/mnras/stx344.

638. T.-h. Shin, J. Clampitt, B. Jain *et al.*, The ellipticity of galaxy cluster haloes from satellite galaxies and weak lensing, *Mon. Not. Roy. Astron. Soc.* **475**, 2421–2437 (2018). doi: 10.1093/mnras/stx3366.

639. M. Shirasaki, E. T. Lau and D. Nagai, Modelling baryonic effects on galaxy cluster mass profiles, *Mon. Not. Roy. Astron. Soc.* **477**, 2804–2814 (2018). doi: 10.1093/mnras/sty763.

640. L. J. King, P. Schneider and V. Springel, Cluster mass profiles from weak lensing: The influence of substructure, *Astron. Astrophys.* **378**, 748–755 (2001). doi: 10.1051/0004-6361:20011178.

641. A. E. Schulz, J. Hennawi and M. White, Characterizing the shapes of galaxy clusters using moments of the gravitational lensing shear, *Astropart. Phys.* **24**, 409–419 (2005). doi: 10.1016/j.astropartphys.2005.09.003.

642. M. Meneghetti, M. Bartelmann, A. Jenkins *et al.*, The effects of ellipticity and substructure on estimates of cluster density profiles based on lensing and kinematics, *Mon. Not. Roy. Astron. Soc.* **381**, 171–186 (2007). doi: 10.1111/j.1365-2966.2007.12225.x.

643. M. D. Schneider, C. S. Frenk and S. Cole, The shapes and alignments of dark matter halos, *J. Cosmo. Astropart. Phys.* **5**, 030 (2012). doi: 10.1088/1475-7516/2012/05/030.

644. M. Manolopoulou and M. Plionis, Galaxy cluster's rotation, *Mon. Not. Roy. Astron. Soc.* **465**, 2616–2633 (2017). doi: 10.1093/mnras/stw2870.

645. P. M. Ricker and C. L. Sarazin. Off-axis cluster mergers. In eds. J. C. Wheeler and H. Martel, *20th Texas Symposium on Relativistic Astrophysics*, Vol. 586, American Institute of Physics Conference Series, pp. 152–154 (2001). doi: 10.1063/1.1419547.

646. E. T. Hamden, C. M. Simpson, K. V. Johnston *et al.*, Measuring transverse motions for nearby galaxy clusters, *Astrophys. J. Lett.* **716**, L205–L208 (2010). doi: 10.1088/2041-8205/716/2/L205.

647. J.-h. He, A. J. Hawken, B. Li *et al.*, Effective dark matter halo catalog in f(R) gravity, *Phys. Rev. Lett.* **115**(7), 071306 (2015). doi: 10.1103/PhysRevLett.115.071306.

648. X. Barcons, K. Nandra, D. Barret *et al.* Athena: The X-ray observatory to study the hot and energetic Universe. In *Journal of Physics Conference Series*, Vol. 610, p. 012008, (2015). doi: 10.1088/1742-6596/610/1/012008.

649. S. Adhikari, N. Dalal and R. T. Chamberlain, Splashback in accreting dark matter

halos, *J. Cosmo. Astropart. Phys.* **11**, 019 (2014). doi: 10.1088/1475-7516/2014/11/019.

650. S. More, B. Diemer and A. V. Kravtsov, The splashback radius as a physical halo boundary and the growth of halo mass, *Astrophys. J.* **810**, 36 (2015). doi: 10.1088/0004-637X/810/1/36.

651. X. Shi, The outer profile of dark matter haloes: An analytical approach, *Mon. Not. Roy. Astron. Soc.* **459**, 3711–3720 (2016). doi: 10.1093/mnras/stw925.

652. S. More, H. Miyatake, M. Takada *et al.*, Detection of the splashback radius and halo assembly bias of massive galaxy clusters, *Astrophys. J.* **825**, 39 (2016). doi: 10.3847/0004-637X/825/1/39.

653. E. Baxter, C. Chang, B. Jain *et al.*, The halo boundary of galaxy clusters in the SDSS, *Astrophys. J.* **841**, 18 (2017). doi: 10.3847/1538-4357/aa6ff0.

654. C. Chang *et al.*, The splashback feature around DES galaxy clusters: Galaxy density and weak lensing profiles, *Astrophys. J.* **864**(1), 83 (2018). doi: 10.3847/1538-4357/aad5e7.

655. S. Chandrasekhar, Brownian motion, dynamical friction, and stellar dynamics, *Rev. Mod. Phys.* **21**, 383–388 (1949). doi: 10.1103/RevModPhys.21.383. URL https://link.aps.org/doi/10.1103/RevModPhys.21.383.

656. J. Binney and S. Tremaine, *Galactic Dynamics: Second Edition.* (Princeton University Press, 2008).

657. S. Adhikari, N. Dalal and J. Clampitt, Observing dynamical friction in galaxy clusters, *J. Cosmo. Astropart. Phys.* **7**, 022 (2016). doi: 10.1088/1475-7516/2016/07/022.

658. S. Adhikari, J. Sakstein, B. Jain *et al.*, Splashback in galaxy clusters as a probe of cosmic expansion and gravity, *J. Cosmo. Astropart. Phys.* **2018**(11), 033 (2018). doi: 10.1088/1475-7516/2018/11/033.

659. G. Dvali, G. Gabadadze and M. Porrati, Metastable gravitons and infinite volume extra dimensions, *Phys. Lett. B.* **484**, 112–118 (2000). doi: 10.1016/S0370-2693(00)00631-6.

660. P. Brax, Screening mechanisms in modified gravity, *Class. Quant. Grav.* **30**(21), 214005 (2013). doi: 10.1088/0264-9381/30/21/214005.

661. A. G. Riess *et al.*, Milky Way Cepheid standards for measuring cosmic distances and application to Gaia DR2: Implications for the Hubble constant, *Astrophys. J.* **861**(2), 126 (2018). doi: 10.3847/1538-4357/aac82e.

662. P. Vielva, E. Martínez-González, R. B. Barreiro *et al.*, Detection of non-Gaussianity in the Wilkinson Microwave Anisotropy Probe first-year data using spherical wavelets, *Astrophys. J.* **609**, 22–34 (2004). doi: 10.1086/421007.

663. M. Cruz, E. Martínez-González, P. Vielva *et al.*, Detection of a non-Gaussian spot in WMAP, *Mon. Not. Roy. Astron. Soc.* **356**, 29–40 (2005). doi: 10.1111/j.1365-2966.2004.08419.x.

664. M. Tegmark, A. de Oliveira-Costa and A. J. Hamilton, High resolution foreground cleaned CMB map from WMAP, *Phys. Rev.* **D68**(12), 123523 (2003). doi: 10.1103/PhysRevD.68.123523.

665. D. J. Schwarz, G. D. Starkman, D. Huterer *et al.*, Is the low-ℓ microwave background cosmic?, *Phys. Rev. Lett.* **93**(22), 221301 (2004). doi: 10.1103/PhysRevLett.93.221301.

666. H. K. Eriksen, A. J. Banday, K. M. Górski *et al.*, Hemispherical power asymmetry

in the third-year Wilkinson Microwave Anisotropy Probe sky maps, *Astrophys. J. Lett.* **660**, L81–L84 (2007). doi: 10.1086/518091.

667. F. K. Hansen, A. J. Banday, K. M. Górski *et al.*, Power asymmetry in cosmic microwave background fluctuations from full sky to sub-degree scales: Is the universe isotropic?, *Astrophys. J.* **704**, 1448–1458 (2009). doi: 10.1088/0004-637X/704/2/1448.

668. Planck Collaboration, P. A. R. Ade, N. Aghanim *et al.*, Planck 2013 results. XXIII. Isotropy and statistics of the CMB, *Astron. Astrophys.* **571**, A23 (2014). doi: 10.1051/0004-6361/201321534.

669. B. Li and J. D. Barrow, The cosmology of f(R) gravity in metric variational approach, *Phys. Rev.* **D75**, 084010 (2007). doi: 10.1103/PhysRevD.75.084010.

670. Y.-C. Cai, N. Padilla and B. Li, Testing gravity using cosmic voids, *Mon. Not. Roy. Astron. Soc.* **451**, 1036–1055 (2015). doi: 10.1093/mnras/stv777.

671. S. D. M. White, The hierarchy of correlation functions and its relation to other measures of galaxy clustering, *Mon. Not. Roy. Astron. Soc.* **186**, 145–154 (1979). doi: 10.1093/mnras/186.2.145.

672. W. C. Saslaw and A. J. S. Hamilton, Thermodynamics and galaxy clustering - Nonlinear theory of high order correlations, *Astrophys. J.* **276**, 13–25 (1984). doi: 10.1086/161589.

673. A. J. S. Hamilton, Galaxy clustering and the method of voids, *Astrophys. J. Lett.* **292**, L35–L39 (1985). doi: 10.1086/184468.

674. J. N. Fry, Cosmological density fluctuations and large-scale structure from N-point correlation functions to the probability distribution, *Astrophys. J.* **289**, 10–17 (1985). doi: 10.1086/162859.

675. J. N. Fry, Statistics of voids in hierarchical universes, *Astrophys. J.* **306**, 358–365 (1986). doi: 10.1086/164348.

676. E. Bertschinger, The self-similar evolution of holes in an Einstein-de Sitter universe, *Astrophys. J. Suppl.* **58**, 1–37 (1985). doi: 10.1086/191027.

677. G. R. Blumenthal, L. N. da Costa, D. S. Goldwirth *et al.*, The largest possible voids, *Astrophys. J.* **388**, 234–241 (1992). doi: 10.1086/171147.

678. V. Demchenko, Y.-C. Cai, C. Heymans *et al.*, Testing the spherical evolution of cosmic voids, *Mon. Not. Roy. Astron. Soc.* **463**, 512–519 (2016). doi: 10.1093/mnras/stw2030.

679. J. M. Bardeen, J. R. Bond, N. Kaiser *et al.*, The statistics of peaks of Gaussian random fields, *Astrophys. J.* **304**, 15–61 (1986). doi: 10.1086/164143.

680. R. van de Weygaert. Voids and the cosmic web: Cosmic depression & spatial complexity. In eds. R. van de Weygaert, S. Shandarin, E. Saar *et al.*, *The Zeldovich Universe: Genesis and Growth of the Cosmic Web*, Vol. 308, *IAU Symposium*, pp. 493–523, (2016). doi: 10.1017/S1743921316010504.

681. A. R. Zentner, The excursion set theory of halo mass functions, halo clustering, and halo growth, *Int. J. Mod. Phys. D.* **16**, 763–815 (2007). doi: 10.1142/S0218271807010511.

682. E. Jennings, Y. Li and W. Hu, The abundance of voids and the excursion set formalism, *Mon. Not. Roy. Astron. Soc.* **434**, 2167–2181 (2013). doi: 10.1093/mnras/stt1169.

683. N. D. Padilla, L. Ceccarelli and D. G. Lambas, Spatial and dynamical properties of

voids in a Λ cold dark matter universe, *Mon. Not. Roy. Astron. Soc.* **363**, 977–990 (2005). doi: 10.1111/j.1365-2966.2005.09500.x.

684. E. Platen, R. van de Weygaert and B. J. T. Jones, A cosmic watershed: The WVF void detection technique, *Mon. Not. Roy. Astron. Soc.* **380**, 551–570 (2007). doi: 10.1111/j.1365-2966.2007.12125.x.

685. M. C. Neyrinck, ZOBOV: A parameter-free void-finding algorithm, *Mon. Not. Roy. Astron. Soc.* **386**, 2101–2109 (2008). doi: 10.1111/j.1365-2966.2008.13180.x.

686. Y. Hoffman, O. Metuki, G. Yepes *et al.*, A kinematic classification of the cosmic web, *Mon. Not. Roy. Astron. Soc.* **425**, 2049–2057 (2012). doi: 10.1111/j.1365-2966.2012.21553.x.

687. D. C. Pan, M. S. Vogeley, F. Hoyle *et al.*, Cosmic voids in Sloan Digital Sky Survey Data Release 7, *Mon. Not. Roy. Astron. Soc.* **421**, 926–934 (2012). doi: 10.1111/j.1365-2966.2011.20197.x.

688. P. M. Sutter, G. Lavaux, B. D. Wandelt *et al.*, A public void catalog from the SDSS DR7 galaxy redshift surveys based on the watershed transform, *Astrophys. J.* **761**, 44 (2012). doi: 10.1088/0004-637X/761/1/44.

689. Y.-C. Cai, M. C. Neyrinck, I. Szapudi *et al.*, A possible cold imprint of voids on the microwave background radiation, *Astrophys. J.* **786**, 110 (2014). doi: 10.1088/0004-637X/786/2/110.

690. S. Nadathur, Testing cosmology with a catalogue of voids in the BOSS galaxy surveys, *Mon. Not. Roy. Astron. Soc.* **461**, 358–370 (2016). doi: 10.1093/mnras/stw1340.

691. Q. Mao, A. A. Berlind, R. J. Scherrer *et al.*, A cosmic void catalog of SDSS DR12 BOSS Galaxies, *Astrophys. J.* **835**, 161 (2017). doi: 10.3847/1538-4357/835/2/161.

692. R. K. Sachs and A. M. Wolfe, Perturbations of a cosmological model and angular variations of the microwave background, *Astrophys. J.* **147**, 73 (1967). doi: 10.1086/148982.

693. B. R. Granett, M. C. Neyrinck and I. Szapudi, An imprint of superstructures on the microwave background due to the integrated Sachs-Wolfe effect, *Astrophys. J. Lett.* **683**, L99 (2008). doi: 10.1086/591670.

694. S. Nadathur, S. Hotchkiss and S. Sarkar, The integrated Sachs-Wolfe imprint of cosmic superstructures: A problem for ΛCDM, *J. Cosmo. Astropart. Phys.* **6**, 042 (2012). doi: 10.1088/1475-7516/2012/06/042.

695. S. Flender, S. Hotchkiss and S. Nadathur, The stacked ISW signal of rare superstructures in ΛCDM, *J. Cosmo. Astropart. Phys.* **2**, 013 (2013). doi: 10.1088/1475-7516/2013/02/013.

696. S. Ilić, M. Langer and M. Douspis, Detecting the integrated Sachs-Wolfe effect with stacked voids, *Astron. Astrophys.* **556**, A51 (2013). doi: 10.1051/0004-6361/201321150.

697. A. Kovács and B. R. Granett, Cold imprint of supervoids in the cosmic microwave background re-considered with Planck and Baryon Oscillation Spectroscopic Survey DR10, *Mon. Not. Roy. Astron. Soc.* **452**, 1295–1302 (2015). doi: 10.1093/mnras/stv1371.

698. Planck Collaboration, P. A. R. Ade, N. Aghanim *et al.*, Planck 2015 results. XXI. The integrated Sachs-Wolfe effect, *Astron. Astrophys.* **594**, A21 (2016). doi: 10.1051/0004-6361/201525831.

699. S. Aiola, A. Kosowsky and B. Wang, Gaussian approximation of peak values in

the integrated Sachs-Wolfe effect, *Phys. Rev.* **D91**(4), 043510 (2015). doi: 10.1103/ PhysRevD.91.043510.

700. S. Nadathur and R. Crittenden, A detection of the integrated Sachs-Wolfe imprint of cosmic superstructures using a matched-filter approach, *Astrophys. J. Lett.* **830**, L19 (2016). doi: 10.3847/2041-8205/830/1/L19.

701. Y.-C. Cai, M. Neyrinck, Q. Mao *et al.*, The lensing and temperature imprints of voids on the cosmic microwave background, *Mon. Not. Roy. Astron. Soc.* **466**, 3364–3375 (2017). doi: 10.1093/mnras/stw3299.

702. C. Alcock and B. Paczynski, An evolution free test for non-zero cosmological constant, *Nature.* **281**, 358 (1979). doi: 10.1038/281358a0.

703. G. Lavaux and B. D. Wandelt, Precision cosmography with stacked voids, *Astrophys. J.* **754**, 109 (2012). doi: 10.1088/0004-637X/754/2/109.

704. P. M. Sutter, A. Pisani, B. D. Wandelt *et al.*, A measurement of the Alcock-Paczyński effect using cosmic voids in the SDSS, *Mon. Not. Roy. Astron. Soc.* **443**, 2983–2990 (2014). doi: 10.1093/mnras/stu1392.

705. Q. Mao, A. A. Berlind, R. J. Scherrer *et al.*, Cosmic voids in the SDSS DR12 BOSS galaxy sample: The Alcock-Paczynski test, *Astrophys. J.* **835**, 160 (2017). doi: 10. 3847/1538-4357/835/2/160.

706. Y.-C. Cai, A. Taylor, J. A. Peacock *et al.*, Redshift-space distortions around voids, *Mon. Not. Roy. Astron. Soc.* **462**, 2465–2477 (2016). doi: 10.1093/mnras/stw1809.

707. N. Hamaus, A. Pisani, P. M. Sutter *et al.*, Constraints on cosmology and gravity from the dynamics of voids, *Phys. Rev. Lett.* **117**(9), 091302 (2016). doi: 10.1103/ PhysRevLett.117.091302.

708. N. Hamaus, M.-C. Cousinou, A. Pisani *et al.*, Multipole analysis of redshift-space distortions around cosmic voids, *J. Cosmo. Astropart. Phys.* **7**, 014 (2017). doi: 10. 1088/1475-7516/2017/07/014.

709. E. Krause, T.-C. Chang, O. Doré *et al.*, The weight of emptiness: The gravitational lensing signal of stacked voids, *Astrophys. J. Lett.* **762**, L20 (2013). doi: 10.1088/ 2041-8205/762/2/L20.

710. Y. Higuchi, M. Oguri and T. Hamana, Measuring the mass distribution of voids with stacked weak lensing, *Mon. Not. Roy. Astron. Soc.* **432**, 1021–1031 (2013). doi: 10.1093/mnras/stt521.

711. P. Melchior, P. M. Sutter, E. S. Sheldon *et al.*, First measurement of gravitational lensing by cosmic voids in SDSS, *Mon. Not. Roy. Astron. Soc.* **440**, 2922–2927 (2014). doi: 10.1093/mnras/stu456.

712. J. Clampitt and B. Jain, Lensing measurements of the mass distribution in SDSS voids, *Mon. Not. Roy. Astron. Soc.* **454**, 3357–3365 (2015). doi: 10.1093/mnras/ stv2215.

713. C. Sanchez *et al.*, Cosmic voids and void lensing in the Dark Energy Survey Science Verification Data, *Mon. Not. Roy. Astron. Soc.* **465**(1), 746–759 (2017). doi: 10.1093/ mnras/stw2745.

714. J. Lee and D. Park, Constraining the dark energy equation of state with cosmic voids, *Astrophys. J. Lett.* **696**, L10–L12 (2009). doi: 10.1088/0004-637X/696/1/L10.

715. G. Lavaux and B. D. Wandelt, Precision cosmology with voids: Definition, methods, dynamics, *Mon. Not. Roy. Astron. Soc.* **403**, 1392–1408 (2010). doi: 10.1111/j. 1365-2966.2010.16197.x.

716. E. G. P. Bos, R. van de Weygaert, K. Dolag *et al.*, The darkness that shaped the void: Dark energy and cosmic voids, *Mon. Not. Roy. Astron. Soc.* **426**, 440–461 (2012). doi: 10.1111/j.1365-2966.2012.21478.x.

717. P. M. Sutter, E. Carlesi, B. D. Wandelt *et al.*, On the observability of coupled dark energy with cosmic voids, *Mon. Not. Roy. Astron. Soc.* **446**, L1–L5 (2015). doi: 10.1093/mnrasl/slu155.

718. B. Li, Voids in coupled scalar field cosmology, *Mon. Not. Roy. Astron. Soc.* **411**, 2615–2627 (2011). doi: 10.1111/j.1365-2966.2010.17867.x.

719. P. Zivick, P. M. Sutter, B. D. Wandelt *et al.*, Using cosmic voids to distinguish f(R) gravity in future galaxy surveys, *Mon. Not. Roy. Astron. Soc.* **451**, 4215–4222 (2015). doi: 10.1093/mnras/stv1209.

720. G. Pollina, M. Baldi, F. Marulli *et al.*, Cosmic voids in coupled dark energy cosmologies: The impact of halo bias, *Mon. Not. Roy. Astron. Soc.* **455**, 3075–3085 (2016). doi: 10.1093/mnras/stv2503.

721. A. Barreira, M. Cautun, B. Li *et al.*, Weak lensing by voids in modified lensing potentials, *J. Cosmo. Astropart. Phys.* **8**, 028 (2015). doi: 10.1088/1475-7516/2015/08/028.

722. E. Massara, F. Villaescusa-Navarro, M. Viel *et al.*, Voids in massive neutrino cosmologies, *J. Cosmo. Astropart. Phys.* **11**, 018 (2015). doi: 10.1088/1475-7516/2015/11/018.

723. L. F. Yang, M. C. Neyrinck, M. A. Aragón-Calvo *et al.*, Warmth elevating the depths: Shallower voids with warm dark matter, *Non. Not. Roy. Astron. Soc.* **451**, 3606–3614 (2015). doi: 10.1093/mnras/stv1087.

724. F.-S. Kitaura, C.-H. Chuang, Y. Liang *et al.*, Signatures of the primordial universe from its emptiness: Measurement of baryon acoustic oscillations from minima of the density field, *Phys. Rev. Lett.* **116**(17), 171301 (2016). doi: 10.1103/PhysRevLett.116.171301.

725. Y. Liang, C. Zhao, C.-H. Chuang *et al.*, Measuring baryon acoustic oscillations from the clustering of voids, *Mon. Not. Roy. Astron. Soc.* **459**, 4020–4028 (2016). doi: 10.1093/mnras/stw884.

726. L. Hui, A. Nicolis and C. Stubbs, Equivalence principle implications of modified gravity models, *Phys. Rev.* **D80**, 104002 (2009). doi: 10.1103/PhysRevD.80.104002.

727. B. Jain, Designing surveys for tests of gravity, *Phil. Trans. Roy. Soc. London Ser. A.* **369**, 5081–5089 (2011). doi: 10.1098/rsta.2011.0286.

728. B. Jain and J. VanderPlas, Tests of modified gravity with dwarf galaxies, *J. Cosmo. Astropart. Phys.* **1110**, 032 (2011). doi: 10.1088/1475-7516/2011/10/032.

729. R. Voivodic, M. Lima, C. Llinares *et al.*, Modeling void abundance in modified gravity, *Phys. Rev.* **D95**(2), 024018 (2017). doi: 10.1103/PhysRevD.95.024018.

730. DESI Collaboration, A. Aghamousa, J. Aguilar *et al.*, The DESI Experiment Part I: Science, targeting, and survey design (2016). arXiv:1611.00036 [astro-ph.IM].

731. M. Sahlén, Í. Zubeldía and J. Silk, Cluster-void degeneracy breaking: Dark energy, planck, and the largest cluster and void, *Astrophys. J. Lett.* **820**, L7 (2016). doi: 10.3847/2041-8205/820/1/L7.

732. M. Sahlén and J. Silk, Cluster-void degeneracy breaking: Modified gravity in the balance (2016). arXiv:1612.06595.

733. E. V. Linder and R. N. Cahn, Parameterized beyond-Einstein growth, *Astropart. Phys.* **28**, 481–488 (2007). doi: 10.1016/j.astropartphys.2007.09.003.
734. G. Pollina, N. Hamaus, K. Paech *et al.*, On the relative bias of void tracers in the Dark Energy Survey (2018). arXiv:1806.06860.
735. D. Gruen *et al.*, Weak lensing by galaxy troughs in DES Science Verification data, *Mon. Not. Roy. Astron. Soc.* **455**(3), 3367–3380 (2016). doi: 10.1093/mnras/stv2506.
736. A. Barreira, S. Bose, B. Li *et al.*, Weak lensing by galaxy troughs with modified gravity, *J. Cosmo. Astropart. Phys.* **2**, 031 (2017). doi: 10.1088/1475-7516/2017/02/031.
737. M. M. Brouwer *et al.*, Studying galaxy troughs and ridges using weak gravitational lensing with the Kilo-Degree Survey, *Mon. Not. Roy. Astron. Soc.* **481**, 5189 (2018). doi: 10.1093/mnras/sty2589.
738. O. Friedrich *et al.*, Density split statistics: Joint model of counts and lensing in cells, *Phys. Rev.* **D98**(2), 023508 (2018). doi: 10.1103/PhysRevD.98.023508.
739. C. T. Davies, M. Cautun and B. Li, Weak lensing by voids in weak lensing maps, *Mon. Not. Roy. Astron. Soc.* **480**, L101–L105 (2018). doi: 10.1093/mnrasl/sly135.
740. T. Chantavat, U. Sawangwit and B. D. Wandelt, Void profile from planck lensing potential map, *Astrophys. J.* **836**, 156 (2017). doi: 10.3847/1538-4357/836/2/156.
741. T. Baker, J. Clampitt, B. Jain *et al.*, Void lensing as a test of gravity, *Phys. Rev.* **D98**(2), 023511 (2018). doi: 10.1103/PhysRevD.98.023511.
742. J. Renk, M. Zumalacárregui, F. Montanari *et al.*, Galileon gravity in light of ISW, CMB, BAO and H_0 data, *J. Cosmo. Astropart. Phys.* **10**, 020 (2017). doi: 10.1088/1475-7516/2017/10/020.
743. E. Paillas, M. Cautun, B. Li *et al.*, The Santiago-Harvard-Edinburgh-Durham void comparison II: Unveiling the Vainshtein screening using weak lensing, *Mon. Not. Roy. Astron. Soc.* **484**(1), 1149–1165 (2019). doi: 10.1093/mnras/stz022.
744. S. Alam, H. Miyatake, S. More *et al.*, Testing gravity on large scales by combining weak lensing with galaxy clustering using CFHTLenS and BOSS CMASS, *Mon. Not. Roy. Astron. Soc.* **465**, 4853–4865 (2017). doi: 10.1093/mnras/stw3056.
745. D. Spolyar, M. Sahlén and J. Silk, Topology and dark energy: Testing gravity in voids, *Phys. Rev. Lett.* **111**(24), 241103 (2013). doi: 10.1103/PhysRevLett.111.241103.
746. N. Hamaus, P. M. Sutter, G. Lavaux *et al.*, Probing cosmology and gravity with redshift-space distortions around voids, *J. Cosmo. Astropart. Phys.* **11**, 036 (2015). doi: 10.1088/1475-7516/2015/11/036.
747. S. Nadathur and W. J. Percival, An accurate linear model for redshift space distortions in the void-galaxy correlation function, *Mon. Not. Roy. Astron. Soc.* **483**(3), 3472–3487 (2019). doi: 10.1093/mnras/sty3372.
748. I. Achitouv, Improved model of redshift-space distortions around voids: Application to quintessence dark energy, *Phys. Rev.* **D96**(8), 083506 (2017). doi: 10.1103/PhysRevD.96.083506.
749. I. Achitouv and Y.-C. Cai, Modeling the environmental dependence of the growth rate of cosmic structure, *Phys. Rev.* **D98**(10), 103502 (2018). doi: 10.1103/PhysRevD.98.103502.
750. I. Achitouv, C. Blake, P. Carter *et al.*, Consistency of the growth rate in different environments with the 6-degree Field Galaxy Survey: Measurement of the void-galaxy

and galaxy-galaxy correlation functions, *Phys. Rev.* **D95**(8), 083502 (2017). doi: 10. 1103/PhysRevD.95.083502.

751. A. J. Hawken *et al.*, The VIMOS Public Extragalactic Redshift Survey: Measuring the growth rate of structure around cosmic voids, *Astron. Astrophys.* **607**, A54 (2017). doi: 10.1051/0004-6361/201629678.

752. A. J. Ross, S. Ho, A. J. Cuesta *et al.*, Ameliorating systematic uncertainties in the angular clustering of galaxies: A study using the SDSS-III, *Mon. Not. Roy. Astron. Soc.* **417**, 1350–1373 (2011). doi: 10.1111/j.1365-2966.2011.19351.x.

753. Y.-S. Song, H. Peiris and W. Hu, Cosmological constraints on f(R) acceleration models, *Phys. Rev.* **D76**(6), 063517 (2007). doi: 10.1103/PhysRevD.76.063517.

754. S. Ho, C. Hirata, N. Padmanabhan *et al.*, Correlation of CMB with large-scale structure. I. Integrated Sachs-Wolfe tomography and cosmological implications, *Phys. Rev.* **D78**(4), 043519 (2008). doi: 10.1103/PhysRevD.78.043519.

755. T. Giannantonio, R. Scranton, R. G. Crittenden *et al.*, Combined analysis of the integrated Sachs-Wolfe effect and cosmological implications, *Phys. Rev.* **D77**(12), 123520 (2008). doi: 10.1103/PhysRevD.77.123520.

756. A. Kovács, The part and the whole: Voids, supervoids, and their ISW imprint, *Mon. Not. Roy. Astron. Soc.* **475**, 1777–1790 (2018). doi: 10.1093/mnras/stx3213.

757. A. J. Ross, W. J. Percival, A. G. Sánchez *et al.*, The clustering of galaxies in the SDSS-III Baryon Oscillation Spectroscopic Survey: Analysis of potential systematics, *Mon. Not. Roy. Astron. Soc.* **424**, 564–590 (2012). doi: 10.1111/j.1365-2966.2012. 21235.x.

758. A. J. Ross *et al.*, The clustering of galaxies in the completed SDSS-III Baryon Oscillation Spectroscopic Survey: Observational systematics and baryon acoustic oscillations in the correlation function, *Mon. Not. Roy. Astron. Soc.* **464**(1), 1168–1191 (2017). doi: 10.1093/mnras/stw2372.

759. H. Desmond, P. G. Ferreira, G. Lavaux *et al.*, Reconstructing the gravitational field of the local universe, *Mon. Not. Roy. Astron. Soc.* **474**(3), 3152–3161 (2018). doi: 10.1093/mnras/stx3062.

760. D. Langlois. Degenerate higher-order scalar-tensor (DHOST) theories. In *Proceedings, 52nd Rencontres de Moriond on Gravitation (Moriond Gravitation 2017)*: La Thuile, Italy, March 25-April 1, 2017, pp. 221–228 (2017).

761. D. F. Mota and D. J. Shaw, Strongly coupled chameleon fields: New horizons in scalar field theory, *Phys. Rev. Lett.* **97**, 151102 (2006). doi: 10.1103/PhysRevLett. 97.151102.

762. D. F. Mota and D. J. Shaw, Evading equivalence principle violations, cosmological and other experimental constraints in scalar field theories with a strong coupling to matter, *Phys. Rev.* **D75**, 063501 (2007). doi: 10.1103/PhysRevD.75.063501.

763. J. K. Bloomfield, C. Burrage and A.-C. Davis, Shape dependence of Vainshtein screening, *Phys. Rev.* **D91**(8), 083510 (2015). doi: 10.1103/PhysRevD.91.083510.

764. R. Brito, A. Terrana, M. Johnson *et al.*, Nonlinear dynamical stability of infrared modifications of gravity, *Phys. Rev.* **D90**, 124035 (2014). doi: 10.1103/PhysRevD. 90.124035.

765. M. Andrews, Y.-Z. Chu and M. Trodden, Galileon forces in the Solar System, *Phys. Rev.* **D88**, 084028 (2013). doi: 10.1103/PhysRevD.88.084028.

766. P. Kanti, N. E. Mavromatos, J. Rizos *et al.*, Dilatonic black holes in higher curvature string gravity, *Phys. Rev.* **D54**, 5049–5058 (1996). doi: 10.1103/PhysRevD.54.5049.

767. T. P. Sotiriou and S.-Y. Zhou, Black hole hair in generalized scalar-tensor gravity, *Phys. Rev. Lett.* **112**, 251102 (2014). doi: 10.1103/PhysRevLett.112.251102.

768. T. P. Sotiriou and S.-Y. Zhou, Black hole hair in generalized scalar-tensor gravity: An explicit example, *Phys. Rev.* **D90**, 124063 (2014). doi: 10.1103/PhysRevD.90.124063.

769. R. Dong, J. Sakstein and D. Stojkovic, Quasinormal modes of black holes in scalar-tensor theories with nonminimal derivative couplings, *Phys. Rev.* **D96**(6), 064048 (2017). doi: 10.1103/PhysRevD.96.064048.

770. H. O. Silva, J. Sakstein, L. Gualtieri *et al.*, Spontaneous scalarization of black holes and compact stars from a Gauss-Bonnet coupling, *Phys. Rev. Lett.* **120**(13), 131104 (2018). doi: 10.1103/PhysRevLett.120.131104.

771. L. Hui and A. Nicolis, No-hair theorem for the Galileon, *Phys. Rev. Lett.* **110**, 241104 (2013). doi: 10.1103/PhysRevLett.110.241104.

772. A.-C. Davis, R. Gregory, R. Jha *et al.*, Astrophysical black holes in screened modified gravity, *J. Cosmo. Astropart. Phys.* **1408**, 033 (2014). doi: 10.1088/1475-7516/2014/08/033.

773. A.-C. Davis, R. Gregory and R. Jha, Black hole accretion discs and screened scalar hair, *J. Cosmo. Astropart. Phys.* **2016**(10), 024 (2016). doi: 10.1088/1475-7516/2016/10/024.

774. R. Kippenhahn and A. Weigert, *Stellar Structure and Evolution* (Springer, 1994).

775. P. Chang and L. Hui, Stellar structure and tests of modified gravity, *Astrophys. J.* **732**, 25 (2011). doi: 10.1088/0004-637X/732/1/25.

776. A.-C. Davis, E. A. Lim, J. Sakstein *et al.*, Modified gravity makes galaxies brighter, *Phys. Rev.* **D85**, 123006 (2012). doi: 10.1103/PhysRevD.85.123006.

777. B. Jain, V. Vikram and J. Sakstein, Astrophysical tests of modified gravity: Constraints from distance indicators in the nearby universe, *Astrophys. J.* **779**, 39 (2013). doi: 10.1088/0004-637X/779/1/39.

778. J. Sakstein, Stellar oscillations in modified gravity, *Phys. Rev.* **D88**(12), 124013 (2013). doi: 10.1103/PhysRevD.88.124013.

779. J. Sakstein. *Astrophysical Tests of Modified Gravity.* PhD thesis, Cambridge U., DAMTP (2014).

780. J. Sakstein and K. Koyama, Testing the Vainshtein mechanism using stars and galaxies, *Int. J. Mod. Phys. D.* **24**(12), 1544021 (2015). doi: 10.1142/S0218271815440216.

781. R. K. Jain, C. Kouvaris and N. G. Nielsen, White dwarf critical tests for modified gravity, *Phys. Rev. Lett.* **116**(15), 151103 (2016). doi: 10.1103/PhysRevLett.116.151103.

782. S. Chandrasekhar, *An Introduction to the Study of Stellar Structure* (University of Chicago Press, 1939).

783. P. Brax, R. Rosenfeld and D. A. Steer, Spherical collapse in chameleon models, *J. Cosmo. Astropart. Phys.* **1008**, 033 (2010). doi: 10.1088/1475-7516/2010/08/033.

784. G. P. Horedt, Topology of the Lane-Emden equation, *Astron. Astrophys.* **177**, 117–130 (1987).

785. G. P. Horedt, Ed., *Polytropes – Applications in Astrophysics and Related Fields*, Vol. 306, Astrophysics and Space Science Library, (Kluwer Academic Publishers, 2004). doi: 10.1007/978-1-4020-2351-4.

786. B. Paxton, L. Bildsten, A. Dotter *et al.*, Modules for Experiments in Stellar Astrophysics (MESA), *Astrophys. J. Suppl.* **192**, 3 (2011). doi: 10.1088/0067-0049/192/1/3.

787. B. Paxton *et al.*, Modules for Experiments in Stellar Astrophysics (MESA): Planets, oscillations, rotation, and massive stars, *Astrophys. J. Suppl.* **208**, 4 (2013). doi: 10.1088/0067-0049/208/1/4.

788. B. Paxton *et al.*, Modules for Experiments in Stellar Astrophysics (MESA): Binaries, pulsations, and explosions, *Astrophys. J. Suppl.* **220**(1), 15 (2015). doi: 10.1088/0067-0049/220/1/15.

789. J. P. Cox, *Theory of Stellar Pulsation* (Princeton University Press, 1980).

790. A. Silvestri, Scalar radiation from Chameleon-shielded regions, *Phys. Rev. Lett.* **106**, 251101 (2011). doi: 10.1103/PhysRevLett.106.251101.

791. A. Upadhye and J. H. Steffen, Monopole radiation in modified gravity (2013). arXiv:1306.6113 [astro-ph.CO].

792. P. Brax, A.-C. Davis and J. Sakstein, Pulsar constraints on screened modified gravity, *Class. Quant. Grav.* **31**, 225001 (2014). doi: 10.1088/0264-9381/31/22/225001.

793. S. Chandrasekhar, The dynamical instability of gaseous masses approaching the Schwarzschild limit in general relativity., *Astrophys. J.* **140**, 417 (1964). doi: 10.1086/147938.

794. J. Sakstein, M. Kenna-Allison and K. Koyama, Stellar pulsations in beyond Horndeski gravity theories, *J. Cosmo. Astropart. Phys.* **2017**(03), 007 (2017). doi: 10.1088/1475-7516/2017/03/007.

795. M. S. R. Delgaty and K. Lake, Physical acceptability of isolated, static, spherically symmetric, perfect fluid solutions of Einstein's equations, *Comput. Phys. Commun.* **115**, 395–415 (1998). doi: 10.1016/S0010-4655(98)00130-1.

796. A. Burrows and J. Liebert, The science of brown dwarfs, *Rev. Mod. Phys.* **65**, 301–336 (1993). doi: 10.1103/RevModPhys.65.301.

797. G. Chabrier, I. Baraffe, J. Leconte *et al.*, The mass-radius relationship from solar-type stars to terrestrial planets: A review, *AIP Conf. Proc.* **1094**, 102–111 (2009). doi: 10.1063/1.3099078.

798. D. Saumon, G. Chabrier and H. M. van Horn, An equation of state for low-mass stars and giant planets, *Astrophys. J. Suppl.* **99**, 713 (1995). doi: 10.1086/192204.

799. D. Segransan, X. Delfosse, T. Forveille *et al.*, Accurate masses of very low mass stars. 3. 16 New or improved masses, *Astron. Astrophys.* **364**, 665 (2000). arXiv:astro-ph/0010585 [astro-ph].

800. E. E. Salpeter, The luminosity function and stellar evolution., *Astrophys. J.* **121**, 161 (1955). doi: 10.1086/145971.

801. R. Kippenhahn, On the thermal behaviour of rotating and magnetic stars, *Astron. Astrophys.* **8**, 50 (1970).

802. T. J. Henry and D. W. McCarthy, Jr., The mass-luminosity relation for stars of mass 1.0 to 0.08 solar mass, *Astron. J.* **106**, 773–789 (1993). doi: 10.1086/116685.

803. S. L. Shapiro and S. A. Teukolsky, *Black holes, White Dwarfs, and Neutron Stars: The Physics of Compact Objects* (Wiley, 1983).

804. J. B. Holberg, T. D. Oswalt and M. A. Barstow, Observational constraints on the degenerate mass-radius relation, *Astron. J.* **143**, 68 (2012). doi: 10.1088/0004-6256/143/3/68.

805. I. Hachisu and M. Kato, A theoretical light-curve model for the recurrent nova v394 coronae austrinae, *Astrophys. J.* **540**, 447 (2000). doi: 10.1086/309338.

806. I. D. Saltas, I. Sawicki and I. Lopes, White dwarfs and revelations, *J. Cosmo. Astropart. Phys.* **1805**(05), 028 (2018). doi: 10.1088/1475-7516/2018/05/028.

807. S. Mereghetti, N. La Palombara, A. Tiengo *et al.*, X-ray and optical observations of the unique binary system HD49798/RXJ0648.0-4418, *Astrophys. J.* **737**, 51 (2011). doi: 10.1088/0004-637X/737/2/51.

808. B. S. Wright and B. Li, Type Ia supernovae, standardizable candles, and gravity, *Phys. Rev.* **D97**(8), 083505 (2018). doi: 10.1103/PhysRevD.97.083505.

809. W. Zhao, B. S. Wright and B. Li, Constraining the time variation of Newton's constant G with gravitational-wave standard sirens and supernovae, *J. Cosmo. Astropart. Phys.* **2018**(10), 052 (2018). doi: 10.1088/1475-7516/2018/10/052.

810. W. L. Freedman and B. F. Madore, The Hubble constant, *Ann. Rev. Astron. Astrophys.* **48**, 673–710 (2010). doi: 10.1146/annurev-astro-082708-101829.

811. R. A. Swaters, R. Sancisi, T. S. van Albada *et al.*, Are dwarf galaxies dominated by dark matter?, *Astrophys. J.* **729**, 118 (2011). doi: 10.1088/0004-637X/729/2/118.

812. V. Vikram, J. Sakstein, C. Davis *et al.*, Astrophysical tests of modified gravity: Stellar and gaseous rotation curves in dwarf galaxies, *Phys. Rev.* **D97**(10), 104055 (2018). doi: 10.1103/PhysRevD.97.104055.

813. V. Vikram, A. Cabré, B. Jain *et al.*, Astrophysical tests of modified gravity: The morphology and kinematics of dwarf galaxies, *J. Cosmo. Astropart. Phys.* **1308**, 020 (2013). doi: 10.1088/1475-7516/2013/08/020.

814. J. Sakstein, Tests of gravity with future space-based experiments, *Phys. Rev.* **D97** (6), 064028 (2018). doi: 10.1103/PhysRevD.97.064028.

815. L. Hui and A. Nicolis, Proposal for an observational test of the Vainshtein mechanism, *Phys. Rev. Lett.* **109**, 051304 (2012). doi: 10.1103/PhysRevLett.109.051304.

816. J. Sakstein, B. Jain, J. S. Heyl *et al.*, Tests of gravity theories using supermassive black holes, *Astrophys. J.* **844**(1), L14 (2017). doi: 10.3847/2041-8213/aa7e26.

817. B. Diemer and A. V. Kravtsov, Dependence of the outer density profiles of halos on their mass accretion rate, *Astrophys. J.* **789**, 1 (2014). doi: 10.1088/0004-637X/789/1/1.

818. A. Asvathaman, J. S. Heyl and L. Hui, Eötvös experiments with supermassive black holes, *Mon. Not. Roy. Astron. Soc.* **465**(3), 3261–3266 (2017). doi: 10.1093/mnras/stw2905.

819. C. Burrage and J. Sakstein, A compendium of chameleon constraints, *J. Cosmo. Astropart. Phys.* **1611**(11), 045 (2016). doi: 10.1088/1475-7516/2016/11/045.

820. P. Brax, A.-C. Davis and R. Jha, Neutron stars in screened modified gravity: Chameleon vs dilaton, *Phys. Rev.* **D95**(8), 083514 (2017). doi: 10.1103/PhysRevD.95.083514.

821. J. Chagoya, K. Koyama, G. Niz *et al.*, Galileons and strong gravity, *J. Cosmo. Astropart. Phys.* **2014**(10), 055 (2014). doi: 10.1088/1475-7516/2014/10/055.

822. E. Berti *et al.*, Testing general relativity with present and future astrophysical observations, *Class. Quant. Grav.* **32**, 243001 (2015). doi: 10.1088/0264-9381/32/24/243001.

823. E. Babichev, C. Charmousis and A. Lehébel, Black holes and stars in Horndeski

theory, *Class. Quant. Grav.* **33**(15), 154002 (2016). doi: 10.1088/0264-9381/33/15/154002.

824. E. Babichev and D. Langlois, Relativistic stars in f(R) gravity, *Phys. Rev.* **D80**, 121501 (2009). doi: 10.1103/PhysRevD.80.121501. [Erratum: *ibid.* **81**, 069901 (2010)].

825. E. Babichev and D. Langlois, Relativistic stars in f(R) and scalar-tensor theories, *Phys. Rev.* **D81**, 124051 (2010). doi: 10.1103/PhysRevD.81.124051.

826. E. Babichev, K. Koyama, D. Langlois *et al.*, Relativistic stars in beyond Horndeski theories, *Class. Quant. Grav.* **33**(23), 235014 (2016). doi: 10.1088/0264-9381/33/23/235014.

827. J. Sakstein, E. Babichev, K. Koyama *et al.*, Towards strong field tests of beyond Horndeski gravity theories, *Phys. Rev.* **D95**(6), 064013 (2017). doi: 10.1103/PhysRevD.95.064013.

828. P. Demorest, T. Pennucci, S. Ransom *et al.*, Shapiro delay measurement of a two solar mass neutron star, *Nature.* **467**, 1081–1083 (2010). doi: 10.1038/nature09466.

829. C. Breu and L. Rezzolla, Maximum mass, moment of inertia and compactness of relativistic stars, *Mon. Not. Roy. Astron. Soc.* **459**(1), 646–656 (2016). doi: 10.1093/mnras/stw575.

830. K. Yagi and N. Yunes, Approximate universal relations for neutron stars and quark stars, *Phys. Rept.* **681**, 1–72 (2017). doi: 10.1016/j.physrep.2017.03.002.

831. J. B. Hartle, Slowly rotating relativistic stars. 1. Equations of structure, *Astrophys. J.* **150**, 1005–1029 (1967). doi: 10.1086/149400.

832. P. Brax, C. Burrage, A.-C. Davis *et al.*, Anomalous coupling of scalars to gauge fields, *Phys. Lett.* **B699**, 5–9 (2011). doi: 10.1016/j.physletb.2011.03.047.

833. C. Burrage, A.-C. Davis and D. J. Shaw, Detecting chameleons: The astronomical polarization produced by chameleon-like scalar fields, *Phys. Rev.* **D79**, 044028 (2009). doi: 10.1103/PhysRevD.79.044028.

834. A.-C. Davis, C. A. O. Schelpe and D. J. Shaw, The chameleonic contribution to the SZ radial profile of the coma cluster, *Phys. Rev.* **D83**, 044006 (2011). doi: 10.1103/PhysRevD.83.044006.

835. A. Avgoustidis, C. Burrage, J. Redondo *et al.*, Constraints on cosmic opacity and beyond the standard model physics from cosmological distance measurements, *J. Cosmo. Astropart. Phys.* **1010**, 024 (2010). doi: 10.1088/1475-7516/2010/10/024.

836. P. Brax, What makes the Universe accelerate? A review on what dark energy could be and how to test it, *Rept. Prog. Phys.* **81**(1), 016902 (2018). doi: 10.1088/1361-6633/aa8e64.

837. J. Martin, Everything you always wanted to know about the cosmological constant problem (but were afraid to ask), *Comptes Rendus Physique* **13**, 566–665 (2012). doi: 10.1016/j.crhy.2012.04.008.

838. D. J. Kapner, T. S. Cook, E. G. Adelberger *et al.*, Tests of the gravitational inverse-square law below the dark-energy length scale, *Phys. Rev. Lett.* **98**, 021101 (2007). doi: 10.1103/PhysRevLett.98.021101.

839. S. K. Lamoreaux, Demonstration of the Casimir force in the 0.6 to 6 micrometers range, *Phys. Rev. Lett.* **78**, 5–8 (1997). doi: 10.1103/PhysRevLett.78.5. [Erratum: *ibid.* **81**, 5475 (1998)].

840. P. Hamilton, M. Jaffe, P. Haslinger *et al.*, Atom-interferometry constraints on dark energy, *Science.* **349**, 849–851 (2015). doi: 10.1126/science.aaa8883.

841. H. Lemmel, P. Brax, A. N. Ivanov *et al.*, Neutron interferometry constrains dark energy chameleon fields, *Phys. Lett.* **B743**, 310–314 (2015). doi: 10.1016/j.physletb. 2015.02.063.

842. V. V. Nesvizhevsky *et al.*, Measurement of quantum states of neutrons in the earth's gravitational field, *Phys. Rev.* **D67**, 102002 (2003). doi: 10.1103/PhysRevD. 67.102002.

843. P. Brax and A.-C. Davis, Casimir, gravitational and neutron tests of dark energy, *Phys. Rev.* **D91**(6), 063503 (2015). doi: 10.1103/PhysRevD.91.063503.

844. T. Damour and A. M. Polyakov, The string dilaton and a least coupling principle, *Nucl. Phys.* **B423**, 532–558 (1994). doi: 10.1016/0550-3213(94)90143-0.

845. M. Pietroni, Dark energy condensation, *Phys. Rev.* **D72**, 043535 (2005). doi: 10. 1103/PhysRevD.72.043535.

846. K. A. Olive and M. Pospelov, Environmental dependence of masses and coupling constants, *Phys. Rev.* **D77**, 043524 (2008). doi: 10.1103/PhysRevD.77.043524.

847. P. Brax, C. Burrage and A.-C. Davis, Screening fifth forces in k-essence and DBI models, *J. Cosmo. Astropart. Phys.* **1301**, 020 (2013). doi: 10.1088/1475-7516/2013/ 01/020.

848. P. Brax and P. Valageas, K-mouflage cosmology: Formation of large-scale structures, *Phys. Rev.* **D90**(2), 023508 (2014). doi: 10.1103/PhysRevD.90.023508.

849. P. Brax, C. van de Bruck, A.-C. Davis *et al.*, Detecting dark energy in orbit: The cosmological chameleon, *Phys. Rev.* **D70**, 123518 (2004). doi: 10.1103/PhysRevD. 70.123518.

850. P. Brax, A.-C. Davis and J. Sakstein, SUPER-Screening, *Phys. Lett.* **B719**, 210–217 (2013). doi: 10.1016/j.physletb.2013.01.044.

851. B. Elder, J. Khoury, P. Haslinger *et al.*, Chameleon dark energy and atom interferometry, *Phys. Rev.* **D94**(4), 044051 (2016). doi: 10.1103/PhysRevD.94.044051.

852. S. Schlögel, S. Clesse and A. Füzfa, Probing modified gravity with atom-interferometry: A numerical approach, *Phys. Rev.* **D93**(10), 104036 (2016). doi: 10.1103/PhysRevD.93.104036.

853. P. Brax, G. Pignol and D. Roulier, Probing strongly coupled chameleons with slow neutrons, *Phys. Rev.* **D88**, 083004 (2013). doi: 10.1103/PhysRevD.88.083004.

854. P. Brax and A.-C. Davis, Atomic interferometry test of dark energy, *Phys. Rev.* **D94** (10), 104069 (2016). doi: 10.1103/PhysRevD.94.104069.

855. C. Burrage, E. J. Copeland and E. A. Hinds, Probing dark energy with atom interferometry, *J. Cosmo. Astropost. Phys.* **2015**(03), 042 (2015). doi: 10.1088/1475-7516/ 2015/03/042.

856. C. Burrage and E. J. Copeland, Using atom interferometry to detect dark energy, *Contemp. Phys.* **57**(2), 164–176 (2016). doi: 10.1080/00107514.2015.1060058.

857. C. Burrage, A. Kuribayashi-Coleman, J. Stevenson *et al.*, Constraining symmetron fields with atom interferometry, *J. Cosmo. Astropost. Phys.* **1612**, 041 (2016). doi: 10.1088/1475-7516/2016/12/041.

858. M. Jaffe, P. Haslinger, V. Xu *et al.*, Testing sub-gravitational forces on atoms from a miniature, in-vacuum source mass, *Nature Phys.* **13**, 938 (2017). doi: 10.1038/ nphys4189.

859. E. G. Adelberger, B. R. Heckel and A. E. Nelson, Tests of the gravitational inverse

square law, *Ann. Rev. Nucl. Part. Sci.* **53**, 77–121 (2003). doi: 10.1146/annurev.nucl. 53.041002.110503.

860. A. Lambrecht, V. V. Nesvizhevsky, R. Onofrio *et al.*, Development of a high-sensitivity torsional balance for the study of the Casimir force in the 1-10 micrometre range, *Class. Quant. Grav.* **22**, 5397–5406 (2005). doi: 10.1088/0264-9381/22/24/012.

861. E. G. Adelberger, B. R. Heckel, S. A. Hoedl *et al.*, Particle physics implications of a recent test of the gravitational inverse sqaure law, *Phys. Rev. Lett.* **98**, 131104 (2007). doi: 10.1103/PhysRevLett.98.131104.

862. A. Upadhye. Particles and forces from chameleon dark energy. In *8th Patras Workshop on Axions, WIMPs and WISPs (AXION-WIMP 2012)* Chicago, Illinois, July 18-22, 2012 (2012).

863. A. Upadhye, Dark energy fifth forces in torsion pendulum experiments, *Phys. Rev.* **D86**, 102003 (2012). doi: 10.1103/PhysRevD.86.102003.

864. A. Upadhye, Symmetron dark energy in laboratory experiments, *Phys. Rev. Lett.* **110**(3), 031301 (2013). doi: 10.1103/PhysRevLett.110.031301.

865. S. K. Lamoreaux and W. T. Buttler, Thermal noise limitations to force measurements with torsion pendulums: Applications to the measurement of the Casimir force and its thermal correction, *Phys. Rev.* **E71**, 036109 (2005). doi: 10.1103/PhysRevE.71. 036109.

866. A. Lambrecht and S. Reynaud. Casimir and short-range gravity tests. In *Proceedings, 46th Rencontres de Moriond on Gravitational Waves and Experimental Gravity:* La Thuile, Italy, March 20-27, 2011 pp. 199–206 (2011).

867. P. Brax, C. van de Bruck, A.-C. Davis *et al.*, Detecting chameleons through Casimir force measurements, *Phys. Rev.* **D76**, 124034 (2007). doi: 10.1103/PhysRevD.76. 124034.

868. P. Brax, C. van de Bruck, A. C. Davis *et al.*, Tuning the mass of chameleon fields in casimir force experiments, *Phys. Rev. Lett.* **104**, 241101 (2010). doi: 10.1103/ PhysRevLett.104.241101.

869. A. Almasi, P. Brax, D. Iannuzzi *et al.*, Force sensor for chameleon and Casimir force experiments with parallel-plate configuration, *Phys. Rev.* **D91**(10), 102002 (2015). doi: 10.1103/PhysRevD.91.102002.

870. H. Abele, T. Jenke, H. Leeb *et al.*, Ramsey's method of separated oscillating fields and its application to gravitationally induced quantum phaseshifts, *Phys. Rev.* **D81**, 065019 (2010). doi: 10.1103/PhysRevD.81.065019.

871. T. Jenke, P. Geltenbort, H. Lemmel *et al.*, Realization of a gravity-resonance-spectroscopy technique, *Nature Phys.* **7**, 468–472 (2011). doi: 10.1038/nphys1970.

872. T. Jenke *et al.*, Gravity resonance spectroscopy constrains dark energy and dark matter scenarios, *Phys. Rev. Lett.* **112**, 151105 (2014). doi: 10.1103/PhysRevLett. 112.151105.

873. V. V. Nesvizhevsky *et al.*, Quantum states of neutrons in the Earth's gravitational field, *Nature.* **415**, 297–299 (2002). doi: 10.1038/415297a.

874. G. Cronenberg, P. Brax, H. Filter *et al.*, Acoustic Rabi oscillations between gravitational quantum states and impact on symmetron dark energy, *Nature Phys.* **14**(10), 1022–1026 (2018). doi: 10.1038/s41567-018-0205-x.

875. A. Westphal, H. Abele, S. Baessler *et al.*, A quantum mechanical description of the

experiment on the observation of gravitationally bound states, *Eur. Phys. J.* **C51**, 367–375 (2007). doi: 10.1140/epjc/s10052-007-0283-x.

876. P. Brax and C. Burrage, Atomic precision tests and light scalar couplings, *Phys. Rev.* **D83**, 035020 (2011). doi: 10.1103/PhysRevD.83.035020.

877. J. Jaeckel and S. Roy, Spectroscopy as a test of Coulomb's law: A probe of the hidden sector, *Phys. Rev.* **D82**, 125020 (2010). doi: 10.1103/PhysRevD.82.125020.

878. C. Schwob, L. Jozefowski, B. de Beauvoir *et al.*, Optical frequency measurement of the S-2- D-12 transitions in hydrogen and deuterium: Rydberg constant and Lamb shift determinations, *Phys. Rev. Lett.* **82**, 4960–4963 (1999). doi: 10.1103/PhysRevLett. 82.4960.

879. G. G. Simon, C. Schmitt, F. Borkowski *et al.*, Absolute electron proton cross-sections at low momentum transfer measured with a high pressure gas target system, *Nucl. Phys.* **A333**, 381–391 (1980). doi: 10.1016/0375-9474(80)90104-9.

880. P. Brax, C. van de Bruck, A.-C. Davis *et al.*, Testing chameleon theories with light propagating through a magnetic field, *Phys. Rev.* **D76**, 085010 (2007). doi: 10.1103/ PhysRevD.76.085010.

881. T. Damour, F. Piazza and G. Veneziano, Runaway dilaton and equivalence principle violations, *Phys. Rev. Lett.* **89**, 081601 (2002). doi: 10.1103/PhysRevLett.89.081601.

882. J. G. Williams, S. G. Turyshev and D. H. Boggs, Lunar laser ranging tests of the equivalence principle, *Class. Quant. Grav.* **29**(18), 184004 (2012). doi: 10.1088/ 0264-9381/29/18/184004.

883. E. Babichev, C. Deffayet and G. Esposito-Farese, Constraints on shift-symmetric scalar-tensor theories with a Vainshtein mechanism from bounds on the time variation of G, *Phys. Rev. Lett.* **107**, 251102 (2011). doi: 10.1103/PhysRevLett.107.251102.

884. P. S. Cowperthwaite *et al.*, The electromagnetic counterpart of the binary neutron star merger LIGO/Virgo GW170817. II. UV, optical, and near-IR light curves and comparison to kilonova models, *Astrophys. J.* **848**(2), L17 (2017). doi: 10.3847/ 2041-8213/aa8fc7.

885. R. S. Decca, D. Lopez, E. Fischbach *et al.*, Novel constraints on light elementary particles and extra-dimensional physics from the Casimir effect, *Eur. Phys. J.* **C51**, 963–975 (2007). doi: 10.1140/epjc/s10052-007-0346-z.

886. Y. J. Chen, W. K. Tham, D. E. Krause *et al.*, Stronger limits on hypothetical Yukawa interactions in the 30–8000 nm range, *Phys. Rev. Lett.* **116**(22), 221102 (2016). doi: 10.1103/PhysRevLett.116.221102.

887. G. Bressi, G. Carugno, R. Onofrio *et al.*, Measurement of the Casimir force between parallel metallic surfaces, *Phys. Rev. Lett.* **88**, 041804 (2002). doi: 10.1103/ PhysRevLett.88.041804.

888. G. Bressi, G. Carugno, A. Galvani *et al.*, Experimental searches for extra-gravitational forces in the submillimetre range, *Class. Quant. Grav.* **17**, 2365–2368 (2000). doi: 10.1088/0264-9381/17/12/308.

889. F. Nitti and F. Piazza, Scalar-tensor theories, trace anomalies and the QCD-frame, *Phys. Rev.* **D86**, 122002 (2012). doi: 10.1103/PhysRevD.86.122002.

890. E. Zavattini *et al.*, New PVLAS results and limits on magnetically induced optical rotation and ellipticity in vacuum, *Phys. Rev.* **D77**, 032006 (2008). doi: 10.1103/ PhysRevD.77.032006.

891. K. Ehret *et al.*, New ALPS results on hidden-sector lightweights, *Phys. Lett.* **B689**, 149–155 (2010). doi: 10.1016/j.physletb.2010.04.066.

892. J. H. Steffen, The CHASE laboratory search for chameleon dark energy, *PoS.* **ICHEP2010**, 446 (2010). arXiv:1011.3802 [hep-ex].

893. P. Brax and K. Zioutas, Solar chameleons, *Phys. Rev.* **D82**, 043007 (2010). doi: 10.1103/PhysRevD.82.043007.

894. P. Brax, A. Lindner and K. Zioutas, Detection prospects for solar and terrestrial chameleons, *Phys. Rev.* **D85**, 043014 (2012). doi: 10.1103/PhysRevD.85.043014.

895. V. Anastassopoulos *et al.*, Search for chameleons with CAST, *Phys. Lett.* **B749**, 172–180 (2015). doi: 10.1016/j.physletb.2015.07.049.

896. C. Llinares, R. Hagala and D. F. Mota, Non-linear phenomenology of disformally coupled quintessence (2019). arXiv:1902.02125.

897. I. Zubeldia and A. Challinor, Cosmological constraints from Planck galaxy clusters with CMB lensing mass bias calibration, *Mon. Not. Roy. Astron. Soc.* **489**, 401–419 (2019). doi: 10.1093/mnras/stz2153.